Ford Super Duty Pick-ups Automotive Repair Manual

by Jeff Killingsworth and John H Haynes

Member of the Guild of Motoring Writers

(36064-10X2)

ABCDE
FGHIJ
KLMNO
PQR

AUTOMOTIVE
PARTS &
ACCESSORIES
ASSOCIATION   MEMBER

Models covered:
Ford Super Duty F-250 and F-350 gasoline and diesel engines
2011 through 2016
Does not include F-450 and F-550 models

Haynes Publishing Group

Sparkford Nr Yeovil
Somerset BA22 7JJ England

Haynes North America, Inc
859 Lawrence Drive
Newbury Park
California 91320 USA
www.haynes.com

Acknowledgements

Technical writers who contributed to this project include Demian Hurst and Scott "Gonzo" Weaver.

© **Haynes North America, Inc. 2017**

With permission from J.H. Haynes & Co. Ltd.

A book in the Haynes Automotive Repair Manual Series

Printed in Malaysia

ISBN-13: 978-1-62092-256-9
ISBN-10: 1-62092-256-8

Library of Congress Control Number: 2017930672

Contents

Haynes photographer and mechanic with a 2013 F-250 4WD, 6.7L diesel crew cab

About this manual

Its purpose

The purpose of this manual is to help you get the best value from your vehicle. It can do so in several ways. It can help you decide what work must be done, even if you choose to have it done by a dealer service department or a repair shop; it provides information and procedures for routine maintenance and servicing; and it offers diagnostic and repair procedures to follow when trouble occurs.

We hope you use the manual to tackle the work yourself. For many simpler jobs, doing it yourself may be quicker than arranging an appointment to get the vehicle into a shop and making the trips to leave it and pick it up. More importantly, a lot of money can be saved by avoiding the expense the shop must pass on to you to cover its labor and overhead costs. An added benefit is the sense of satisfaction and accomplishment that you feel after doing the job yourself.

Using the manual

The manual is divided into Chapters. Each Chapter is divided into numbered Sections, which are headed in bold type between horizontal lines. Each Section consists of consecutively numbered paragraphs.

The reference numbers used in illustration captions pinpoint the pertinent Section and the Step within that Section. That is, illustration 3.2 means the illustration refers to Section 3 and Step (or paragraph) 2 within that Section.

Procedures, once described in the text, are not normally repeated. When it's necessary to refer to another Chapter, the reference will be given as Chapter and Section number. Cross references given without use of the word "Chapter" apply to Sections and/or paragraphs in the same Chapter. For example, "see Section 8" means in the same Chapter.

References to the left or right side of the vehicle assume you are sitting in the driver's seat, facing forward.

Even though we have prepared this manual with extreme care, neither the publisher nor the author can accept responsibility for any errors in, or omissions from, the information given.

NOTE

A **Note** provides information necessary to properly complete a procedure or information which will make the procedure easier to understand.

CAUTION

A **Caution** provides a special procedure or special steps which must be taken while completing the procedure where the Caution is found. Not heeding a Caution can result in damage to the assembly being worked on.

WARNING

A **Warning** provides a special procedure or special steps which must be taken while completing the procedure where the Warning is found. Not heeding a Warning can result in personal injury.

Introduction

These pick-ups are available regular cab, Supercab and Crew Cab, short bed and long bed, 2WD and 4WD models.

Available engines are the 6.2L, 16-valve V8 gasoline engine as well as the new Ford-designed 6.7L, 32-valve turbo diesel V8. All models are equipped with the On Board Diagnostic Second-Generation (OBD-II) computerized engine management system that controls virtually every aspect of engine operation. OBD-II monitors emissions system components for signs of degradation and engine operation for any malfunction that could affect emissions, turning on the CHECK ENGINE light if any faults are detected.

Chassis layout is conventional, with the engine mounted at the front and the power being transmitted through a 6-speed automatic transmission and a driveshaft to the solid rear axle. On 4WD models, a transfer case transmits power to a front differential by way of a driveshaft and then to the front wheels through the axles.

Suspension is independent twin I-beam with coil springs in the front on 2WD models or coil springs and a solid front axle with radius arms on 4WD models. The rear suspension features semi-elliptical leaf springs. All models use telescopic shock absorbers at each corner and a front and rear stabilizer bar.

The brakes are disc-type on all four wheels, with an Anti-Lock Brake System (ABS) as standard equipment.

The power-assisted recirculating ball-type steering gear is mounted on the chassis frame rail to the left of the engine.

Most models have a power assisted disc-type front and rear brake system with an Anti-lock Brake System (ABS) as standard equipment.

Vehicle identification numbers

Modifications are a continuing and unpublicized process in vehicle manufacturing. Since spare parts lists and manuals are compiled on a numerical basis, the individual vehicle numbers are necessary to correctly identify the component required.

Vehicle Identification Number (VIN)

This very important identification number is stamped on a plate attached to the dashboard inside the windshield on the driver's side of the vehicle (see illustration). The VIN also appears on the Vehicle Certificate of Title and Registration. It contains information such as where and when the vehicle was manufactured, the model year and the body style.

VIN engine and model year codes

Two particularly important pieces of information found in the VIN are the engine code and the model year code. Counting from the left, the engine code letter designation is the 8th digit and the model year code letter designation is the 10th digit.

On the models covered by this manual the engine codes are:

6   6.2L 16-valve, Overhead camshaft (OHC), V8 gasoline

T   6.7L V8 turbo diesel

On the models covered by this manual the model year codes are:

B   2011
C   2012
D   2013
E   2014
F   2015
G   2016

Vehicle Certification Label

The Vehicle Certification Label is attached to the driver's side door pillar (see illustration). Information on this label includes the name of the manufacturer, the month and year of production, as well as information on the options with which it is equipped. This label is especially useful for matching the color and type of paint for repair work.

Engine identification number

Labels containing the engine code, engine number and build date can be found on the valve cover (6.2L engine) or the crankcase vent oil separator (6.7L diesel) (see illustration). The engine number is also stamped onto a machined pad on the external surface of the engine block.

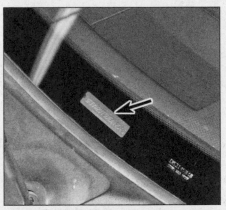

The VIN is visible through the windshield on the driver's side

The Vehicle Certification Label is affixed to the driver's side door pillar

The engine identification label is affixed to the crankcase vent oil separator on 6.7L diesel models

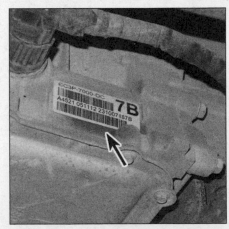

The automatic transmission identification tag is located on the left (driver's) side of the case

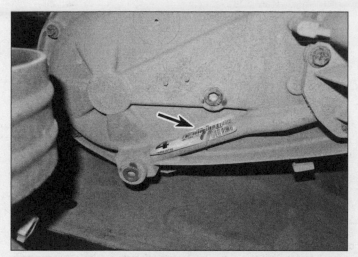

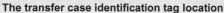

The transfer case identification tag location

The VECI label is affixed to the underside of the hood

Automatic transmission identification number

The automatic transmission ID number is on a label affixed to the left side of the case (see illustration).

The transmission code is also shown on the Vehicle Certification label, under "TR."

CODE	Transmission type
P	6R140, 6-speed automatic
W	6R140, 6-speed automatic (diesel models)

Transfer case identification number

10   The transfer case ID number is stamped on a tag which is fastened to the rear cover (see illustration).

Differential identification number

11   The differential/axle ID number is located on the Vehicle Certification Label. Some axles are also equipped with a tag on which this information is stamped.

CODE	GEAR RATIO/AXLE TYPE
31	3.31, non limited-slip
35	3.55, non limited-slip
37	3.73 non limited-slip
3E	3.73 electronic locking differential
3H	3.31 electronic locking differential
3J	3.55 electronic locking differential
3L	3.73 limited slip
41	4.10 non limited-slip
43	4.30 non limited-slip
48	4.88 non limited-slip
4E	4.10 electronic locking differential
4L	4.30 limited slip
4M	4.30 electronic locking differential
4N	4.10 limited slip
8L	4.88 limited slip

Vehicle Emissions Control Information label

12   This label is found in the engine compartment (see illustration). See Chapter 6A or Chapter 6B   for more information on this label.

Service parts identification or Regular Production Option (RPO) code label

This label is located in the glove box (see illustration). It lists the VIN, paint number, options and other information specific to the vehicle. You may sometimes need to refer to this label when you order parts.

Recall information

Vehicle recalls are carried out by the manufacturer in the rare event of a possible safety-related defect. The vehicle's registered owner is contacted at the address on file at the Department of Motor Vehicles and given the details of the recall. Remedial work is carried out free of charge at a dealer service department.

If you are the new owner of a used vehicle which was subject to a recall and you want to be sure that the work has been carried out, it's best to contact a dealer service department and ask about your individual vehicle - you'll need to furnish them your Vehicle Identification Number (VIN).

The table below is based on information provided by the National Highway Traffic Safety Administration (NHTSA), the body which oversees vehicle recalls in the United States. The recall database is updated constantly. For the latest information on vehicle recalls, check the NHTSA website at www.nhtsa.gov, www.safercar.gov, or call the NHTSA hotline at 1-888-327-4236.

Recall date	Recall campaign number	Model(s) affected	Concern
April 17, 2014	14V204000	2013, 2014 - F-250, F-350 Super Duty	Ford is recalling certain vehicles equipped with 6.2L gasoline or 6.7L diesel engines and 6R140 transmissions. An error with the transmission control software may be such that while shifting from "Park" to "Drive" the vehicle may display "Drive" but possibly engage "Reverse" for 1.5 seconds before engaging "Drive." Therefore, the vehicles do not conform to Federal Motor Vehicle Safety Standard (FMVSS) number 102, "Transmission Shift Lever Sequence, Starter Interlock, and Transmission Braking Effect." If the vehicle is in "Reverse" while the while the gear selection shows "Drive," unexpected movement could occur, increasing the risk of a crash.
March 25, 2015	15V175000	2011, 2012, 2013, 2014, 2015 - F-250 Super Duty, F-350 Super Duty	Ford is recalling certain vehicles equipped with ambulance or fire engine preparation packages. The affected vehicles may detect an incorrect Exhaust Gas Temperature Sensor (EGT) fault indicating that the vehicle is too hot, causing the engine management system to shut down the engine and prevent its immediate restart. Engine shut down and a subsequent cool down period may delay medical treatment or assistance to those needing it, increasing their risk of injury.
April 26, 2016	16V246000	2016 - F-250 Super Duty, F-350 Super Duty	Ford is recalling certain vehicles where the tires installed during assembly of the affected vehicles may have sustained damage to the inner sidewall, possibly resulting in a loss of air or a tire rupture. A rapid loss of air or a tire rupture while the vehicle is being driven may cause a loss of vehicle control, increasing the risk of a crash.

Buying parts

Replacement parts are available from many sources, which generally fall into one of two categories - authorized dealer parts departments and independent retail auto parts stores. Our advice concerning these parts is as follows:

Retail auto parts stores: Good auto parts stores will stock frequently needed components which wear out relatively fast, such as clutch components, exhaust systems, brake parts, tune-up parts, etc. These stores often supply new or reconditioned parts on an exchange basis, which can save a considerable amount of money. Discount auto parts stores are often very good places to buy materials and parts needed for general vehicle maintenance such as oil, grease, filters, spark plugs, belts, touch-up paint, bulbs, etc. They also usually sell tools and general accessories, have convenient hours, charge lower prices and can often be found not far from home.

Authorized dealer parts department: This is the best source for parts which are unique to the vehicle and not generally available elsewhere (such as major engine parts, transmission parts, trim pieces, etc.).

Warranty information: If the vehicle is still covered under warranty, be sure that any replacement parts purchased - regardless of the source - do not invalidate the warranty!

To be sure of obtaining the correct parts, have engine and chassis numbers available and, if possible, take the old parts along for positive identification.

Maintenance techniques, tools and working facilities

Maintenance techniques

There are a number of techniques involved in maintenance and repair that will be referred to throughout this manual. Application of these techniques will enable the home mechanic to be more efficient, better organized and capable of performing the various tasks properly, which will ensure that the repair job is thorough and complete.

Fasteners

Fasteners are nuts, bolts, studs and screws used to hold two or more parts together. There are a few things to keep in mind when working with fasteners. Almost all of them use a locking device of some type, either a lockwasher, locknut, locking tab or thread adhesive. All threaded fasteners should be clean and straight, with undamaged threads and undamaged corners on the hex head where the wrench fits. Develop the habit of replacing all damaged nuts and bolts with new ones. Special locknuts with nylon or fiber inserts can only be used once. If they are removed, they lose their locking ability and must be replaced with new ones.

Rusted nuts and bolts should be treated with a penetrating fluid to ease removal and prevent breakage. Some mechanics use turpentine in a spout-type oil can, which works quite well. After applying the rust penetrant, let it work for a few minutes before trying to loosen the nut or bolt. Badly rusted fasteners may have to be chiseled or sawed off or removed with a special nut breaker, available at tool stores.

If a bolt or stud breaks off in an assembly, it can be drilled and removed with a special tool commonly available for this purpose. Most automotive machine shops can perform this task, as well as other repair procedures, such as the repair of threaded holes that have been stripped out.

Flat washers and lockwashers, when removed from an assembly, should always be replaced exactly as removed. Replace any damaged washers with new ones. Never use a lockwasher on any soft metal surface (such as aluminum), thin sheet metal or plastic.

Fastener sizes

For a number of reasons, automobile manufacturers are making wider and wider use of metric fasteners. Therefore, it is important to be able to tell the difference between standard (sometimes called U.S. or SAE) and metric hardware, since they cannot be interchanged.

All bolts, whether standard or metric, are sized according to diameter, thread pitch and length. For example, a standard 1/2 - 13 x 1 bolt is 1/2 inch in diameter, has 13 threads per inch and is 1 inch long. An M12 - 1.75 x 25 metric bolt is 12 mm in diameter, has a thread pitch of 1.75 mm (the distance between threads) and is 25 mm long. The two bolts are nearly identical, and easily confused, but they are not interchangeable.

In addition to the differences in diameter, thread pitch and length, metric and standard bolts can also be distinguished by examining the bolt heads. To begin with, the distance across the flats on a standard bolt head is measured in inches, while the same dimension on a metric bolt is sized in millimeters (the same is true for nuts). As a result, a standard wrench should not be used on a metric bolt and a metric wrench should not be used on a standard bolt. Also, most standard bolts have slashes radiating out from the center of the head to denote the grade or strength of the bolt, which is an indication of the amount of torque that can be applied to it. The greater the number of slashes, the greater the strength of the bolt. Grades 0 through 5 are commonly used on automobiles. Metric bolts have a property class (grade) number, rather than a slash, molded into their heads to indicate bolt strength. In this case, the higher the number, the stronger the bolt. Property class numbers 8.8, 9.8 and 10.9 are commonly used on automobiles.

Strength markings can also be used to distinguish standard hex nuts from metric hex nuts. Many standard nuts have dots stamped into one side, while metric nuts are marked with a number. The greater the number of dots, or the higher the number, the greater the strength of the nut.

Metric studs are also marked on their ends according to property class (grade). Larger studs are numbered (the same as metric bolts), while smaller studs carry a geometric code to denote grade.

It should be noted that many fasteners, especially Grades 0 through 2, have no distinguishing marks on them. When such is the case, the only way to determine whether it is standard or metric is to measure the thread pitch or compare it to a known fastener of the same size.

Standard fasteners are often referred to as SAE, as opposed to metric. However, it should be noted that SAE technically refers to a non-metric fine thread fastener only. Coarse thread non-metric fasteners are referred to as USS sizes.

Since fasteners of the same size (both standard and metric) may have different strength ratings, be sure to reinstall any bolts, studs or nuts removed from your vehicle in their original locations. Also, when replacing a fastener with a new one, make sure that the new one has a strength rating equal to or greater than the original.

Tightening sequences and procedures

Most threaded fasteners should be tightened to a specific torque value (torque is the twisting force applied to a threaded component such as a nut or bolt). Overtightening the fastener can weaken it and cause it to break, while undertightening can cause it to eventually come loose. Bolts, screws and studs, depending on the material they are made of and their thread diameters, have specific torque values, many of which are noted in the Specifications at the beginning of each Chapter. Be sure to follow the torque recommen-

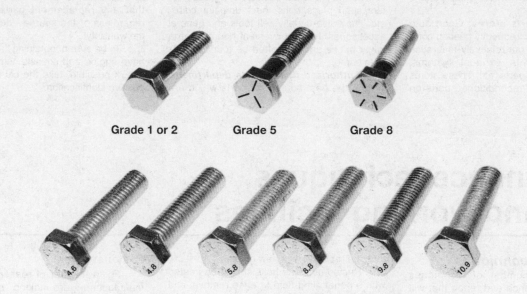

Grade 1 or 2   Grade 5   Grade 8

Bolt strength marking (standard/SAE/USS; bottom - metric)

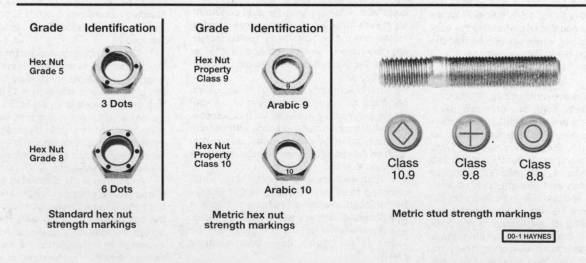

Grade	Identification
Hex Nut Grade 5	3 Dots
Hex Nut Grade 8	6 Dots

Standard hex nut strength markings

Grade	Identification
Hex Nut Property Class 9	Arabic 9
Hex Nut Property Class 10	Arabic 10

Metric hex nut strength markings

Class 10.9   Class 9.8   Class 8.8

Metric stud strength markings

00-1 HAYNES

dations closely. For fasteners not assigned a specific torque, a general torque value chart is presented here as a guide. These torque values are for dry (unlubricated) fasteners threaded into steel or cast iron (not aluminum). As was previously mentioned, the size and grade of a fastener determine the amount of torque that can safely be applied to it. The figures listed here are approximate for Grade 2 and Grade 3 fasteners. Higher grades can tolerate higher torque values.

Fasteners laid out in a pattern, such as cylinder head bolts, oil pan bolts, differential cover bolts, etc., must be loosened or tightened in sequence to avoid warping the component. This sequence will normally be shown in the appropriate Chapter. If a specific pattern is not given, the following procedures can be used to prevent warping.

Initially, the bolts or nuts should be assembled finger-tight only. Next, they should be tightened one full turn each, in a criss-cross or diagonal pattern. After each one has been tightened one full turn, return to the first one and tighten them all one-half turn, following the same pattern. Finally, tighten each of them one-quarter turn at a time until each fastener has been tightened to the proper torque. To loosen and remove the fasteners, the procedure would be reversed.

Metric thread sizes

	Ft-lbs	Nm
M-6	6 to 9	9 to 12
M-8	14 to 21	19 to 28
M-10	28 to 40	38 to 54
M-12	50 to 71	68 to 96
M-14	80 to 140	109 to 154

Pipe thread sizes

1/8	5 to 8	7 to 10
1/4	12 to 18	17 to 24
3/8	22 to 33	30 to 44
1/2	25 to 35	34 to 47

U.S. thread sizes

1/4 - 20	6 to 9	9 to 12
5/16 - 18	12 to 18	17 to 24
5/16 - 24	14 to 20	19 to 27
3/8 - 16	22 to 32	30 to 43
3/8 - 24	27 to 38	37 to 51
7/16 - 14	40 to 55	55 to 74
7/16 - 20	40 to 60	55 to 81
1/2 - 13	55 to 80	75 to 108

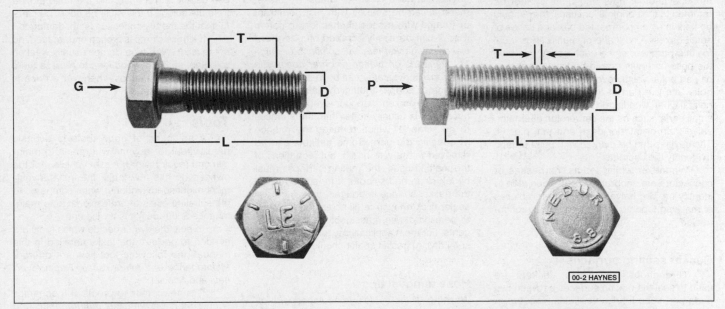

Standard (SAE and USS) bolt dimensions/grade marks

G   Grade marks (bolt strength)
L   Length (in inches)
T   Thread pitch (number of threads per inch)
D   Nominal diameter (in inches)

Metric bolt dimensions/grade marks

P   Property class (bolt strength)
L   Length (in millimeters)
T   Thread pitch (distance between threads in millimeters)
D   Diameter

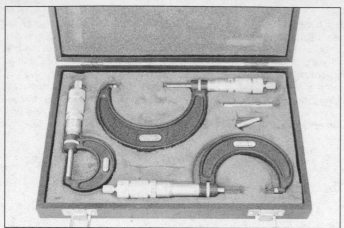

Micrometer set

Dial indicator set

Component disassembly

Component disassembly should be done with care and purpose to help ensure that the parts go back together properly. Always keep track of the sequence in which parts are removed. Make note of special characteristics or marks on parts that can be installed more than one way, such as a grooved thrust washer on a shaft. It is a good idea to lay the disassembled parts out on a clean surface in the order that they were removed. It may also be helpful to make sketches or take instant photos of components before removal.

When removing fasteners from a component, keep track of their locations. Sometimes threading a bolt back in a part, or putting the washers and nut back on a stud, can prevent mix-ups later. If nuts and bolts cannot be returned to their original locations, they should be kept in a compartmented box or a series of small boxes. A cupcake or muffin tin is ideal for this purpose, since each cavity can hold the bolts and nuts from a particular area (i.e. oil pan bolts, valve cover bolts, engine mount bolts, etc.). A pan of this type is especially helpful when working on assemblies with very small parts, such as the carburetor, alternator, valve train or interior dash and trim pieces. The cavities can be marked with paint or tape to identify the contents.

Whenever wiring looms, harnesses or connectors are separated, it is a good idea to identify the two halves with numbered pieces of masking tape so they can be easily reconnected.

Gasket sealing surfaces

Throughout any vehicle, gaskets are used to seal the mating surfaces between two parts and keep lubricants, fluids, vacuum or pressure contained in an assembly.

Many times these gaskets are coated with a liquid or paste-type gasket sealing compound before assembly. Age, heat and pressure can sometimes cause the two parts to stick together so tightly that they are very difficult to separate. Often, the assembly can

be loosened by striking it with a soft-face hammer near the mating surfaces. A regular hammer can be used if a block of wood is placed between the hammer and the part. Do not hammer on cast parts or parts that could be easily damaged. With any particularly stubborn part, always recheck to make sure that every fastener has been removed.

Avoid using a screwdriver or bar to pry apart an assembly, as they can easily mar the gasket sealing surfaces of the parts, which must remain smooth. If prying is absolutely necessary, use an old broom handle, but keep in mind that extra clean up will be necessary if the wood splinters.

After the parts are separated, the old gasket must be carefully scraped off and the gasket surfaces cleaned. Stubborn gasket material can be soaked with rust penetrant or treated with a special chemical to soften it so it can be easily scraped off. **Caution:** *Never use gasket removal solutions or caustic chemicals on plastic or other composite components.* A scraper can be fashioned from a piece of copper tubing by flattening and sharpening one end. Copper is recommended because it is usually softer than the surfaces to be scraped, which reduces the chance of gouging the part. Some gaskets can be removed with a wire brush, but regardless of the method used, the mating surfaces must be left clean and smooth. If for some reason the gasket surface is gouged, then a gasket sealer thick enough to fill scratches will have to be used during reassembly of the components. For most applications, a non-drying (or semi-drying) gasket sealer should be used.

Hose removal tips

Warning: *If the vehicle is equipped with air conditioning, do not disconnect any of the A/C hoses without first having the system depressurized by a dealer service department or a service station.*

Hose removal precautions closely parallel gasket removal precautions. Avoid scratching or gouging the surface that the

hose mates against or the connection may leak. This is especially true for radiator hoses. Because of various chemical reactions, the rubber in hoses can bond itself to the metal spigot that the hose fits over. To remove a hose, first loosen the hose clamps that secure it to the spigot. Then, with slip-joint pliers, grab the hose at the clamp and rotate it around the spigot. Work it back and forth until it is completely free, then pull it off. Silicone or other lubricants will ease removal if they can be applied between the hose and the outside of the spigot. Apply the same lubricant to the inside of the hose and the outside of the spigot to simplify installation.

As a last resort (and if the hose is to be replaced with a new one anyway), the rubber can be slit with a knife and the hose peeled from the spigot. If this must be done, be careful that the metal connection is not damaged.

If a hose clamp is broken or damaged, do not reuse it. Wire-type clamps usually weaken with age, so it is a good idea to replace them with screw-type clamps whenever a hose is removed.

Tools

A selection of good tools is a basic requirement for anyone who plans to maintain and repair his or her own vehicle. For the owner who has few tools, the initial investment might seem high, but when compared to the spiraling costs of professional auto maintenance and repair, it is a wise one.

To help the owner decide which tools are needed to perform the tasks detailed in this manual, the following tool lists are offered: *Maintenance and minor repair, Repair/overhaul* and *Special.*

The newcomer to practical mechanics should start off with the *maintenance and minor repair* tool kit, which is adequate for the simpler jobs performed on a vehicle. Then, as confidence and experience grow, the owner can tackle more difficult tasks, buying additional tools as they are needed. Eventually the basic kit will be expanded into the *repair and overhaul* tool set. Over a period of time, the

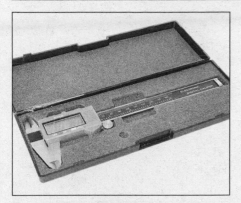

Dial caliper

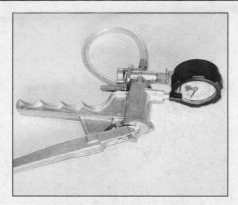

Hand-operated vacuum pump

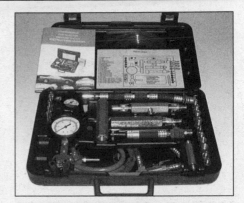

Fuel pressure gauge set

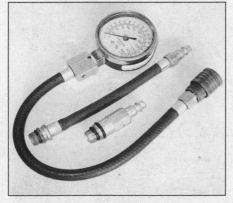

Compression gauge with spark plug hole adapter

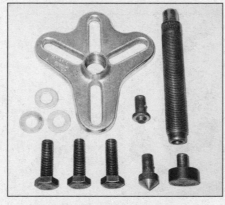

Damper/steering wheel puller

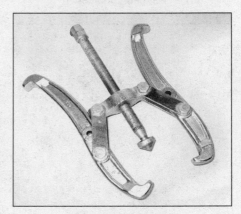

General purpose puller

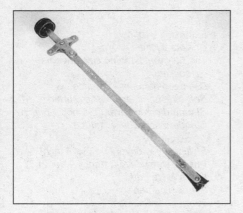

Hydraulic lifter removal tool

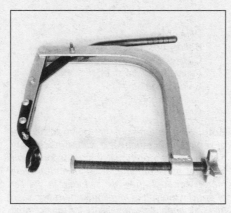

Valve spring compressor

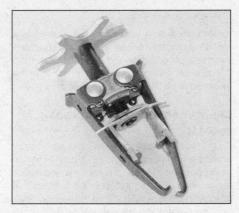

Valve spring compressor

Ridge reamer

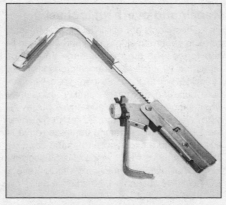

Piston ring groove cleaning tool

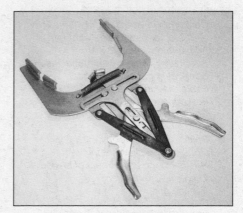

Ring removal/installation tool

Ring compressor

Cylinder hone

Brake hold-down spring tool

Torque angle gauge

Clutch plate alignment tool

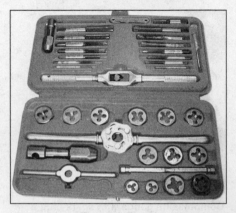

Tap and die set

experienced do-it-yourselfer will assemble a tool set complete enough for most repair and overhaul procedures and will add tools from the special category when it is felt that the expense is justified by the frequency of use.

Maintenance and minor repair tool kit

The tools in this list should be considered the minimum required for performance of routine maintenance, servicing and minor repair work. We recommend the purchase of combination wrenches (box-end and open-end combined in one wrench). While more expensive than open end wrenches, they offer the advantages of both types of wrench.

Combination wrench set (1/4-inch to
* 1 inch or 6 mm to 19 mm)*
Adjustable wrench, 8 inch
Spark plug wrench with rubber insert
Spark plug gap adjusting tool
Feeler gauge set
Brake bleeder wrench
Standard screwdriver (5/16-inch x
* 6 inch)*
Phillips screwdriver (No. 2 x 6 inch)
Combination pliers - 6 inch
Hacksaw and assortment of blades
Tire pressure gauge
Grease gun
Oil can
Fine emery cloth

Wire brush
Battery post and cable cleaning tool
Oil filter wrench
Funnel (medium size)
Safety goggles
Jackstands (2)
Drain pan

Note: *If basic tune-ups are going to be part of routine maintenance, it will be necessary to purchase a good quality stroboscopic timing light and combination tachometer/dwell meter. Although they are included in the list of special tools, it is mentioned here because they are absolutely necessary for tuning most vehicles properly.*

Repair and overhaul tool set

These tools are essential for anyone who plans to perform major repairs and are in addition to those in the maintenance and minor repair tool kit. Included is a comprehensive set of sockets which, though expensive, are invaluable because of their versatility, especially when various extensions and drives are available. We recommend the 1/2-inch drive over the 3/8-inch drive. Although the larger drive is bulky and more expensive, it has the capacity of accepting a very wide range of large sockets. Ideally, however, the mechanic should have a 3/8-inch drive set and a 1/2-inch drive set.

Socket set(s)
Reversible ratchet

Extension - 10 inch
Universal joint
Torque wrench (same size drive as
* sockets)*
Ball peen hammer - 8 ounce
Soft-face hammer (plastic/rubber)
Standard screwdriver (1/4-inch x 6 inch)
Standard screwdriver (stubby -
* 5/16-inch)*
Phillips screwdriver (No. 3 x 8 inch)
Phillips screwdriver (stubby - No. 2)
Pliers - vise grip
Pliers - lineman's
Pliers - needle nose
Pliers - snap-ring (internal and external)
Cold chisel - 1/2-inch
Scribe
Scraper (made from flattened copper
* tubing)*
Centerpunch
Pin punches (1/16, 1/8, 3/16-inch)
Steel rule/straightedge - 12 inch
Allen wrench set (1/8 to 3/8-inch or
* 4 mm to 10 mm)*
A selection of files
Wire brush (large)
Jackstands (second set)
Jack (scissor or hydraulic type)

Note: *Another tool which is often useful is an electric drill with a chuck capacity of 3/8-inch and a set of good quality drill bits.*

Special tools

The tools in this list include those which are not used regularly, are expensive to buy, or which need to be used in accordance with their manufacturer's instructions. Unless these tools will be used frequently, it is not very economical to purchase many of them. A consideration would be to split the cost and use between yourself and a friend or friends. In addition, most of these tools can be obtained from a tool rental shop on a temporary basis.

This list primarily contains only those tools and instruments widely available to the public, and not those special tools produced by the vehicle manufacturer for distribution to dealer service departments. Occasionally, references to the manufacturer's special tools are included in the text of this manual. Generally, an alternative method of doing the job without the special tool is offered. However, sometimes there is no alternative to their use. Where this is the case, and the tool cannot be purchased or borrowed, the work should be turned over to the dealer service department or an automotive repair shop.

Valve spring compressor
Piston ring groove cleaning tool
Piston ring compressor
Piston ring installation tool
Cylinder compression gauge
Cylinder ridge reamer
Cylinder surfacing hone
Cylinder bore gauge
Micrometers and/or dial calipers
Hydraulic lifter removal tool
Balljoint separator
Universal-type puller
Impact screwdriver
Dial indicator set
Stroboscopic timing light (inductive pick-up)
Hand operated vacuum/pressure pump
Tachometer/dwell meter
Universal electrical multimeter
Cable hoist
Brake spring removal and installation tools
Floor jack

Buying tools

For the do-it-yourselfer who is just starting to get involved in vehicle maintenance and repair, there are a number of options available when purchasing tools. If maintenance and minor repair is the extent of the work to be done, the purchase of individual tools is satisfactory. If, on the other hand, extensive work is planned, it would be a good idea to purchase a modest tool set from one of the large retail chain stores. A set can usually be bought at a substantial savings over the individual tool prices, and they often come with a tool box. As additional tools are needed, add-on sets, individual tools and a larger tool box can be purchased to expand the tool selection. Building a tool set gradually allows the cost of the tools to be spread over a longer period of time and gives the mechanic the freedom to choose only those tools that will actually be used.

Tool stores will often be the only source of some of the special tools that are needed, but regardless of where tools are bought, try to avoid cheap ones, especially when buying screwdrivers and sockets, because they won't last very long. The expense involved in replacing cheap tools will eventually be greater than the initial cost of quality tools.

Care and maintenance of tools

Good tools are expensive, so it makes sense to treat them with respect. Keep them clean and in usable condition and store them properly when not in use. Always wipe off any dirt, grease or metal chips before putting them away. Never leave tools lying around in the work area. Upon completion of a job, always check closely under the hood for tools that may have been left there so they won't get lost during a test drive.

Some tools, such as screwdrivers, pliers, wrenches and sockets, can be hung on a panel mounted on the garage or workshop wall, while others should be kept in a tool box or tray. Measuring instruments, gauges, meters, etc. must be carefully stored where they cannot be damaged by weather or impact from other tools.

When tools are used with care and stored properly, they will last a very long time. Even with the best of care, though, tools will wear out if used frequently. When a tool is damaged or worn out, replace it. Subsequent jobs will be safer and more enjoyable if you do.

How to repair damaged threads

Sometimes, the internal threads of a nut or bolt hole can become stripped, usually from overtightening. Stripping threads is an all-too-common occurrence, especially when working with aluminum parts, because aluminum is so soft that it easily strips out.

Usually, external or internal threads are only partially stripped. After they've been cleaned up with a tap or die, they'll still work. Sometimes, however, threads are badly damaged. When this happens, you've got three choices:

1) *Drill and tap the hole to the next suitable oversize and install a larger diameter bolt, screw or stud.*

2) *Drill and tap the hole to accept a threaded plug, then drill and tap the plug to the original screw size. You can also buy a plug already threaded to the original size. Then you simply drill a hole to the specified size, then run the threaded plug into the hole with a bolt and jam nut. Once the plug is fully seated, remove the jam nut and bolt.*

3) *The third method uses a patented thread repair kit like Heli-Coil or Slimsert. These easy-to-use kits are designed to repair damaged threads in straight-through holes and blind holes. Both are available as kits which can handle a variety of sizes and thread patterns. Drill the hole, then tap it with the special included tap. Install the Heli-Coil and the hole is back to its original diameter and thread pitch.*

Regardless of which method you use, be sure to proceed calmly and carefully. A little impatience or carelessness during one of these relatively simple procedures can ruin your whole day's work and cost you a bundle if you wreck an expensive part.

Working facilities

Not to be overlooked when discussing tools is the workshop. If anything more than routine maintenance is to be carried out, some sort of suitable work area is essential.

It is understood, and appreciated, that many home mechanics do not have a good workshop or garage available, and end up removing an engine or doing major repairs outside. It is recommended, however, that the overhaul or repair be completed under the cover of a roof.

A clean, flat workbench or table of comfortable working height is an absolute necessity. The workbench should be equipped with a vise that has a jaw opening of at least four inches.

As mentioned previously, some clean, dry storage space is also required for tools, as well as the lubricants, fluids, cleaning solvents, etc. which soon become necessary.

Sometimes waste oil and fluids, drained from the engine or cooling system during normal maintenance or repairs, present a disposal problem. To avoid pouring them on the ground or into a sewage system, pour the used fluids into large containers, seal them with caps and take them to an authorized disposal site or recycling center. Plastic jugs, such as old antifreeze containers, are ideal for this purpose.

Always keep a supply of old newspapers and clean rags available. Old towels are excellent for mopping up spills. Many mechanics use rolls of paper towels for most work because they are readily available and disposable. To help keep the area under the vehicle clean, a large cardboard box can be cut open and flattened to protect the garage or shop floor.

Whenever working over a painted surface, such as when leaning over a fender to service something under the hood, always cover it with an old blanket or bedspread to protect the finish. Vinyl covered pads, made especially for this purpose, are available at auto parts stores.

Booster battery (jump) starting

1   Observe these precautions when using a booster battery to start a vehicle:

a) *Before connecting the booster battery, make sure the ignition switch is in the Off position.*

b) *Turn off the lights, heater and other electrical loads.*

c) *Your eyes should be shielded. Safety goggles are a good idea.*

d) *Make sure the booster battery is the same voltage as the dead one in the vehicle.*

e) *The two vehicles MUST NOT TOUCH each other!*

f) *Make sure the transmission is in Park (automatic).*

g) *If the booster battery is not a maintenance-free type, remove the vent caps and lay a cloth over the vent holes.*

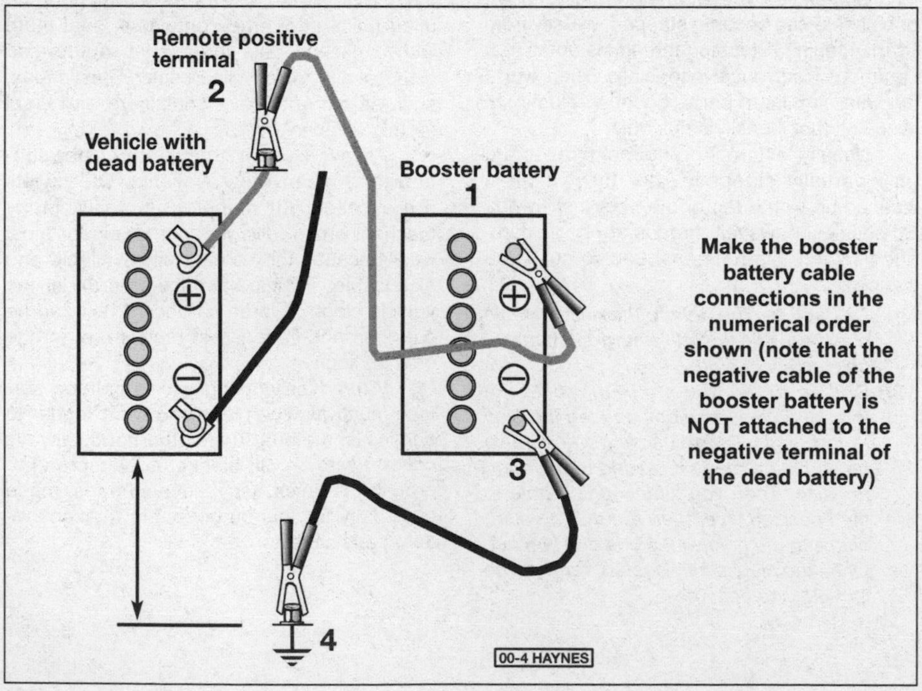

2   The main battery on these vehicles (some models have an optional second battery) is located in the right-rear corner of the engine compartment.

3   Connect the red jumper cable to the positive (+) terminals of each battery.

Note: *The positive terminal on the battery in these vehicles is located under a protective red cover which must be flipped up.*

4   Connect one end of the black cable to the negative (-) terminal of the booster battery. The other end of this cable should be connected to a good ground on the engine, such as a bracket, or a bolt on the body (see illustration). Make sure the cable will not come into contact with the fan, drivebelts or other moving parts of the engine.

5   Start the engine using the booster battery, then run the booster vehicle at a fast idle for a few minutes to instill some charge in the dead battery. Let the engine idle, then disconnect the jumper cables in the reverse order of connection. The vehicle with the dead battery may have to be driven for 20 minutes or more to sufficiently recharge the battery for independent starting.

Remote positive terminal

2

Vehicle with dead battery

Booster battery

1

Make the booster battery cable connections in the numerical order shown (note that the negative cable of the booster battery is NOT attached to the negative terminal of the dead battery)

3

4

00-4 HAYNES

Jacking and towing

Jacking

Warning: The *jack supplied with the vehicle should only be used for changing a tire or placing jackstands under the frame. Never work under the vehicle or start the engine while this jack is being used as the only means of support.*

1   The vehicle should be on level ground. Place the shift lever in Park, if you have an automatic, or Reverse if you have a manual transmission. Block the wheel diagonally opposite the wheel being changed. Set the parking brake.

2   Remove the spare tire and jack from stowage. Remove the wheel cover and trim ring (if so equipped) with the tapered end of the lug nut wrench by inserting and twisting the handle and then prying against the back of the wheel cover. Loosen the wheel lug nuts about 1/4-to-1/2 turn each.

3   Place the jack under the vehicle in the indicated position (see illustrations). Turn the jack handle clockwise until the tire clears the ground. Remove the lug nuts and pull the wheel off. Replace it with the spare.

4   Install the lug nuts with the beveled edges facing in. Tighten them snugly. Don't attempt to tighten them completely until the vehicle is lowered or it could slip off the jack. Turn the jack handle counterclockwise to lower the vehicle. Remove the jack and tighten the lug nuts in a diagonal pattern.

5   Install the cover and be sure it's snapped into place all the way around.

6   Stow the tire, jack and wrench. Unblock the wheels.

Towing

7   We recommend these vehicles (except four-wheel drive models) be towed from the rear, with the rear wheels off the ground. Vehicles with four-wheel drive must only be towed with all four wheels off the ground.

8   Equipment specifically designed for towing should be used. It must be attached to the main structural members of the vehicle, not the bumpers or brackets.

9   Safety is a major consideration when towing and all applicable state and local laws must be obeyed. A safety chain must be used at all times.

10   The parking brake must be released and the transmission must be in Neutral. The steering must be unlocked (ignition switch in the Off position). Remember that power steering and power brakes won't work with the engine off.

7.3a Left (driver's side) front jacking location - 4WD models (the jack head must engage the flat area to the left of the differential housing)

7.3b Right (passenger's side) front jacking location - 4WD models

7.3c Front jacking locations - 2WD models

7.3d Rear jacking location - all models

Automotive chemicals and lubricants

A number of automotive chemicals and lubricants are available for use during vehicle maintenance and repair. They include a wide variety of products ranging from cleaning solvents and degreasers to lubricants and protective sprays for rubber, plastic and vinyl.

Cleaners

Carburetor cleaner and choke cleaner is a strong solvent for gum, varnish and carbon. Most carburetor cleaners leave a dry-type lubricant film which will not harden or gum up. Because of this film it is not recommended for use on electrical components.

Brake system cleaner is used to remove brake dust, grease and brake fluid from the brake system, where clean surfaces are absolutely necessary. It leaves no residue and often eliminates brake squeal caused by contaminants.

Electrical cleaner removes oxidation, corrosion and carbon deposits from electrical contacts, restoring full current flow. It can also be used to clean spark plugs, carburetor jets, voltage regulators and other parts where an oil-free surface is desired.

Demoisturants remove water and moisture from electrical components such as alternators, voltage regulators, electrical connectors and fuse blocks. They are non-conductive and non-corrosive.

Degreasers are heavy-duty solvents used to remove grease from the outside of the engine and from chassis components. They can be sprayed or brushed on and, depending on the type, are rinsed off either with water or solvent.

Lubricants

Motor oil is the lubricant formulated for use in engines. It normally contains a wide variety of additives to prevent corrosion and reduce foaming and wear. Motor oil comes in various weights (viscosity ratings) from 0 to 50. The recommended weight of the oil depends on the season, temperature and the demands on the engine. Light oil is used in cold climates and under light load conditions. Heavy oil is used in hot climates and where high loads are encountered. Multi-viscosity oils are designed to have characteristics of both light and heavy oils and are available in a number of weights from 0W-20 to 20W-50.

Gear oil is designed to be used in differentials, manual transmissions and other areas where high-temperature lubrication is required.

Chassis and wheel bearing grease is a heavy grease used where increased loads and friction are encountered, such as for wheel bearings, balljoints, tie-rod ends and universal joints.

High-temperature wheel bearing grease is designed to withstand the extreme temperatures encountered by wheel bearings in disc brake equipped vehicles. It usually contains molybdenum disulfide (moly), which is a dry-type lubricant.

White grease is a heavy grease for metal-to-metal applications where water is a problem. White grease stays soft under both low and high temperatures (usually from -100 to +190-degrees F), and will not wash off or dilute in the presence of water.

Assembly lube is a special extreme pressure lubricant, usually containing moly, used to lubricate high-load parts (such as main and rod bearings and cam lobes) for initial start-up of a new engine. The assembly lube lubricates the parts without being squeezed out or washed away until the engine oiling system begins to function.

Silicone lubricants are used to protect rubber, plastic, vinyl and nylon parts.

Graphite lubricants are used where oils cannot be used due to contamination problems, such as in locks. The dry graphite will lubricate metal parts while remaining uncontaminated by dirt, water, oil or acids. It is electrically conductive and will not foul electrical contacts in locks such as the ignition switch.

Moly penetrants loosen and lubricate frozen, rusted and corroded fasteners and prevent future rusting or freezing.

Heat-sink grease is a special electrically non-conductive grease that is used for mounting electronic ignition modules where it is essential that heat is transferred away from the module.

Sealants

RTV sealant is one of the most widely used gasket compounds. Made from silicone, RTV is air curing, it seals, bonds, waterproofs, fills surface irregularities, remains flexible, doesn't shrink, is relatively easy to remove, and is used as a supplementary sealer with almost all low and medium temperature gaskets.

Anaerobic sealant is much like RTV in that it can be used either to seal gaskets or to form gaskets by itself. It remains flexible, is solvent resistant and fills surface imperfections. The difference between an anaerobic sealant and an RTV-type sealant is in the curing. RTV cures when exposed to air, while an anaerobic sealant cures only in the absence of air. This means that an anaerobic sealant cures only after the assembly of parts, sealing them together.

Thread and pipe sealant is used for sealing hydraulic and pneumatic fittings and vacuum lines. It is usually made from a Teflon compound, and comes in a spray, a paint-on liquid and as a wrap-around tape.

Chemicals

Anti-seize compound prevents seizing, galling, cold welding, rust and corrosion in fasteners. High-temperature ant-seize, usually made with copper and graphite lubricants, is used for exhaust system and exhaust manifold bolts.

Anaerobic locking compounds are used to keep fasteners from vibrating or working loose and cure only after installation, in the absence of air. Medium strength locking compound is used for small nuts, bolts and screws that may be removed later. High-strength locking compound is for large nuts, bolts and studs which aren't removed on a regular basis.

Oil additives range from viscosity index improvers to chemical treatments that claim to reduce internal engine friction. It should be noted that most oil manufacturers caution against using additives with their oils.

Gas additives perform several functions, depending on their chemical makeup. They usually contain solvents that help dissolve gum and varnish that build up on carburetor, fuel injection and intake parts. They also serve to break down carbon deposits that form on the inside surfaces of the combustion chambers. Some additives contain upper cylinder lubricants for valves and piston rings, and others contain chemicals to remove condensation from the gas tank.

Miscellaneous

Brake fluid is specially formulated hydraulic fluid that can withstand the heat and pressure encountered in brake systems. Care must be taken so this fluid does not come in contact with painted surfaces or plastics. An opened container should always be resealed to prevent contamination by water or dirt.

Weatherstrip adhesive is used to bond weatherstripping around doors, windows and trunk lids. It is sometimes used to attach trim pieces.

Undercoating is a petroleum-based, tar-like substance that is designed to protect metal surfaces on the underside of the vehicle from corrosion. It also acts as a sound-deadening agent by insulating the bottom of the vehicle.

Waxes and polishes are used to help protect painted and plated surfaces from the weather. Different types of paint may require the use of different types of wax and polish. Some polishes utilize a chemical or abrasive cleaner to help remove the top layer of oxidized (dull) paint on older vehicles. In recent years many non-wax polishes that contain a wide variety of chemicals such as polymers and silicones have been introduced. These non-wax polishes are usually easier to apply and last longer than conventional waxes and polishes.

Conversion factors

Length (distance)
Inches (in)	X	25.4	= Millimeters (mm)	X 0.0394	= Inches (in)
Feet (ft)	X	0.305	= Meters (m)	X 3.281	= Feet (ft)
Miles	X	1.609	= Kilometers (km)	X 0.621	= Miles

Volume (capacity)
Cubic inches (cu in; in^3)	X	16.387	= Cubic centimeters (cc; cm^3)	X 0.061	= Cubic inches (cu in; in^3)
Imperial pints (Imp pt)	X	0.568	= Liters (l)	X 1.76	= Imperial pints (Imp pt)
Imperial quarts (Imp qt)	X	1.137	= Liters (l)	X 0.88	= Imperial quarts (Imp qt)
Imperial quarts (Imp qt)	X	1.201	= US quarts (US qt)	X 0.833	= Imperial quarts (Imp qt)
US quarts (US qt)	X	0.946	= Liters (l)	X 1.057	= US quarts (US qt)
Imperial gallons (Imp gal)	X	4.546	= Liters (l)	X 0.22	= Imperial gallons (Imp gal)
Imperial gallons (Imp gal)	X	1.201	= US gallons (US gal)	X 0.833	= Imperial gallons (Imp gal)
US gallons (US gal)	X	3.785	= Liters (l)	X 0.264	= US gallons (US gal)

Mass (weight)
Ounces (oz)	X	28.35	= Grams (g)	X 0.035	= Ounces (oz)
Pounds (lb)	X	0.454	= Kilograms (kg)	X 2.205	= Pounds (lb)

Force
Ounces-force (ozf; oz)	X	0.278	= Newtons (N)	X 3.6	= Ounces-force (ozf; oz)
Pounds-force (lbf; lb)	X	4.448	= Newtons (N)	X 0.225	= Pounds-force (lbf; lb)
Newtons (N)	X	0.1	= Kilograms-force (kgf; kg)	X 9.81	= Newtons (N)

Pressure
Pounds-force per square inch (psi; lbf/in^2; lb/in^2)	X	0.070	= Kilograms-force per square centimeter (kgf/cm^2; kg/cm^2)	X 14.223	= Pounds-force per square inch (psi; lbf/in^2; lb/in^2)
Pounds-force per square inch (psi; lbf/in^2; lb/in^2)	X	0.068	= Atmospheres (atm)	X 14.696	= Pounds-force per square inch (psi; lbf/in^2; lb/in^2)
Pounds-force per square inch (psi; lbf/in^2; lb/in^2)	X	0.069	= Bars	X 14.5	= Pounds-force per square inch (psi; lbf/in^2; lb/in^2)
Pounds-force per square inch (psi; lbf/in^2; lb/in^2)	X	6.895	= Kilopascals (kPa)	X 0.145	= Pounds-force per square inch (psi; lbf/in^2; lb/in^2)
Kilopascals (kPa)	X	0.01	= Kilograms-force per square centimeter (kgf/cm^2; kg/cm^2)	X 98.1	= Kilopascals (kPa)

Torque (moment of force)
Pounds-force inches (lbf in; lb in)	X	1.152	= Kilograms-force centimeter (kgf cm; kg cm)	X 0.868	= Pounds-force inches (lbf in; lb in)
Pounds-force inches (lbf in; lb in)	X	0.113	= Newton meters (Nm)	X 8.85	= Pounds-force inches (lbf in; lb in)
Pounds-force inches (lbf in; lb in)	X	0.083	= Pounds-force feet (lbf ft; lb ft)	X 12	= Pounds-force inches (lbf in; lb in)
Pounds-force feet (lbf ft; lb ft)	X	0.138	= Kilograms-force meters (kgf m; kg m)	X 7.233	= Pounds-force feet (lbf ft; lb ft)
Pounds-force feet (lbf ft; lb ft)	X	1.356	= Newton meters (Nm)	X 0.738	= Pounds-force feet (lbf ft; lb ft)
Newton meters (Nm)	X	0.102	= Kilograms-force meters (kgf m; kg m)	X 9.804	= Newton meters (Nm)

Vacuum
Inches mercury (in. Hg)	X	3.377	= Kilopascals (kPa)	X 0.2961	= Inches mercury
Inches mercury (in. Hg)	X	25.4	= Millimeters mercury (mm Hg)	X 0.0394	= Inches mercury

Power
Horsepower (hp)	X	745.7	= Watts (W)	X 0.0013	= Horsepower (hp)

Velocity (speed)
Miles per hour (miles/hr; mph)	X	1.609	= Kilometers per hour (km/hr; kph)	X 0.621	= Miles per hour (miles/hr; mph)

Fuel consumption*
Miles per gallon, Imperial (mpg)	X	0.354	= Kilometers per liter (km/l)	X 2.825	= Miles per gallon, Imperial (mpg)
Miles per gallon, US (mpg)	X	0.425	= Kilometers per liter (km/l)	X 2.352	= Miles per gallon, US (mpg)

Temperature
Degrees Fahrenheit = (°C x 1.8) + 32

Degrees Celsius (Degrees Centigrade; °C) = (°F - 32) x 0.56

*It is common practice to convert from miles per gallon (mpg) to liters/100 kilometers (l/100km), where mpg (Imperial) x l/100 km = 282 and mpg (US) x l/100 km = 235

DECIMALS to MILLIMETERS

Decimal	mm	Decimal	mm
0.001	0.0254	0.500	12.7000
0.002	0.0508	0.510	12.9540
0.003	0.0762	0.520	13.2080
0.004	0.1016	0.530	13.4620
0.005	0.1270	0.540	13.7160
0.006	0.1524	0.550	13.9700
0.007	0.1778	0.560	14.2240
0.008	0.2032	0.570	14.4780
0.009	0.2286	0.580	14.7320
		0.590	14.9860
0.010	0.2540		
0.020	0.5080		
0.030	0.7620		
0.040	1.0160	0.600	15.2400
0.050	1.2700	0.610	15.4940
0.060	1.5240	0.620	15.7480
0.070	1.7780	0.630	16.0020
0.080	2.0320	0.640	16.2560
0.090	2.2860	0.650	16.5100
		0.660	16.7640
0.100	2.5400	0.670	17.0180
0.110	2.7940	0.680	17.2720
0.120	3.0480	0.690	17.5260
0.130	3.3020		
0.140	3.5560		
0.150	3.8100		
0.160	4.0640	0.700	17.7800
0.170	4.3180	0.710	18.0340
0.180	4.5720	0.720	18.2880
0.190	4.8260	0.730	18.5420
		0.740	18.7960
0.200	5.0800	0.750	19.0500
0.210	5.3340	0.760	19.3040
0.220	5.5880	0.770	19.5580
0.230	5.8420	0.780	19.8120
0.240	6.0960	0.790	20.0660
0.250	6.3500		
0.260	6.6040		
0.270	6.8580	0.800	20.3200
0.280	7.1120	0.810	20.5740
0.290	7.3660	0.820	20.8280
		0.830	21.0820
0.300	7.6200	0.840	21.3360
0.310	7.8740	0.850	21.5900
0.320	8.1280	0.860	21.8440
0.330	8.3820	0.870	22.0980
0.340	8.6360	0.880	22.3520
0.350	8.8900	0.890	22.6060
0.360	9.1440		
0.370	9.3980		
0.380	9.6520		
0.390	9.9060	0.900	22.8600
0.400	10.1600	0.910	23.1140
0.410	10.4140	0.920	23.3680
0.420	10.6680	0.930	23.6220
0.430	10.9220	0.940	23.8760
0.440	11.1760	0.950	24.1300
0.450	11.4300	0.960	24.3840
0.460	11.6840	0.970	24.6380
0.470	11.9380	0.980	24.8920
0.480	12.1920	0.990	25.1460
0.490	12.4460	1.000	25.4000

FRACTIONS to DECIMALS to MILLIMETERS

Fraction	Decimal	mm	Fraction	Decimal	mm
1/64	0.0156	0.3969	33/64	0.5156	13.0969
1/32	0.0312	0.7938	17/32	0.5312	13.4938
3/64	0.0469	1.1906	35/64	0.5469	13.8906
1/16	0.0625	1.5875	9/16	0.5625	14.2875
5/64	0.0781	1.9844	37/64	0.5781	14.6844
3/32	0.0938	2.3812	19/32	0.5938	15.0812
7/64	0.1094	2.7781	39/64	0.6094	15.4781
1/8	0.1250	3.1750	5/8	0.6250	15.8750
9/64	0.1406	3.5719	41/64	0.6406	16.2719
5/32	0.1562	3.9688	21/32	0.6562	16.6688
11/64	0.1719	4.3656	43/64	0.6719	17.0656
3/16	0.1875	4.7625	11/16	0.6875	17.4625
13/64	0.2031	5.1594	45/64	0.7031	17.8594
7/32	0.2188	5.5562	23/32	0.7188	18.2562
15/64	0.2344	5.9531	47/64	0.7344	18.6531
1/4	0.2500	6.3500	3/4	0.7500	19.0500
17/64	0.2656	6.7469	49/64	0.7656	19.4469
9/32	0.2812	7.1438	25/32	0.7812	19.8438
19/64	0.2969	7.5406	51/64	0.7969	20.2406
5/16	0.3125	7.9375	13/16	0.8125	20.6375
21/64	0.3281	8.3344	53/64	0.8281	21.0344
11/32	0.3438	8.7312	27/32	0.8438	21.4312
23/64	0.3594	9.1281	55/64	0.8594	21.8281
3/8	0.3750	9.5250	7/8	0.8750	22.2250
25/64	0.3906	9.9219	57/64	0.8906	22.6219
13/32	0.4062	10.3188	29/32	0.9062	23.0188
27/64	0.4219	10.7156	59/64	0.9219	23.4156
7/16	0.4375	11.1125	15/16	0.9375	23.8125
29/64	0.4531	11.5094	61/64	0.9531	24.2094
15/32	0.4688	11.9062	31/32	0.9688	24.6062
31/64	0.4844	12.3031	63/64	0.9844	25.0031
1/2	0.5000	12.7000	1	1.0000	25.4000

Safety first!

Regardless of how enthusiastic you may be about getting on with the job at hand, take the time to ensure that your safety is not jeopardized. A moment's lack of attention can result in an accident, as can failure to observe certain simple safety precautions. The possibility of an accident will always exist, and the following points should not be considered a comprehensive list of all dangers. Rather, they are intended to make you aware of the risks and to encourage a safety conscious approach to all work you carry out on your vehicle.

Essential DOs and DON'Ts

DON'T rely on a jack when working under the vehicle. Always use approved jackstands to support the weight of the vehicle and place them under the recommended lift or support points.

DON'T attempt to loosen extremely tight fasteners (i.e. wheel lug nuts) while the vehicle is on a jack - it may fall.

DON'T start the engine without first making sure that the transmission is in Neutral (or Park where applicable) and the parking brake is set.

DON'T remove the radiator cap from a hot cooling system - let it cool or cover it with a cloth and release the pressure gradually.

DON'T attempt to drain the engine oil until you are sure it has cooled to the point that it will not burn you.

DON'T touch any part of the engine or exhaust system until it has cooled sufficiently to avoid burns.

DON'T siphon toxic liquids such as gasoline, antifreeze and brake fluid by mouth, or allow them to remain on your skin.

DON'T inhale brake lining dust - it is potentially hazardous (see *Asbestos* below).

DON'T allow spilled oil or grease to remain on the floor - wipe it up before someone slips on it.

DON'T use loose fitting wrenches or other tools which may slip and cause injury.

DON'T push on wrenches when loosening or tightening nuts or bolts. Always try to pull the wrench toward you. If the situation calls for pushing the wrench away, push with an open hand to avoid scraped knuckles if the wrench should slip.

DON'T attempt to lift a heavy component alone - get someone to help you.

DON'T rush or take unsafe shortcuts to finish a job.

DON'T allow children or animals in or around the vehicle while you are working on it.

DO wear eye protection when using power tools such as a drill, sander, bench grinder, etc. and when working under a vehicle.

DO keep loose clothing and long hair well out of the way of moving parts.

DO make sure that any hoist used has a safe working load rating adequate for the job.

DO get someone to check on you periodically when working alone on a vehicle.

DO carry out work in a logical sequence and make sure that everything is correctly assembled and tightened.

DO keep chemicals and fluids tightly capped and out of the reach of children and pets.

DO remember that your vehicle's safety affects that of yourself and others. If in doubt on any point, get professional advice.

Steering, suspension and brakes

These systems are essential to driving safety, so make sure you have a qualified shop or individual check your work. Also, compressed suspension springs can cause injury if released suddenly - be sure to use a spring compressor.

Airbags

Airbags are explosive devices that can **CAUSE** injury if they deploy while you're working on the vehicle. Follow the manufacturer's instructions to disable the airbag whenever you're working in the vicinity of airbag components.

Asbestos

Certain friction, insulating, sealing, and other products - such as brake linings, brake bands, clutch linings, torque converters, gaskets, etc. - may contain asbestos or other hazardous friction material. Extreme care must be taken to avoid inhalation of dust from such products, since it is hazardous to health. If in doubt, assume that they do contain asbestos.

Fire

Remember at all times that gasoline is highly flammable. Never smoke or have any kind of open flame around when working on a vehicle. But the risk does not end there. A spark caused by an electrical short circuit, by two metal surfaces contacting each other, or even by static electricity built up in your body under certain conditions, can ignite gasoline vapors, which in a confined space are highly explosive. Do not, under any circumstances, use gasoline for cleaning parts. Use an approved safety solvent.

Always disconnect the battery ground (-) cable at the battery before working on any part of the fuel system or electrical system. Never risk spilling fuel on a hot engine or exhaust component. It is strongly recommended that a fire extinguisher suitable for use on fuel and electrical fires be kept handy in the garage or workshop at all times. Never try to extinguish a fuel or electrical fire with water.

Fumes

Certain fumes are highly toxic and can quickly cause unconsciousness and even death if inhaled to any extent. Gasoline vapor falls into this category, as do the vapors from some cleaning solvents. Any draining or pouring of such volatile fluids should be done in a well ventilated area.

When using cleaning fluids and solvents, read the instructions on the container carefully. Never use materials from unmarked containers.

Never run the engine in an enclosed space, such as a garage. Exhaust fumes contain carbon monoxide, which is extremely poisonous. If you need to run the engine, always do so in the open air, or at least have the rear of the vehicle outside the work area.

The battery

Never create a spark or allow a bare light bulb near a battery. They normally give off a certain amount of hydrogen gas, which is highly explosive.

Always disconnect the battery ground (-) cable at the battery before working on the fuel or electrical systems.

If possible, loosen the filler caps or cover when charging the battery from an external source (this does not apply to sealed or maintenance-free batteries). Do not charge at an excessive rate or the battery may burst.

Take care when adding water to a non maintenance-free battery and when carrying a battery. The electrolyte, even when diluted, is very corrosive and should not be allowed to contact clothing or skin.

Always wear eye protection when cleaning the battery to prevent the caustic deposits from entering your eyes.

Household current

When using an electric power tool, inspection light, etc., which operates on household current, always make sure that the tool is correctly connected to its plug and that, where necessary, it is properly grounded. Do not use such items in damp conditions and, again, do not create a spark or apply excessive heat in the vicinity of fuel or fuel vapor.

Secondary ignition system voltage

A severe electric shock can result from touching certain parts of the ignition system (such as the spark plug wires) when the engine is running or being cranked, particularly if components are damp or the insulation is defective. In the case of an electronic ignition system, the secondary system voltage is much higher and could prove fatal.

Hydrofluoric acid

This extremely corrosive acid is formed when certain types of synthetic rubber, found in some O-rings, oil seals, fuel hoses, etc. are exposed to temperatures above 750-degrees F (400-degrees C). The rubber changes into a charred or sticky substance containing the acid. *Once formed, the acid remains dangerous for years. If it gets onto the skin, it may be necessary to amputate the limb concerned.*

When dealing with a vehicle which has suffered a fire, or with components salvaged from such a vehicle, wear protective gloves and discard them after use.

Troubleshooting

Contents

This section provides an easy reference guide to the more common problems which may occur during the operation of your vehicle. These problems and their possible causes are grouped under headings denoting various components or systems, such as Engine, Cooling system, etc. They also refer you to the chapter and/or section which deals with the problem.

Remember that successful troubleshooting is not a mysterious black art practiced only by professional mechanics. It is simply the result of the right knowledge combined with an intelligent, systematic approach to the problem. Always work by a process of elimination, starting with the simplest solution and working through to the most complex - and never overlook the obvious. Anyone can run the gas tank dry or leave the lights on overnight, so don't assume that you are exempt from such oversights.

Finally, always establish a clear idea of why a problem has occurred and take steps to ensure that it doesn't happen again. If the electrical system fails because of a poor connection, check the other connections in the system to make sure that they don't fail as well. If a particular fuse continues to blow, find out why - don't just replace one fuse after another. Remember, failure of a small component can often be indicative of potential failure or incorrect functioning of a more important component or system.

Engine

1 Engine will not rotate when attempting to start

1 Battery terminal connections loose or corroded. Check the cable terminals at the battery. Tighten the cable or remove corrosion as necessary.
2 Battery discharged or faulty. If the cable connections are clean and tight on the battery posts, turn the key to the On position and switch on the headlights and/or windshield wipers. If they fail to function, the battery is discharged.
3 Automatic transmission not completely engaged in Park.
4 Broken, loose or disconnected wiring in the starting circuit. Inspect all wiring and connectors at the battery, starter solenoid and ignition switch.
5 Starter motor pinion jammed in driveplate ring gear. Remove starter and inspect pinion and driveplate (Chapter 5).
6 Starter solenoid faulty (Chapter 5).
7 Starter motor faulty (Chapter 5).
8 Ignition switch faulty (Chapter 12).
9 Starter relay faulty.
10 Body Control Module (BCM) or Powertrain Control Module (PCM) faulty.

2 Engine rotates but will not start

1 Fuel tank empty, fuel filter plugged or fuel line restricted.
2 Fault in the fuel injection system (Chapter 4).
3 Battery discharged (engine rotates slowly). Check the operation of electrical components as described in the previous Section.
4 Battery terminal connections loose or corroded (see previous Section).
5 Fuel pump faulty (Chapter 4).
6 Ignition system faulty (see Chapter 5).
7 Worn, faulty or incorrectly gapped spark plugs (Chapter 1).
8 Broken, loose or disconnected wires at the ignition coils (Chapter 5).

3 Starter motor operates without rotating engine

1 Starter pinion sticking. Remove the starter (Chapter 5) and inspect.
2 Starter pinion or driveplate teeth worn or broken. Remove the driveplate access cover and inspect.

4 Engine hard to start when cold

1 Discharged or low battery. Check as described in Section 2.
2 Fault in the fuel or ignition systems (Chapters 4 and 5).
3 Injector(s) leaking (Chapter 4).

5 Engine hard to start when hot

1 Air filter clogged (Chapter 1).
2 Fault in the fuel or ignition systems (Chapters 4 and 5).
3 Fuel not reaching the injectors (see Chapter 4).
4 Low cylinder compression (Chapter 2A).

6 Starter motor noisy or excessively rough in engagement

1 Pinion or flywheel gear teeth worn or broken. Remove the cover at the rear of the engine (if equipped) and inspect.
2 Starter motor mounting bolts loose or missing.

7 Engine starts but stops immediately

1 Fault in the fuel or ignition systems (Chapters 4 and 5).
2 Vacuum leak at the gasket surfaces of the intake manifold or throttle body. Make sure all mounting bolts/nuts are tightened securely and all vacuum hoses connected to the manifold are positioned properly and in good condition.

3 Restricted intake or exhaust systems (Chapter 4).

8 Engine lopes while idling or idles erratically

1 Vacuum leakage. Check the mounting bolts/nuts at the throttle body and intake manifold for tightness. Make sure all vacuum hoses are connected and in good condition. Use a stethoscope or a length of fuel hose held against your ear to listen for vacuum leaks while the engine is running. A hissing sound will be heard. A soapy water solution will also detect leaks.
2 Fault in the fuel or ignition systems (Chapters 4 and 5).
3 Plugged PCV valve or hose (see Chapter 1).
4 Air filter clogged (Chapter 1).
5 Fuel pump not delivering sufficient fuel to the fuel injectors (see Chapter 4).
6 Leaking head gasket. Perform a compression check (Chapter 2C).
7 Camshaft lobes worn (Chapter 2A).
8 Faulty valve lifter (Chapter 2A).

9 Engine misses at idle speed

1 Spark plugs worn, fouled or not gapped properly (Chapter 1).
2 Fault in the fuel or ignition systems (Chapters 4 and 5).
3 Vacuum leaks at intake manifold or hose connections. Check as described in Section 9.
4 Uneven or low cylinder compression. Check compression as described in Chapter 2C.

10 Engine misses throughout driving speed range

1 Fuel filter clogged and/or impurities in the fuel system (Chapter 1).
2 Faulty or incorrectly gapped spark plugs (Chapter 1).
3 Fault in the fuel or ignition systems (Chapters 4 and 5).

4   Faulty emissions system components (Chapter 6).

5   Low or uneven cylinder compression pressures. Remove the spark plugs and test the compression with a gauge (Chapter 2C).

6   Vacuum leaks at the throttle body, intake manifold or vacuum hoses (see Section 9).

11   Engine stalls

1   Fuel filter clogged and/or water and impurities in the fuel system (Chapter 1).

2   Fault in the fuel system or sensors (Chapters 4 and 6).

3   Faulty emissions system components (Chapter 6).

4   Faulty or incorrectly gapped spark plugs (Chapter 1).

5   Vacuum leak at the throttle body, intake manifold or vacuum hoses. Check as described in Section 9.

12   Engine lacks power

1   Fault in the fuel or ignition systems (Chapters 4 and 5).

2   Faulty or incorrectly gapped spark plugs (Chapter 1).

3   Faulty coils (Chapter 5).

4   Brakes binding (Chapter 1).

5   Automatic transmission fluid level incorrect (Chapter 1).

6   Fuel filter clogged and/or impurities in the fuel system (Chapter 1).

7   Emissions control system not functioning properly (Chapter 6).

8   Use of substandard fuel. Fill the tank with the proper fuel.

9   Low or uneven cylinder compression pressures. Test with a compression tester, which will detect leaking valves and/or a blown head gasket (Chapter 2A).

10   Restriction in the intake or exhaust system (Chapter 4).

13   Engine backfires

1   Emissions system not functioning properly (Chapter 6).

2   Fault in the fuel or ignition systems (Chapters 4 and 5).

3   Vacuum leak at the throttle body, intake manifold or vacuum hoses. Check as described in Section 9.

4   Valves sticking (Chapter 2A).

14   Pinging or knocking engine sounds during acceleration or uphill

1   Incorrect grade of fuel. Fill the tank with fuel of the proper octane rating.

2   Fault in the fuel or ignition systems (Chapters 4 and 5).

3   Improper spark plugs. Check the plug type against the VECI label located in the engine compartment. Also check the plugs for damage (Chapter 1).

4   Faulty emissions system (Chapter 6).

5   Vacuum leak. Check as described in Section 10.

15   Engine continues to run after switching off

Faulty ignition switch (Chapter 12).

Engine electrical system

16   Battery will not hold a charge

1   Drivebelt or tensioner defective (Chapter 1).

2   Electrolyte level low or battery discharged (Chapter 1).

3   Battery terminals loose or corroded (Chapter 1).

4   Alternator not charging properly (Chapter 5).

5   Loose, broken or faulty wiring in the charging circuit (Chapter 5).

6   Battery defective internally.

17   Alternator light fails to go out

1   Fault in the alternator or charging circuit (Chapter 5).

2   Drivebelt or tensioner defective (Chapter 1).

18   Alternator light fails to come on when key is turned on

1   Instrument cluster warning light bulb defective (Chapter 12).

2   Alternator faulty (Chapter 5).

3   Fault in the instrument cluster printed circuit, dashboard wiring or bulb holder (Chapter 12).

Fuel system

19   Excessive fuel consumption

1   Dirty or clogged air filter element (Chapter 1).

2   Emissions system not functioning properly (Chapter 6).

3   Fault in the fuel or ignition systems (Chapters 4 and 5).

4   Low tire pressure or incorrect tire size (Chapter 1).

5   Restricted exhaust system (Chapter 4).

6   Brakes binding (Chapter 9).

20   Fuel leakage and/or fuel odor

1   Leak in a fuel feed or vent line (Chapter 4).

2   Tank overfilled. Fill only to automatic shut-off.

3   Evaporative emissions system canister clogged (Chapter 6).

4   Vapor leaks from system lines or injectors (Chapter 4).

Cooling system

21   Overheating

1   Insufficient coolant in the system (Chapter 1).

2   Drivebelt or tensioner defective (Chapter 1).

3   Radiator core blocked or radiator grille dirty and restricted (see Chapter 3).

4   Thermostat faulty (Chapter 3).

5   Fan blades broken or cracked (Chapter 3).

6   Expansion tank cap not maintaining proper pressure. Have the cap pressure tested by a gas station or repair shop.

7   Fault in electrical circuit for cooling fans (Chapter 3).

22   Overcooling

1   Thermostat faulty (Chapter 3).

2   Inaccurate temperature gauge (Chapter 12).

3   Fault in electrical circuit for cooling fans (Chapter 3).

23   External coolant leakage

1   Deteriorated or damaged hoses or loose clamps. Replace hoses and/or tighten the clamps at the hose connections (Chapter 1).

2   Water pump seals defective. If this is the case, water will drip from the weep hole in the water pump body (Chapter 3).

3   Leakage from the radiator core or side tank(s). This will require the radiator to be professionally repaired (see Chapter 3 for removal procedures).

4   Engine drain plug(s) leaking (Chapter 1) or water jacket core plugs leaking.

5   Leakage at the heater core. Signs of leakage should show up on interior carpeting (Chapter 3).

24 Internal coolant leakage

Note: *Internal coolant leaks can usually be detected by examining the oil. Check the dipstick and inside of the valve cover for water deposits and an oil consistency like that of a milkshake.*

1 Leaking cylinder head gasket. Have the cooling system pressure tested.

2 Cracked cylinder bore or cylinder head. Remove the head(s) and inspect (Chapter 2A).

3 Leaking intake manifold gasket (Chapter 2A).

25 Coolant loss

1 Too much coolant in the system (Chapter 1).

2 Coolant boiling away due to overheating (Chapter 3).

3 External or internal leakage (see Sections 23 and 24 above).

4 Faulty expansion tank cap. Have the cap pressure tested.

26 Poor coolant circulation

1 Inoperative water pump. A quick test is to pinch the top radiator hose closed with your hand while the engine is idling, then let it loose. You should feel the surge of coolant if the pump is working properly (see Chapter 1).

2 Restriction in the cooling system. Drain, flush and refill the system (Chapter 1). If necessary, remove the radiator (Chapter 3) and have it reverse flushed.

3 Drivebelt or tensioner defective (Chapter 1).

4 Thermostat sticking (Chapter 3).

5 Drivebelt incorrectly routed, causing the pump to turn backward (Chapter 1).

Automatic transmission

27 General shift mechanism problems

1 Chapter 7A deals with checking and adjusting the shift cable on automatic transmissions. Common problems that may be attributed to poorly adjusted cable are:

a) Engine starting in gears other than Park or Neutral.

b) Indicator on shifter pointing to a gear other than the one actually being selected.

c) Vehicle moves when in Park.

2 Refer to Chapter 7A to adjust the linkage.

3 Problem with the electronic shift solenoid. Check for Diagnostic Trouble Codes (Chapter 6).

28 Transmission will not downshift with accelerator pedal pressed to the floor

Transmission pressure control solenoid valve faulty. Check for Diagnostic Trouble Codes (Chapter 6).

29 Transmission slips, shifts rough, is noisy or has no drive in forward or reverse gears

1 Of the many probable causes for the above problems, the home mechanic should be concerned with only one possibility - fluid level.

2 Before taking the vehicle to a repair shop, check the level and condition of the fluid as described in Chapter 1. Correct fluid level as necessary or change the fluid and filter if needed. If the problem persists, have a professional diagnose the probable cause.

3 If the transmission shifts late and the shifts are harsh, suspect a faulty transmission pressure control solenoid valve. Check for Diagnostic Trouble Codes (Chapter 6).

30 Fluid leakage

1 Automatic transmission fluid is a deep red color. Fluid leaks should not be confused with engine oil, which can easily be blown by airflow to the transmission.

2 To pinpoint a leak, first remove all built-up dirt and grime from around the transmission. Degreasing agents and/or steam cleaning will achieve this. With the underside clean, drive the vehicle at low speeds so airflow will not blow the leak far from its source. Raise the vehicle and determine where the leak is coming from. Common areas of leakage are:

a) Pan: Tighten the mounting bolts and/or replace the pan gasket as necessary (see Chapter 1).

b) Filler pipe: Replace the rubber seal where the pipe enters the transmission case.

c) Transmission oil lines: Tighten the connectors where the lines enter the transmission case and/or replace the lines.

d) Vent pipe: Transmission overfilled and/or water in fluid (Chapter 1).

e) Speed sensor connector: Replace the O-ring where the vehicle speed sensor enters the transmission case (Chapter 6).

Transfer case

31 Transfer case is difficult to shift into the desired range

1 Speed may be too great to permit engagement. Stop the vehicle and shift into the desired range.

2 Shift linkage loose, bent or binding. Check the linkage for damage or wear and replace or lubricate as necessary (Chapter 7B).

3 If the vehicle has been driven on a paved surface for some time, the driveline torque can make shifting difficult. Stop and shift into two-wheel drive on paved or hard surfaces.

4 Insufficient or incorrect grade of lubricant. Drain and refill the transfer case with the specified lubricant (Chapter 1).

5 Worn or damaged internal components. Disassembly and overhaul of the transfer case, by a qualified shop, may be necessary.

6 Fault in the electrical system of the front axle or automatic transfer case. Check for Diagnostic Trouble Codes (Chapter 6).

32 Transfer case noisy in all gears

Insufficient or incorrect grade of lubricant. Drain and refill (Chapter 1).

33 Noisy or jumps out of four-wheel drive Low range

1 Transfer case not fully engaged. Stop the vehicle, shift into Neutral and then engage 4L.

2 Shift linkage loose, worn or binding. Tighten, repair or lubricate linkage as necessary.

3 Shift fork cracked, inserts worn or fork binding on the rail. Disassemble and repair as necessary (Chapter 7B).

4 Fault in the electrical system of the front axle or automatic transfer case. Check for Diagnostic Trouble Codes (Chapter 6).

34 Lubricant leaks from the vent or output shaft seals

1 Transfer case is overfilled. Drain to the proper level (Chapter 1).

2 Vent is clogged or jammed closed. Clear or replace the vent.

3 Output shaft seal incorrectly installed or damaged. Replace the seal and check contact surfaces for nicks and scoring.

Driveshaft

35 Oil leak at seal end of driveshaft

Defective transmission or transfer case oil seal. See Chapter 7A for replacement procedures. While this is done, check the splined yoke for burrs or a rough condition that may be damaging the seal. Burrs can be removed with crocus cloth or a fine whetstone.

36 Knock or clunk when the transmission is under initial load (just after transmission is put into gear)

1 Loose or disconnected rear suspension components. Check all mounting bolts, nuts and bushings (see Chapter 10).
2 Loose driveshaft bolts. Inspect all bolts and nuts and tighten them to the specified torque.
3 Worn or damaged universal joint bearings. Check for wear (see Chapter 8).

37 Metallic grinding sound consistent with vehicle speed

Pronounced wear in the universal joint bearings. Check as described in Chapter 8.

38 Vibration

Note: *Before assuming that the driveshaft is at fault, make sure the tires are perfectly balanced and perform the following test.*
1 Install a tachometer inside the vehicle to monitor engine speed as the vehicle is driven. Drive the vehicle and note the engine speed at which the vibration (roughness) is most pronounced. Now shift the transmission to a different gear and bring the engine speed to the same point.
2 If the vibration occurs at the same engine speed (rpm) regardless of which gear the transmission is in, the driveshaft is NOT at fault since the driveshaft speed varies.
3 If the vibration decreases or is eliminated when the transmission is in a different gear at the same engine speed, refer to the following probable causes.
4 Bent or dented driveshaft. Inspect and replace as necessary (see Chapter 8).
5 Undercoating or built-up dirt, etc., on the driveshaft. Clean the shaft thoroughly and recheck.
6 Worn universal joint bearings (see Chapter 8).
7 Driveshaft and/or companion flange out of balance. Check for missing weights on the shaft. Remove the driveshaft (see Chapter 8) and reinstall 180-degrees from original position, then retest. Have the driveshaft professionally balanced if the problem persists.

Axles

39 Noise

1 Road noise. No corrective procedures available.
2 Tire noise. Inspect tires and check tire pressures (Chapter 1).
3 Rear wheel bearings worn or damaged (Chapter 8).

40 Vibration

See probable causes under Driveshaft. Proceed under the guidelines listed for the driveshaft. If the problem persists, check the rear wheel bearings by raising the rear of the vehicle and spinning the rear wheels by hand. Listen for evidence of rough (noisy) bearings. Remove and inspect (see Chapter 8).

41 Oil leakage

1 Pinion seal damaged (see Chapter 8).
2 Axleshaft oil seals damaged (see Chapter 8).
3 Differential inspection cover leaking. Tighten the bolts or replace the gasket as required (see Chapter 8).

Driveaxles (4WD)

42 Clicking noise on turns

Worn or damaged outboard CV joints (Chapter 8).

43 Shudder or vibration during acceleration

1 Excessive toe-in. Have alignment checked.
2 Incorrect spring heights (Chapter 10).
3 Worn or damaged inboard or outboard CV joints (Chapter 8).
4 Sticking inboard CV joint assembly (Chapter 8).

44 Vibration at highway speeds

1 Out-of-balance front wheels and/or tires (Chapters 1 and 10).
2 Out-of-round front tires (Chapters 1 and 10).
3 Worn CV joints (Chapter 8).

Brakes

45 Vehicle pulls to one side during braking

1 Defective, damaged or oil contaminated disc brake pads on one side. Inspect as described in Chapter 9.
2 Excessive wear of brake pad material or disc on one side. Inspect and correct as necessary.
3 Loose or disconnected front suspension components. Inspect and tighten all bolts to the specified torque (Chapter 10).
4 Defective brake caliper assembly. Remove the caliper and inspect for a stuck piston or other damage (Chapter 9).
5 Inadequate lubrication of front brake caliper slide pins. Remove caliper and lubricate slide pins (Chapter 9).

46 Noise (grinding or high-pitched squeal with the brakes applied)

1 Brake pads worn out. Replace the pads with new ones immediately (Chapter 9).
2 Linings contaminated with dirt or grease. Replace pads.
3 Scored disc or drum (Chapter 9).

47 Excessive brake pedal travel

1 Partial brake system failure. Inspect the entire system (Chapter 9) and correct as required.
2 Insufficient fluid in the master cylinder. Check (Chapter 1), add fluid and bleed the system if necessary (Chapter 9).

48 Brake pedal feels spongy when depressed

1 Air in the hydraulic lines. Bleed the brake system (Chapter 9).
2 Faulty flexible hoses. Inspect all system hoses and lines. Replace parts as necessary.
3 Master cylinder mounting bolts/nuts loose.
4 Master cylinder defective (Chapter 9).

49 Excessive effort required to stop vehicle

1 Power brake booster not operating properly (see check in Chapter 1, repairs in Chapter 9).
2 Excessively worn pads or shoes. Inspect and replace if necessary (Chapter 9).
3 One or more caliper or wheel cylinder pistons seized or sticking. Inspect and replace as required (Chapter 9).

4   Brake pads contaminated with oil or grease. Inspect and replace as required (Chapter 9).
5   New pads installed and not yet seated. It will take a while for the new material to seat against the disc.

50   Pedal travels to the floor with little resistance

1   Little or no fluid in the master cylinder reservoir caused by leaking caliper piston(s), loose, damaged or disconnected brake lines. Inspect the entire system and correct as necessary.
2   Faulty master cylinder (Chapter 9).

51   Brake pedal pulsates during brake application

Disc(s) warped. Check for excessive lateral runout and parallelism. Have the discs resurfaced or replace them with new ones (Chapter 9).

Suspension and steering systems

52   Vehicle pulls to one side

1   Tire pressures uneven or tires mismatched (Chapter 1).
2   Defective tire (Chapter 1).
3   Excessive wear in suspension or steering components (Chapter 10).
4   Front end in need of alignment.
5   Front brakes dragging. Inspect the brakes as described in Chapter 9.

53   Shimmy, shake or vibration

1   Tire or wheel out-of-balance or out-of-round. Have professionally balanced.

2   Loose, worn or out-of-adjustment front wheel bearings (Chapter 1).
3   Shock absorbers and/or suspension components worn or damaged (Chapter 10).

54   Excessive pitching and/or rolling around corners or during braking

1   Defective shock absorbers. Replace as a set (Chapter 10).
2   Broken or weak springs and/or suspension components. Inspect as described in Chapters 1 and 10.

55   Excessively stiff steering

1   Lack of fluid in power steering fluid reservoir (Chapter 1).
2   Incorrect tire pressures (Chapter 1).
3   Lack of lubrication at steering joints (see Chapter 1).
4   Front end out of alignment.
5   Lack of power assistance (Chapter 10).

56   Excessive play in steering

1   Worn front wheel bearings (Chapter 10).
2   Excessive wear in suspension or steering components (Chapter 10).
3   Steering gear worn (Chapter 10).

57   Lack of power assistance

1   Drivebelt or tensioner faulty (Chapter 1).
2   Fluid level low (Chapter 1).
3   Hoses or lines restricted. Inspect and replace parts as necessary.
4   Air in power steering system. Bleed the system (Chapter 10).

58   Excessive tire wear (not specific to one area)

1   Incorrect tire pressures (Chapter 1).
2   Tires out-of-balance. Have professionally balanced.
3   Wheels damaged. Inspect and replace as necessary.
4   Suspension or steering components excessively worn (Chapter 10).

59   Excessive tire wear on outside edge

1   Inflation pressures incorrect (Chapter 1).
2   Excessive speed in turns.
3   Front-end alignment incorrect. Have the front end professionally aligned.
4   Suspension arm bent (Chapter 10).

60   Excessive tire wear on inside edge

1   Inflation pressures incorrect (Chapter 1).
2   Front-end alignment incorrect. Have the front end professionally aligned.
3   Loose or damaged steering components (Chapter 10).

61   Tire tread worn in one place

1   Tires out-of-balance.
2   Damaged or buckled wheel. Inspect and replace if necessary.
3   Defective tire (Chapter 1).

Notes

Chapter 1
Tune-up and routine maintenance

Contents

Specifications

Note: *Listed here are manufacturer recommendations at the time this manual was written. Manufacturers occasionally upgrade their fluid and lubricant specifications, so check with your local auto parts store for current recommendations.*

Recommended lubricants and fluids

Engine oil
  Type
      Gasoline engine .... API "certified for gasoline engines"
      Diesel engine .... API grade CJ-4 or CH-4/SJ/SH
  Viscosity
    Gasoline engine
      2015 and earlier models .... 5W-20 Synthetic blend (USA), 5W-20 premium oil (Canada)
      2016 models .... 5W-30 Synthetic blend (USA), 5W-30 premium oil (Canada)
    Diesel engine
      Above 30-degrees F .... 15W-40 Super duty diesel motor oil
      -10-degrees to 30-degrees F .... 10W-30 Super duty diesel motor oil
      Consistently below -10-degrees F .... 0W-30 Super all season motor oil (XO-0W30-LAS) or
                                        5W-40 Full synthetic diesel motor oil
Power steering fluid .... MERCON V automatic transmission fluid
Brake fluid .... DOT 3 heavy duty brake fluid
Automatic transmission fluid
  2015 and earlier models .... MERCON V automatic transmission fluid.
  2016 models .... MERCON LV automatic transmission fluid.

Caution: *Do not use MERCON V or dual-usage MERCON/MERCON V automatic transmission fluid. Do not mix types.*

Recommended lubricants and fluids (continued)

Transfer case lubricant	Motorcraft transfer case luid ESP-M2C166-H or equivalent
Coolant	Motorcraft Premium Orange Antifreeze/coolant VC-3-B (USA); CVC-3-B2 (Canada) or equivalent.
Front wheel bearing grease..	NLGI No. 2 lithium base grease containing polyethylene and molybdenum disulfide
Chassis grease..	NLGI No. 2 lithium base grease containing polyethylene and molybdenum disulfide
Differential lubricant	
Front axle	Motorcraft SAE 80W-90 Premium rear axle lubricant (USA); Motorcraft SAE 80W-90
Premium axle lubricant (Canada)	
Rear axle	
Ford 10.50 inch axles	SAE 75W-140 QL synthetic gear lubricant
Dana 80 axles	SAE 75W-90 QL synthetic gear lubricant
Dana S110 and S130 axles	SAE 75W-140 QL synthetic gear lubricant*
Dana full floating axle	Motorcraft Premium long-life grease (XG-1-E1), or equivalent
Ford full floating axle	Motorcraft Premium long-life grease (XG-1-E1), or equivalent

Note: *The lubricant capacity will vary depending upon the angle at which the axle housing is mounted. The lubricant level must be level with the fill plug hole opening (slightly coming out of the hole).*

Capacities*

Engine oil (with filter change)	
Gasoline engine	7.0 quarts
Diesel engine	13 quarts
Automatic transmission (capacity with torque converter drained)	
TorqShift..	17.5 quarts
TorqShift6 (6R140)	
2011 early build	
Diesel engine..	18.1 quarts
Gas engine	19.1 quarts
2011 late build and all later models	
Diesel engine..	16.3 quarts
Gasoline engine	17.3 quarts
Transfer case	2.0 quarts
Differential	
Front	5.8 pints
Rear	
10.5 Inch axle	6.9 pints
DANA M80 axle	8.5 pints

Note: *The best way to determine the amount of fluid to add during a routine fluid change is to measure the amount drained. It is important not to overfill the transmission. After draining the transmission, begin the refilling procedure by initially adding 6-1/2 quarts, then adding 1/2-pint at a time until the level is correct on the dipstick.*

Cooling system

Gasoline engine	25.4 quarts
Diesel engine	
2015 and earlier models	29.4 quarts
2016 models	30.3 quarts

Note: *All capacities approximate. Add as necessary to bring to the appropriate level.*

Brakes

Disc brake pad thickness (minimum)..	1/8-inch
Parking brake shoe lining thickness (minimum)	1/32-inch above rivet heads

Ignition system

Spark plug
  Type .... Motorcraft SP-526 (12405) or equivalent
  Gap .... 0.042 to 0.046 inch
Ignition timing .... 10-degrees BTDC (base timing - not adjustable)
Firing order
  Gasoline engine 1-5-4-8-6-3-7-2
  Diesel engine 1-3-7-2-6-5-4-8

6.2L V8 ENGINE
2015 and earlier models
1-3-7-2-6-5-4-8
2016 models
1-5-4-8-6-3-7-2
36064-01-00.00b HAYNES

DIESEL ENGINE
1-3-7-2-6-5-4-8
36064-01-00.01c HAYNES

Cylinder location diagram - 6.2L gasoline engine

Cylinder location diagram - 6.7L diesel engine

Torque specifications

Ft-lbs (unless otherwise indicated)

Note: *One foot-pound (ft-lb) of torque is equivalent to 12 inch-pounds (in-lbs) of torque. Torque values below approximately 15 foot-pounds are expressed in inch-pounds, because most foot-pound torque wrenches are not accurate at these smaller values.*

Automatic transmission
  Fluid pan bolts
    TorqShift .... 177 in-lbs
    TorqShift6 (6R140) .... 80 in-lbs
  Pan drain plug .... 159 in-lbs
Front hub adjusting nut (2WD models)
  Step 1 21
  Step 2 .... Back off 1/2-turn
  Step 3 .... 18 in-lbs
Spark plugs .... 106 in-lbs
Wheel lug nuts* 165

Note: * *The manufacturer recommends that wheel lug nuts be retightened 100 miles after being installed. On dual-rear wheels, the lug nuts should have a second retightening at 500 miles.*

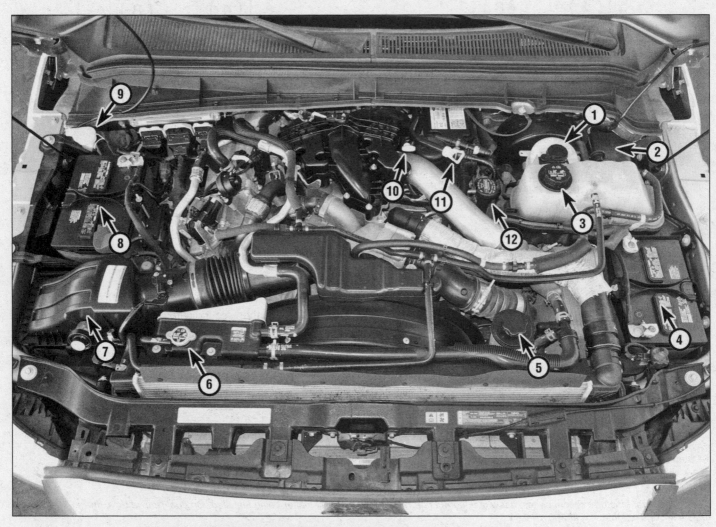

Typical engine compartment components (6.7L diesel engine)

1	Brake fluid reservoir	5	Power steering fluid reservoir	8	Battery (primary)
2	Underhood fuse/relay block	6	Secondary coolant expansion tank	9	Windshield washer fluid reservoir
3	Engine coolant expansion tank		- Charge Air Cooler (CAC)/Exhaust	10	Engine oil dipstick
	(primary cooling system)		Gas Recirculation (EGR) system	11	Automatic transmission fluid dipstick
4	Battery (secondary)	7	Air filter housing	12	Engine oil filler cap

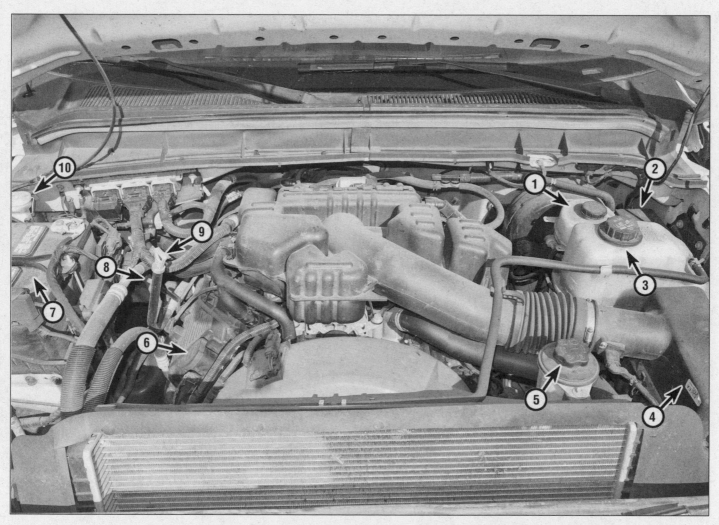

Typical engine compartment components (6.2L gasoline engine)

1	Brake fluid reservoir	5	Power steering fluid reservoir	8	Engine oil dipstick
2	Underhood fuse/relay block	6	Engine oil filler cap	9	Automatic transmission fluid dipstick
3	Engine coolant expansion tank	7	Battery	10	Windshield washer fluid reservoir
4	Air filter housing				

Typical engine underside components (6.7L diesel 4WD model shown)

1	Lower radiator hose	5	Front axle housing	8	Engine oil drain plug
2	Lower radiator hose	6	Front disc brake caliper	9	Exhaust pipe
3	Steering damper	7	Engine oil filter	10	Automatic transmission fluid drain plug
4	Front axle housing check/fill plug				

Typical rear underside components

1 *Fuel tank*	4 *Shock absorber*	6 *Differential cover*
2 *Exhaust pipe*	5 *Parking brake cable*	7 *Rear disc brake caliper*
3 *Leaf spring*		

1 Maintenance schedule

The following maintenance intervals are based on the assumption that the vehicle owner will be doing the maintenance or service work, as opposed to having a dealer service department or other repair shop do the work. Although the time/mileage intervals are loosely based on factory recommendations, most have been shortened to ensure, for example, that such items as lubricants and fluids are checked/changed at intervals that promote maximum engine/driveline service life. Also, subject to the preference of the individual owner interested in keeping his or her vehicle in peak condition at all times, and with the vehicle's ultimate resale in mind, many of the maintenance procedures may be performed more often than recommended in the following schedule. We encourage such owner initiative.

When the vehicle is new it should be serviced initially by a factory authorized dealer service department to protect the factory warranty. In many cases the initial maintenance check is done at no cost to the owner (check with your dealer service department for more information).

Every 250 miles or weekly, whichever comes first

Check the engine oil level (Section 4)
Check the engine coolant level (Section 4)
Check the windshield washer fluid level (Section 4)
Check the brake fluid level (Section 4)
Check the tires and tire pressures (Section 5)
On diesel models, check the air filter restriction gauge (Section 24)

Every 3000 miles or 3 months, whichever comes first

All items listed above, plus:
Check the power steering fluid level (Section 6)
Check the automatic transmission fluid level (Section 7)
Check and drain if necessary, the diesel engine fuel/water separator (Section 8)
Change the engine oil and oil filter (Section 9)
Lubricate the chassis (Section 10)
Check the engine drivebelt (Section 29)

Every 6000 miles or 6 months, whichever comes first

All items listed above, plus:
Check and service the battery (Section 11)
Inspect and replace, if necessary, the windshield wiper blades (Section 12)
Rotate the tires (Section 13)
Inspect the exhaust system (Section 14)
Check the seat belt operation (Section 15)

Every 15,000 miles or 12 months, whichever comes first

All items listed above, plus:
Inspect and replace, if necessary, all underhood hoses (Section 16)

Inspect the cooling system (Section 17)
Check the fuel system (Section 18)
Replace the fuel filter (Section 19)
Inspect the steering and suspension components (Section 20)
Inspect the brakes (Section 21)
Check the transfer case lubricant level (Section 22)
Check the differential lubricant level (Section 23)

Every 30,000 miles or 24 months, whichever comes first

Replace the air filter (Section 24)*
Service the cooling system (drain, flush and refill) (Section 25)
Change the automatic transmission fluid and filter (Section 26)**
Inspect and repack the front wheel bearings (2WD models) (Section 27)
Change the brake fluid (Section 28)

Every 60,000 miles or 48 months, whichever comes first

Check the engine drivebelt (Section 29)
Check the PCV valve (Section 30)
Replace the spark plugs (Section 31)
Inspect and replace, if necessary, the ignition coils (Section 32)

Every 100,000 miles or 60 months, whichever comes first

Change the transfer case lubricant (Section 34)
Change the differential lubricant (Section 33)
Check the spark plug wires (Section 35)

Replace more often if is the vehicle is driven in dusty areas

**If the vehicle is operated in continuous stop-and-go driving or in mountainous areas, change at 15,000 miles*

2 Introduction

1   This Chapter is designed to help the home mechanic maintain the Ford Super Duty F-250 and F-350 with the goals of maximum performance, economy, safety and reliability in mind.

2   Included is a master maintenance schedule, followed by procedures dealing specifically with each item on the schedule. Visual checks, adjustments, component replacement and other helpful items are included. Refer to the accompanying illustrations of the engine compartment and the underside of the vehicle for the locations of various components.

3   Servicing the vehicle, in accordance with the mileage/time maintenance schedule and the step-by-step procedures will result in a planned maintenance program that should produce a long and reliable service life. Keep in mind that it is a comprehensive plan, so maintaining some items but not others at the specified intervals will not produce the same results.

4   As you service the vehicle, you will discover that many of the procedures can - and should - be grouped together because of the nature of the particular procedure you're performing or because of the close proximity of two otherwise unrelated components to one another.

5   For example, if the vehicle is raised for chassis lubrication, you should inspect the exhaust, suspension, steering and fuel systems while you're under the vehicle. When you're rotating the tires, it makes good sense to check the brakes since the wheels are already removed. Finally, let's suppose you have to borrow or rent a torque wrench. Even if you only need it to tighten the spark plugs, you might as well check the torque of as many critical fasteners as time allows.

6   The first step in this maintenance program is to prepare yourself before the actual work begins. Read through all the procedures you're planning to do, then gather up all the parts and tools needed. If it looks like you might run into problems during a particular job, seek advice from a mechanic or an experienced do-it-yourselfer.

Owner's Manual and VECI label information

7   Your vehicle owner's manual was written for your year and model and contains very specific information on component locations, specifications, fuse ratings, part numbers, etc. The owner's manual is an important resource for the do-it-yourselfer to have; if one was not supplied with your vehicle, it can generally be ordered from a dealer parts department.

8   Among other important information, the Vehicle Emissions Control Information (VECI) label contains specifications and procedures for applicable tune-up adjustments (if applicable) and, in some instances, spark plug replacement (see Chapter 6A for more information on the VECI label). The information

on this label is the exact maintenance data recommended by the manufacturer. This data often varies by intended altitude, local emissions regulations, month of manufacture, etc.

9   This Chapter contains procedural details, safety information and more ambitious maintenance intervals than you might find in manufacturer's literature. However, you may find procedures or specifications in your owner's manual or VECI label can be considered correct, since it is specific to your particular vehicle.

3 Tune-up general information

1   The term tune-up is used in this manual to represent a combination of individual operations rather than one specific procedure.

2   If, from the time the vehicle is new, the routine maintenance schedule is followed closely and frequent checks are made of fluid levels and high wear items, as suggested throughout this manual, the engine will be kept in relatively good running condition and the need for additional work will be minimized.

3   More likely than not, however, there will be times when the engine is running poorly due to lack of regular maintenance. This is even more likely if a used vehicle, which has not received regular and frequent maintenance checks, is purchased. In such cases, an engine tune-up will be needed outside of the regular routine maintenance intervals.

4   The first step in any tune-up or diagnostic procedure to help correct a poor running engine is a cylinder compression check. A compression check (see Chapter 2C) will help determine the condition of internal engine components and should be used as a guide for tune-up and repair procedures. If, for instance, a compression check indicates serious internal engine wear, a conventional tune-up will not improve the performance of the engine and would be a waste of time and money. Because of its importance, the compression check should be done by someone with the right equipment and the knowledge to

use it properly.

5   The following procedures are those most often needed to bring a generally poor running engine back into a proper state of tune.

Minor tune-up
Check all engine related fluids (Section 4)
Clean, inspect and test the battery (Section 11)
Check all underhood hoses (Section 16)
Check the cooling system (Section 17)
Check the fuel system (Section 18)
Check the air filter (Section 24)

Major tune-up
All items listed under Minor tune-up, plus . . .
Replace the fuel filter (Section 19)
Replace the air filter (Section 24)
Check the drivebelt (Section 29)
Replace the PCV valve (Section 30)
Replace the spark plugs (Section 31)
Check the charging system (Chapter 5)

4 Fluid level checks (every 250 miles or weekly)

1   Fluids are an essential part of the lubrication, cooling, brake and windshield washer systems. Because the fluids gradually become depleted and/or contaminated during normal operation of the vehicle, they must be periodically replenished. See *Recommended lubricants and fluids* at the beginning of this Chapter before adding fluid to any of the following components.

Note: *The vehicle must be on level ground when fluid levels are checked.*

Engine oil

2   The oil level is checked with a dipstick, which is located on the right-side of gasoline models or (see illustrations) on the left-side on diesel models. The dipstick extends through a metal tube down into the oil pan.

3   The oil level should be checked before the vehicle has been driven, or about five minutes after the engine has been shut off. If the oil is

4.2a Engine oil dipstick location - gasoline engine models

4.2b Engine oil dipstick location - diesel engine models

4.4 The oil level should be at or near the MAX area on the dipstick - if it isn't, add enough oil to bring the level to near the MAX mark

4.6a On diesel models, the oil filler cap is located on the tube connected to the left valve cover

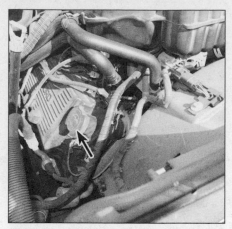

4.6b On gasoline models, the oil filler cap is located on the right valve cover

4.9a The coolant expansion tank is located at the rear of the engine compartment, in front of the battery on all models - keep the level above the COLD FILL mark - DO NOT remove the cap until the engine has cooled completely

4.9b Diesel engine models have a secondary coolant expansion tank located at the front of the engine compartment, mounted to the fan shroud - keep the level near the MAX mark - DO NOT remove the cap until the engine has cooled completely

checked immediately after driving the vehicle, some of the oil will remain in the upper part of the engine, resulting in an inaccurate reading on the dipstick. On diesel engines, wait at least 20 minutes after the engine is shut off before checking the oil level because it takes longer for the greater quality of oil in the upper part to drain into the oil pan.

4   Pull the dipstick out of the tube and wipe all the oil from the end with a clean rag or paper towel. Insert the clean dipstick all the way back into the tube and pull it out again. Note the oil at the end of the dipstick. At its highest point, the level on gasoline engines should be between the MIN and MAX marks on the dipstick (see illustration). On diesel engines the level should be between the ADD and OPERATING RANGE marks on the dipstick.

5   On gasoline engines it takes one quart of oil to raise the level from the MIN mark to the MAX mark on the dipstick. Do not allow the

level to drop below the MIN mark or oil starvation may cause engine damage. Conversely, overfilling the engine (adding oil above the MAX mark) may cause oil fouled spark plugs, oil leaks or oil seal failures. On diesel engines it takes two quarts of oil to raise the level from the ADD to the OPERATING RANGE mark on the dipstick. Maintaining the oil level above the OPERATING RANGE mark can cause excessive oil consumption.

6   To add oil, remove the filler cap from the valve cover (see illustrations). After adding oil, wait a few minutes to allow the level to stabilize, then pull out the dipstick and check the level again. Add more oil if required. Install the filler cap and tighten it by hand only.

7   Checking the oil level is an important preventive maintenance step. A consistently low oil level indicates oil leakage through damaged seals, defective gaskets or past worn rings or valve guides. If the oil looks milky in color or has water droplets in it, the cylinder

head gasket(s) may be blown or the head(s) or block may be cracked. The engine should be checked immediately. The condition of the oil should also be checked. Whenever you check the oil level, slide your thumb and index finger up the dipstick before wiping off the oil. If you see small dirt or metal particles clinging to the dipstick, the oil should be changed (see Section 9).

Engine coolant

Warning: *Do not allow antifreeze to come in contact with your skin or painted surfaces of the vehicle. Flush contaminated areas immediately with plenty of water. Don't store new coolant or leave old coolant lying around where it's accessible to children or pets - they're attracted by its sweet smell. Ingestion of even a small amount of coolant can be fatal! Wipe up garage floor and drip pan spills immediately. Keep antifreeze containers covered and repair cooling system leaks as soon as they're noticed.*

8   All vehicles covered by this manual are equipped with a pressurized coolant recovery system. A plastic expansion tank located at the left rear corner of the engine compartment is connected by a hose to the radiator. As the engine heats up during operation, the expanding coolant fills the tank. Diesel engine models are equipped with a secondary system for the Charge Air Cooler (CAC) and Exhaust Gas Recirculation (EGR) system; the secondary expansion tank is mounted to the radiator fan shroud.

Warning: *Do not remove the expansion tank cap to check the coolant level when the engine is warm!*

9   The coolant level in the tank should be checked regularly. The level in the tank varies with the temperature of the engine. When the engine is cold, the coolant level should be at or slightly above the COLD FILL mark on the reservoir. If it isn't, remove the cap from the tank and add a 50/50 mixture of ethylene glycol based antifreeze and water (see illustrations).

4.15 The brake fluid level should be kept between the MIN and MAX marks on the translucent plastic reservoir

4.22 The windshield washer reservoir is located in the engine compartment, at the right rear corner behind the battery

10   Drive the vehicle and recheck the coolant level. Don't use rust inhibitors or additives. If only a small amount of coolant is required to bring the system up to the proper level, water can be used. However, repeated additions of water will dilute the antifreeze and water solution. In order to maintain the proper ratio of antifreeze and water, always top up the coolant level with the correct mixture. An empty plastic milk jug or bleach bottle makes an excellent container for mixing coolant.

11   On the main cooling system, if the coolant level drops consistently, inspect the radiator, hoses, filler cap, drain plugs and water pump (see Section 17); another possibility is a blown head gasket. On the secondary system, if the coolant level drops consistently, there may be a leak at the Charge Air Cooler, turbocharger or conections (see Chapter 4B), or in the EGR system. If no leaks are noted, have the expansion tank cap pressure tested by a service station.

12   If you have to remove the expansion tank cap, wait until the engine has cooled completely, then wrap a thick cloth around the cap and unscrew it slowly, stopping if you hear a hissing noise. If coolant or steam escapes, let the engine cool down longer, then remove the cap.

13   Check the condition of the coolant as well. It should be relatively clear. If it's brown or rust colored, the system should be drained, flushed and refilled. Even if the coolant appears to be normal, the corrosion inhibitors wear out, so it must be replaced at the specified intervals.

Brake fluid

14   The brake master cylinder is mounted on the front of the power booster unit in the engine compartment.

15   To check the fluid level of the brake cylinder, simply look at the MAX and MIN marks on the reservoir (see illustration), the level should be between the marks.

16   If the level is low, wipe the top of the reservoir and cap with a clean rag to prevent contamination of the brake system before unscrewing the cap.

17   Add only the specified brake fluid to the brake reservoir (refer to *Recommended lubricants and fluids* at the front of this Chapter or to your owner's manual). Mixing different types of brake fluid can damage the system. Fill the brake master cylinder reservoir only to the MAX line.
Warning: *Use caution when filling the reservoir - brake fluid can harm your eyes and damage painted surfaces. Do not use brake fluid that is more than one year old or has been left open. Brake fluid absorbs moisture from the air. Excess moisture can cause a dangerous loss of braking.*

18   While the reservoir cap is removed, inspect the master cylinder reservoir for contamination. If deposits, dirt particles or water droplets are present, the system should be drained and refilled.

19   After filling the reservoir to the proper level, make sure the cap is tightened securely to prevent fluid leakage.

20   The fluid in the brake master cylinder will drop slightly as the brake pads at each wheel wear down during normal operation. If the master cylinder requires repeated replenishing to keep it at the proper level, this is an indication of leakage in the brake system, which should be corrected immediately. If the brake system shows an indication of leakage check all brake lines and connections, along with the calipers, wheel cylinders and booster (see Section 21 for more information).

21   If, upon checking the brake master cylinder fluid level, you discover the reservoir is empty or nearly empty, the system should be bled (see Chapter 9).

Windshield washer fluid

22   Fluid for the windshield washer system is stored in a plastic reservoir located at the right rear corner of the engine compartment (see illustration).

23   In milder climates, plain water can be used in the reservoir, but it should be kept no more than 2/3 full to allow for expansion if the water freezes. In colder climates, use windshield washer system antifreeze, available at any auto parts store, to lower the freezing point of the fluid. Mix the antifreeze with water in accordance with the manufacturer's directions on the container.
Caution: *Do not use cooling system antifreeze - it will damage the vehicle's paint.*

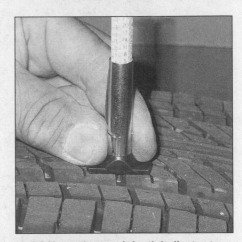

5.2 Use a tire tread depth indicator to monitor tire wear - they are available at auto parts stores and service stations and cost very little

5   Tire and tire pressure checks (every 250 miles or weekly)

1   Periodic inspection of the tires may spare you the inconvenience of being stranded with a flat tire. It can also provide you with vital information regarding possible problems in the steering and suspension systems before major damage occurs.

2   The original tires on this vehicle are equipped with 1/2-inch wide bands that will appear when tread depth reaches 1/16-inch, at which point they can be considered worn out. Tread wear can be monitored with a simple, inexpensive device known as a tread depth indicator (see illustration).

UNDERINFLATION

CUPPING

Cupping may be caused by:
- Underinflation and/or mechanical irregularities such as out-of-balance condition of wheel and/or tire, and bent or damaged wheel.
- Loose or worn steering tie-rod or steering idler arm.
- Loose, damaged or worn front suspension parts.

OVERINFLATION

INCORRECT TOE-IN OR EXTREME CAMBER

FEATHERING DUE TO MISALIGNMENT

5.3 This chart will help you determine the condition of the tires, the probable cause(s) of abnormal wear and the corrective action necessary

3   Note any abnormal tread wear (see illustration). Tread pattern irregularities such as cupping, flat spots and more wear on one side than the other are indications of front end alignment and/or balance problems. If any of these conditions are noted, take the vehicle to a tire shop or service station to correct the problem.

4   Look closely for cuts, punctures and embedded nails or tacks. Sometimes a tire will hold air pressure for a short time or leak down very slowly after a nail has embedded itself in the tread. If a slow leak persists, check the valve stem core to make sure it is tight (see illustration). Examine the tread for an object that may have embedded itself in the tire or for a "plug" that may have begun to leak (radial tire punctures are repaired with a plug that is installed in a puncture). If a puncture is suspected, it can be easily verified by spraying a solution of soapy water onto the puncture area (see illustration). The soapy solution will bubble if there is a leak. Unless the puncture is unusually large, a tire shop or service station can usually repair the tire.

5   Carefully inspect the inner sidewall of each tire for evidence of brake fluid leakage. If you see any, inspect the brakes immediately.

5.4a If a tire loses air on a steady basis, check the valve stem core first to make sure it's snug (special inexpensive wrenches are commonly available at auto parts stores)

5.4b If the valve stem core is tight, raise the corner of the vehicle with the low tire and spray a soapy water solution onto the tread as the tire is turned slowly - leaks will cause small bubbles to appear

5.8 To extend the life of the tires, check the air pressure at least once a week with an accurate gauge (don't forget the spare!)

6.3 The power steering fluid level should be kept between the MIN and MAX marks on the translucent plastic reservoir

6   Correct air pressure adds miles to the life span of the tires, improves mileage and enhances overall ride quality. Tire pressure cannot be accurately estimated by looking at a tire, especially if it's a radial. A tire pressure gauge is essential. Keep an accurate gauge in the glove compartment. The pressure gauges attached to the nozzles of air hoses at gas stations are often inaccurate.

7   Always check tire pressure when the tires are cold. Cold, in this case, means the vehicle has not been driven over a mile in the three hours preceding a tire pressure check. A pressure rise of four to eight pounds is not uncommon once the tires are warm.

8   Unscrew the valve cap protruding from the wheel or hubcap and push the gauge firmly onto the valve stem (see illustration). Note the reading on the gauge and compare the figure to the recommended tire pressure shown on the tire placard on the driver's side door. Be sure to reinstall the valve cap to keep dirt and moisture out of the valve stem mechanism. Check all four tires and, if necessary, add enough air to bring them up to the recommended pressure.

9   Don't forget to keep the spare tire inflated to the specified pressure (refer to the pressure molded into the tire sidewall).

6   Power steering fluid level check (every 3000 miles or 3 months)

1   Check the power steering fluid level periodically to avoid steering system problems, such as damage to the pump. **Caution:** *DO NOT hold the steering wheel against either stop (extreme left or right turn) for more than five seconds. If you do, the power steering pump could be damaged.*

2   The fluid reservoir for the power steering pump is mounted to the left side of the radiator.

3   To check the power steering fluid level, simply look at the MAX and MIN marks on the side of the reservoir (see illustration). The

7.4a The automatic transmission fluid dipstick is located on the driver's side of the engine on diesel models . . .

level should be between the marks.

4   Add small amounts of fluid until the level is correct. **Caution:** *Do not overfill the reservoir. If too much fluid is added, remove the excess with a clean syringe or suction pump.*

5   If the reservoir requires frequent fluid additions, all power steering hoses, hose connections, the power steering pump and the steering gear assembly should be carefully checked for leaks.

7   Automatic transmission fluid level check (every 3000 miles or 3 months)

1   The automatic transmission fluid level should be carefully maintained. Low fluid level can lead to slipping or loss of drive, while overfilling can cause foaming and loss of fluid. Either condition can cause transmission damage.

2   Since transmission fluid expands as it heats up, the fluid level should be checked

7.4b . . . and on the passenger's side of the engine on gasoline powered models

when the transmission is warm (at normal operating temperature). If the vehicle has just been driven over 20 miles (32 km), the transmission can be considered warm. Caution: If the vehicle has just been driven for a long time at high speed or in city traffic in hot weather, or if it has been pulling a trailer, an accurate fluid level reading cannot be obtained. Allow the transmission to cool down for about 30 minutes. You can also check the transmission fluid level when the transmission is cold. If the vehicle has not been driven for over five hours and the fluid is about room temperature (70 to 95-degrees F), the transmission is cold. However, the fluid level is normally checked with the transmission warm to ensure accurate results.

3   Immediately after driving the vehicle, park it on a level surface, set the parking brake and start the engine. While the engine is idling, depress the brake pedal and move the selector lever through all the gear ranges, beginning and ending in Park.

4   Locate the automatic transmission dipstick tube in the engine compartment (see illustrations).

7.6 Check the fluid with the transmission at normal operating temperature - the level should be kept in the HOT range in the cross-hatched area (don't add fluid if the level is anywhere in the cross-hatched area)

8.3 Turn the knob (1) counterclockwise and allow the accumulated water and fuel to flow out from the drain (2)

5   With the engine still idling, pull the dipstick from the tube, wipe it off with a clean rag, push it all the way back into the tube and withdraw it again, then note the fluid level.

6   If the transmission is cold, the level

should be in the "COLD" range on the dipstick; if it's warm to hot, the fluid level should be in the "HOT" range (in the cross-hatched area) (see illustration). If the level is low, add the specified automatic transmission fluid through the dipstick tube - use a funnel to prevent spills.

7   Add just enough of the recommended fluid to fill the transmission to the proper level. It takes about one pint to raise the level from the low mark to the high mark when the fluid is hot, so add the fluid a little at a time and keep checking the level until it's correct.

8   The condition of the fluid should also be checked along with the level. If the fluid is black or a dark reddish-brown color, or if it smells burned, it should be changed (see Section 26). If you are in doubt about its condition, purchase some new fluid and compare the two for color and smell.

has collected).

2   Place a small container under the filter drain (diesel fuel can damage asphalt paving).

3   With the engine off, turn the knob (see illustration) on the filter housing counterclockwise and allow the accumulated water to drain out, then close the valve by turning the knob clockwise.

4   Remove the container and dispose of the fuel/water mixture properly.

---

9   Engine oil and filter change (every 3000 miles or 3 months)

1   Frequent oil changes are the most important preventive maintenance procedures that can be done by the home mechanic. As engine oil ages, it becomes diluted and contaminated, which leads to premature engine wear.

2   Make sure that you have all the necessary tools before you begin this procedure (see illustration). You should also have plenty of rags or newspapers handy for mopping up oil spills.

3   Access to the oil drain plug and filter will be improved if the vehicle can be lifted on a hoist, driven onto ramps or supported by jackstands.

Warning: *Do not work under a vehicle supported only by a jack - always use jackstands!*

4   If you haven't changed the oil on this vehicle before, get under it and locate the oil drain plug and the oil filter. The exhaust components will be warm as you work, so note how they are routed to avoid touching them when you are under the vehicle.

5   Start the engine and allow it to reach normal operating temperature - oil and sludge will flow out more easily when warm. If new oil, a filter or tools are needed, use the vehicle to go get them and warm up the engine/oil at the same time. Park on a level surface and shut off the engine when it's warmed up. Remove the oil filler cap from the valve cover.

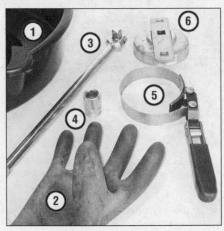

9.2 These tools are required when changing the engine oil and filter

1   **Drain pan** - *It should be fairly shallow in depth, but wide to prevent spills*

2   **Rubber gloves** - *When removing the drain plug and filter, you will get oil on your hands (the gloves will prevent burns)*

3   **Breaker bar** - *Sometimes the oil drain plug is tight, and a long breaker bar is needed to loosen it*

4   **Socket** – *To be used with the breaker bar or a ratchet (must be the correct size to fit the drain plug - six-point preferred)*

5   **Filter wrench** - *This is a metal bandtype wrench, which requires clearance around the filter to be effective*

6   **Filter wrench** - *This type fits on the bottom of the filter and can be turned with a ratchet or breaker bar (different-size wrenches are available for different types of filters)*

---

8   Fuel filter/water separator draining (diesel models) (every 3000 miles or 3 months)

Warning: *Diesel fuel isn't as volatile as gasoline, but it is flammable, so take extra precautions when you work on any part of the fuel system. Don't smoke or allow open flames or bare light bulbs near the work area. Don't work in a garage or other enclosed space where there is a gas-type appliance (such as a water heater or clothes dryer). Finally, when you perform any work on the fuel system, wear safety glasses, latex gloves and have a Class B type fire extinguisher on hand. If you spill any diesel fuel on your skin, rinse it off immediately with soap and water.*

1   The diesel engine fuel filter incorporates a water separator that removes and traps water in the fuel. It's called the Diesel Fuel Conditioner Module (DFCM), located on the driver's side frame-rail. The water must be drained from the filter monthly and at the specified interval or when the "Water In Fuel" light is on (indicating at least 100 cc of water

9.7 Use a proper size box-end wrench or socket to remove the oil drain plug and avoid rounding it off

9.12 The oil filter is usually on very tight and will require a special oil filter wrench to remove it - DO NOT use the wrench to tighten the new filter

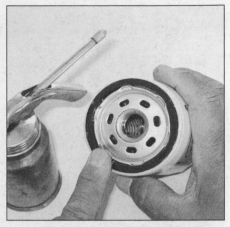

9.15 Lubricate the oil filter gasket with clean engine oil before installing the filter on the engine

6   Raise the vehicle and support it on jackstands. Make sure it is safely supported!

7   Being careful not to touch the hot exhaust components, position a drain pan under the plug in the bottom of the engine, then remove the plug (see illustration). It's a good idea to wear a rubber glove while unscrewing the plug the final few turns to avoid being scalded by hot oil.

8   It may be necessary to move the drain pan slightly as oil flow slows to a trickle. Inspect the old oil for the presence of metal particles.

9   After all the oil has drained, wipe off the drain plug with a clean rag. Any small metal particles clinging to the plug would immediately contaminate the new oil.

10   Clean the area around the drain plug opening, reinstall the plug and tighten it securely, but don't strip the threads.

11   Move the drain pan into position under the oil filter.

12   Loosen the oil filter by turning it counterclockwise with a filter wrench (see illustration).

13   Once the filter is loose, use your hands to unscrew it from the block.

14   Using a clean rag, wipe off the mounting surface on the block. Also, make sure that none of the old gasket remains stuck to the mounting surface. It can be removed with a scraper if necessary.

15   Compare the old filter with the new one to make sure they are the same type. Smear some engine oil on the rubber gasket of the new filter and screw it into place (see illustration). Overtightening the filter will damage the gasket, so don't use a filter wrench. Most filter manufacturers recommend tightening the filter by hand only. Normally they should be tightened 3/4-turn (on diesel models 1-1/4 to 2-turns) after the gasket contacts the block, but be sure to follow the directions on the filter or container.

16   Remove all tools and materials from under the vehicle, being careful not to spill the oil in the drain pan, then lower the vehicle.

17   Add new oil to the engine through the oil filler cap. Use a funnel to prevent oil from spilling onto the top of the engine. Add the specified amount of oil, minus one quart. Wait a few minutes to allow the oil to drain into the pan, then check the level on the dipstick (see Section 4 if necessary). If the oil level is in the OK range, install the filler cap.

18   Start the engine and run it for about a minute. While the engine is running, look under the vehicle and check for leaks at the oil pan drain plug and around the oil filter. If either one is leaking, stop the engine and tighten the plug or filter slightly.

19   Wait a few minutes, then recheck the level on the dipstick. Add oil as necessary to bring the level into the OK range.

20   During the first few trips after an oil change, make it a point to check frequently for leaks and proper oil level.

21   The old oil drained from the engine cannot be reused in its present state and should be disposed of. Check with your local auto parts store, disposal facility or environmental agency to see if they will accept the oil for recycling. After the oil has cooled it can be drained into a container (capped plastic jugs, topped bottles, milk cartons, etc.) for transport to one of these disposal sites. Don't dispose of the oil by pouring it on the ground or down a drain!

10   Chassis lubrication (every 3000 miles or 3 months)

1   Refer to *Recommended lubricants and fluids* at the front of this Chapter to obtain the necessary grease, etc. You will also need a grease gun (see illustration). Occasionally plugs will be installed rather than grease fittings. If so, grease fittings will have to be purchased and installed.

2   Look under the vehicle for grease fittings or plugs on the steering, suspension, and driveline components. They are normally found on the balljoints, tie-rod ends and uni-

versal joints. If there are plugs, remove them and install grease fittings, which will thread into the component. An automotive parts store will be able to supply the correct fittings. Straight, as well as angled, fittings are available.

3   For easier access under the vehicle, raise it with a jack and place jackstands under the frame. Make sure it is safely supported by the stands. If the wheels are to be removed at this interval for tire rotation or brake inspection, loosen the lug nuts slightly while the

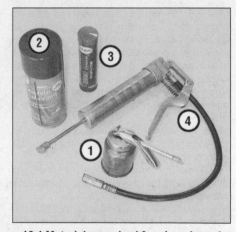

10.1 Materials required for chassis and body lubrication

1   **Engine oil** - *Light engine oil in a can like this can be used for door and hood hinges*

2   **Graphite spray** - *Used to lubricate lock cylinders*

3   **Grease** - *Grease, in a variety of types and weights, is available for use in a grease gun. Check the Specifications for your requirements*

4   **Grease gun** - *A common grease gun, shown here with a detachable hose and nozzle, is needed for chassis lubrication. After use, clean it thoroughly!*

10.6 On the steering joints, pump grease into the fittings until the rubber seals are firm to the touch

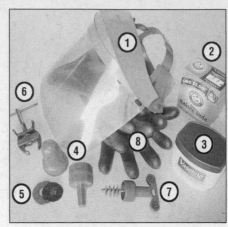

11.1 Tools and materials required for battery maintenance

1   *Face shield/safety goggles - When removing corrosion with a brush, the acidic particles can easily fly up into your eyes*
2   *Baking soda - A solution of baking soda and water can be used to neutralize corrosion*
3   *Petroleum jelly - A layer of this on the battery posts will help prevent corrosion*
4   *Battery post/cable cleaner - This wire brush cleaning tool will remove all traces of corrosion from the battery posts and cable clamps*
5   *Treated felt washers - Placing one of these on each post, directly under the cable clamps, will help prevent corrosion*
6   *Puller - Sometimes the cable clamps are very difficult to pull off the posts, even after the nut/bolt has been completely loosened. This tool pulls the clamp straight up and off the post without damage*
7   *Battery post/cable cleaner - Here is another cleaning tool which is a slightly different version of Number 4 above, but it does the same thing*
8   *Rubber gloves - Another safety item to consider when servicing the battery; remember that's acid inside the battery!*

vehicle is still on the ground.

4   Before beginning, force a little grease out of the nozzle to remove any dirt from the end of the gun. Wipe the nozzle clean with a rag.

5   With the grease gun and plenty of clean rags, crawl under the vehicle and begin lubricating the components.

6   Wipe the balljoint grease fitting clean and push the nozzle firmly over it. Squeeze the trigger on the grease gun to force grease into the component. The balljoints should be lubricated until the rubber seal is firm to the touch (see illustration). Do not pump too much grease into the fittings as it could rupture the seal. For all other suspension and steering components, continue pumping grease into the fitting until it oozes out of the joint between the two components. If it escapes around the grease gun nozzle, the fitting is clogged or the nozzle is not completely seated on the fitting. Resecure the gun nozzle to the fitting and try again. If necessary, replace the fitting with a new one.

Note: *Not all models have grease fittings for the steering joints. They are factory sealed and do not need lubrication.*

7   Wipe the excess grease from the components and the grease fitting. Repeat the procedure for the remaining fittings.

8   On models equipped with an automatic transmission, lubricate the shift linkage with a little clean engine oil.

9   On 4WD models, lubricate the transfer case shift mechanism contact surfaces with clean engine oil.

10   Lubricate the driveshaft slip-joints.

Note: *It may be necessary to disassemble the driveshaft to lubricate the slip joint (see Chapter 8).*

11   Lubricate conventional universal joints until grease can be seen coming out of the contact points.

12   While you are under the vehicle, clean and lubricate the parking brake cable along with the cable guides and levers. This can be done by smearing some chassis grease onto the cable and its related parts with your fingers.

13   Open the hood and smear a little chassis grease on the hood latch mechanism. Have an assistant pull the hood release lever from inside the vehicle as you lubricate the cable at the latch.

14   Lubricate all the hinges (door, hood, etc.) with engine oil to keep them in proper working order.

15   The key lock cylinders can be lubricated with spray-on graphite or silicone lubricant, which is available at auto parts stores.

16   Lubricate the door weatherstripping with silicone spray. This will reduce chafing and retard wear.

11   Battery check, maintenance and charging (every 6000 miles or 6 months)

Warning: *Certain precautions must be followed when checking and servicing the battery. Hydrogen gas, which is highly flammable, is always present in the battery cells, so keep lighted tobacco and all other open flames and sparks away from the battery. The electrolyte inside the battery is actually diluted sulfuric acid, which will cause injury if splashed on your skin or in your eyes. It will also ruin clothes and painted surfaces. When removing the battery cables, always detach the negative cable first and hook it up last!*

1   A routine preventive maintenance program for the battery(ies) in your vehicle is the only way to ensure quick and reliable starts. But before performing any battery maintenance, make sure that you have the proper equipment necessary to work safely around the battery (see illustration).

2   There are also several precautions that should be taken whenever battery maintenance is performed. Before servicing the battery, always turn the engine and all accessories off and disconnect the cable(s) from the negative terminal(s) of the battery(ies) (see Chapter 5).

3   The battery produces hydrogen gas, which is both flammable and explosive. Never

create a spark, smoke or light a match around the battery. Always charge the battery in a ventilated area.

4   Electrolyte contains poisonous and corrosive sulfuric acid. Do not allow it to get in your eyes, on your skin or your clothes. Never ingest it. Wear protective safety glasses when working near the battery. Keep children away from the battery.

5   Note the external condition of the battery. If the positive terminal and cable clamp on your vehicle's battery is equipped with a rubber protector, make sure that it's not torn or damaged. It should completely cover the terminal. Look for any corroded or loose con-

11.6a Battery terminal corrosion usually appears as light, fluffy powder

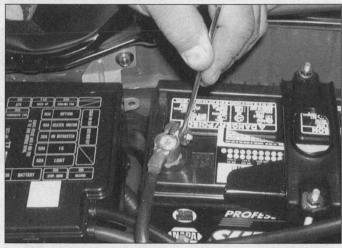

11.6b Removing the cable from a battery post with a wrench - sometimes special battery pliers are required for this procedure if corrosion has caused deterioration of the nut hex (always remove the ground cable first and hook it uplast!)

11.7a When cleaning the cable clamps, all corrosion must be removed (the inside of the clamp is tapered to match the taper on the post, so don't remove too much material)

11.7b Regardless of the type of tool used on the battery posts, a clean, shiny surface should be the result

nections, cracks in the case or cover or loose hold-down clamps. Also check the entire length of each cable for cracks and frayed conductors.

6   If corrosion, which looks like white, fluffy deposits (see illustration) is evident, particularly around the terminals, the battery should be removed for cleaning. Loosen the cable clamp bolts with a wrench, being careful to remove the ground cable first, and slide them off the terminals (see illustration). Then disconnect the hold-down clamp bolt and nut, remove the clamp and lift the battery from the engine compartment.

7   Clean the cable clamps thoroughly with a battery brush or a terminal cleaner and a solution of warm water and baking soda (see illustration). Wash the terminals and the top of the battery case with the same solution but make sure that the solution doesn't get into the battery. When cleaning the cables, terminals and battery top, wear safety goggles and

rubber gloves to prevent any solution from coming in contact with your eyes or hands. Wear old clothes too - even diluted, sulfuric acid splashed onto clothes will burn holes in them. If the terminals have been extensively corroded, clean them up with a terminal cleaner (see illustration). Thoroughly wash all cleaned areas with plain water.

8   Make sure that the battery tray is in good condition and the hold-down clamp bolts are tight. If the battery is removed from the tray, make sure no parts remain in the bottom of the tray when the battery is reinstalled. When reinstalling the hold-down clamp bolts, do not overtighten them.

9   Information on removing and installing the battery(ies) can be found in Chapter 5. Information on jump starting can be found at the front of this manual. For more detailed battery checking procedures, refer to the *Haynes Automotive Electrical Manual.*

Cleaning
10   Corrosion on the hold-down components, battery case and surrounding areas can be removed with a solution of water and baking soda. Thoroughly rinse all cleaned areas with plain water.

11   Any metal parts of the vehicle damaged by corrosion should be covered with a zinc-based primer, then painted.

Charging
Warning: *When batteries are being charged, hydrogen gas, which is very explosive and flammable, is produced. Do not smoke or allow open flames near a charging or a recently charged battery. Wear eye protection when near the battery during charging. Also, make sure the charger is unplugged before connecting or disconnecting the battery from the charger.*

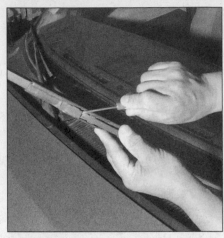

12.4 Pry the cover open to access the release lever

12.5 Press on the release tab and push the blade assembly down out of the hook in the arm

12   Slow-rate charging is the best way to restore a battery that's discharged to the point where it will not start the engine. It's also a good way to maintain the battery charge in a vehicle that's only driven a few miles between starts. Maintaining the battery charge is particularly important in the winter when the battery must work harder to start the engine and electrical accessories that drain the battery are in greater use.

13   It's best to use a one or two-amp battery charger (sometimes called a "trickle" charger). They are the safest and put the least strain on the battery. They are also the least expensive. For a faster charge, you can use a higher amperage charger, but don't use one rated more than 1/10th the amp/hour rating of the battery. Rapid boost charges that claim to restore the power of the battery in one to two hours are hardest on the battery and can

damage batteries not in good condition. This type of charging should only be used in emergency situations.

14   The average time necessary to charge a battery should be listed in the instructions that come with the charger. As a general rule, a trickle charger will charge a battery in 12 to 16 hours.

---

12   Windshield wiper blade inspection and replacement (every 6000 miles or 6 months)

---

1   The windshield wiper and blade assembly should be inspected periodically for damage, loose components and cracked or worn blade elements.

2   Road film can build up on the wiper

blades and affect their efficiency, so they should be washed regularly with a mild detergent solution.

3   If the wiper blade elements are cracked, worn or warped, or no longer clean adequately, they should be replaced with new ones.

4   Lift the arm assembly away from the glass for clearance, use a small screwdriver to release the cover (see illustration).

5   Press on the release lever, then slide the wiper blade assembly out of the hook in the end of the arm (see illustration).

6   Use needle-nose pliers to compress the blade element, then slide the element out of the frame and discard it (see illustration).

7   Installation is the reverse of removal.

---

13   Tire rotation (every 6000 miles or 6 months)

Caution: *Some vehicles have different front and rear tire pressures. On models with different pressures, when the tires are rotated the tire pressures must be adjusted to the correct pressure and the tire pressure sensors must be relearned or "trained" using a tire monitor activation tool. The tire light will flash and a "TRAIN LF/RF TIRE" message will be displayed on the instrument panel display. To relearn or train the sensor, a tire pressure monitor activation tool must be used. This is a procedure best performed at a dealership.*

1   The tires should be rotated at the specified intervals and whenever uneven wear is noticed. Since the vehicle will be raised and the tires removed anyway, check the brakes also (see Section 21).

2   Radial tires must be rotated in a specific pattern (see illustrations).

3   Refer to the information in *Jacking and towing* at the front of this manual for the proper procedure to follow when raising the vehicle and changing a tire.

4   The vehicle must be raised on a hoist or supported on jackstands to get all four wheels off the ground. Make sure the vehicle is safely supported!

5   After the rotation procedure is finished, check and adjust the tire pressures as necessary and be sure to check the lug nut tightness.

---

14   Exhaust system check (every 6000 miles or 6 months)

---

1   With the engine cold (at least three hours after the vehicle has been driven), check the complete exhaust system from the engine to the end of the tailpipe. Ideally, the inspection should be done with the vehicle on a hoist to permit unrestricted access. If a hoist isn't available, raise the vehicle and support it securely on jackstands.

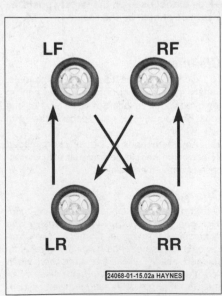

13.2a The recommended four-tire rotation pattern for non-directional tires

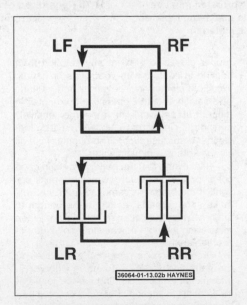

13.2b The recommended tire rotation pattern for vehicles with dual rear wheels

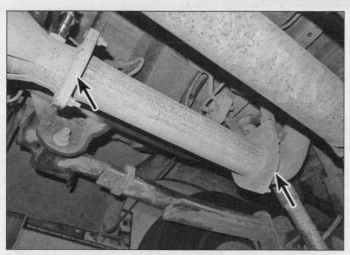

14.2a Check the connections for exhaust leaks - also check that the clamp retaining nuts are securely tightened

14.2b Check the exhaust system hangers for damage and cracks

2   Check the exhaust pipes and connections for evidence of leaks, severe corrosion and damage. Make sure that all brackets and hangers are in good condition and tight (see illustrations).

3   At the same time, inspect the underside of the body for holes, corrosion, open seams, etc. which may allow exhaust gases to enter the passenger compartment. Seal all body openings with silicone or body putty.

4   Rattles and other noises can often be traced to the exhaust system, especially the mounts and hangers. Try to move the pipes, muffler and catalytic converter. If the components can come in contact with the body or suspension parts, secure the exhaust system with new mounts.

5   Check the running condition of the engine by inspecting inside the end of the tailpipe. The exhaust deposits here are an indication of engine state-of-tune. If the pipe is black and sooty or coated with white deposits, the engine may need a tune-up, including a thorough fuel system inspection and adjustment.

15   Seat belt check (every 6000 miles or 6 months)

1   Check seat belts, buckles, latch plates and guide loops for obvious damage and signs of wear.

2   See if the seat belt reminder light comes on when the key is turned to the Run or Start position. A chime should also sound.

3   The seat belts are designed to lock up during a sudden stop or impact, yet allow free movement during normal driving. Make sure the retractors return the belt against your chest while driving and rewind the belt fully when the buckle is unlatched.

4   If any of the above checks reveal problems with the seat belt system, replace parts as necessary.

16   Underhood hose check and replacement (every 15,000 miles or 12 months)

Warning: *Replacement of air conditioning hoses must be left to a dealer service department or air conditioning shop that has the equipment to depressurize the system safely. Never remove air conditioning components or hoses until the system has been depressurized.*

General

1   High temperatures under the hood can cause deterioration of the rubber and plastic hoses used for engine, accessory and emission systems operation. Periodic inspection should be made for cracks, loose clamps, material hardening and leaks.

2   Information specific to the cooling system hoses can be found in Section 17.

3   Most (but not all) hoses are secured to the fittings with clamps. Where clamps are used, check to be sure they haven't lost their tension, allowing the hose to leak. If clamps aren't used, make sure the hose has not expanded and/or hardened where it slips over the fitting, allowing it to leak.

PCV system hose

4   To reduce hydrocarbon emissions, crankcase blow-by gas is vented through the PCV valve in the rocker arm cover to the intake manifold via a rubber hose on most models. The blow-by gases mix with incoming air in the intake manifold before being burned in the combustion chambers.

5   Check the PCV hose for cracks, leaks and other damage. Disconnect it from the valve cover and the intake manifold and check the inside for obstructions. If it's clogged, clean it out with solvent.

Vacuum hoses

6   It's quite common for vacuum hoses, especially those in the emissions system, to be color coded or identified by colored stripes molded into them. Various systems require hoses with different wall thickness, collapse resistance and temperature resistance. When replacing hoses, be sure the new ones are made of the same material.

7   Often the only effective way to check a hose is to remove it completely from the vehicle. If more than one hose is removed, be sure to label the hoses and fittings to ensure correct installation.

8   When checking vacuum hoses, be sure to include any plastic T-fittings in the check. Inspect the fittings for cracks and the hose where it fits over each fitting for distortion, which could cause leakage.

9   A small piece of vacuum hose (1/4-inch inside diameter) can be used as a stethoscope to detect vacuum leaks. Hold one end of the hose to your ear and probe around vacuum hoses and fittings, listening for the "hissing" sound characteristic of a vacuum leak.

Warning: *When probing with the vacuum hose stethoscope, be careful not to come into contact with moving engine components such as drivebelts, the cooling fan, etc.*

Fuel hose

Warning: *Gasoline and diesel fuels are flammable, so take extra precautions when you work on any part of the fuel system. Don't smoke or allow open flames or bare light bulbs near the work area, and don't work in a garage where a gas-type appliance (such as a water heater or clothes dryer) is present. Since fuel is carcinogenic, wear latex gloves when there's a possibility of being exposed to fuel, and, if you spill any fuel on your skin, rinse it off immediately with soap and water. Mop up any spills immediately and do not store fuel-soaked rags where they could ignite. The fuel system is under constant pressure, so, if*

Check for a chafed area that could fail prematurely.

Check for a soft area indicating the hose has deteriorated inside.

Overtightening the clamp on a hardened hose will damage the hose and cause a leak.

Check each hose for swelling and oil-soaked ends. Cracks and breaks can be located by squeezing the hose.

17.4 Hoses, like drivebelts, have a habit of failing at the worst possible time - to prevent the inconvenience of a blown radiator or heater hose, inspect them carefully as shown here

any fuel lines are to be disconnected, the fuel pressure in the system must be relieved first (see Chapter 4A for more information). When you perform any kind of work on the fuel system, wear safety glasses and have a Class B type fire extinguisher on hand.

10   The fuel lines are usually under pressure, so if any fuel lines are to be disconnected be prepared to catch spilled fuel.

Warning: Your vehicle is equipped with fuel injection and you must relieve the fuel system pressure before servicing the fuel lines. Refer to Chapter 4A for the fuel system pressure relief procedure.

11   Check all flexible fuel lines for deterioration and chafing. Check especially for cracks in areas where the hose bends and just before fittings, such as where a hose attaches to the fuel pump, fuel filter and fuel injection unit.

12   When replacing a hose, use only hose that is specifically designed for your fuel injec-

tion system.

13   Spring-type clamps are sometimes used on fuel return or vapor lines. These clamps often lose their tension over a period of time, and can be "sprung" during removal. Replace all spring-type clamps with screw clamps whenever a hose is replaced. Some fuel lines use spring-lock type couplings, which require a special tool to disconnect. See Chapter 4A for more information on this type of couplings.

Metal lines

14   Sections of metal line are often used for fuel line between the fuel pump and the fuel injection unit. Check carefully to make sure the line isn't bent, crimped or cracked.

15   If a section of metal fuel line must be replaced, use seamless steel tubing only, since copper and aluminum tubing do not have the strength necessary to withstand vibration caused by the engine.

16   Check the metal brake lines where they enter the master cylinder and brake proportioning unit (if used) for cracks in the lines and loose fittings. Any sign of brake fluid leakage calls for an immediate thorough inspection of the brake system.

17   Cooling system check (every 15,000 miles or 12 months)

1   Many major engine failures can be attributed to a faulty cooling system. The cooling system also plays an important role in prolonging transmission life because it cools the fluid.

2   The engine should be cold for the cooling system check, so perform the following procedure before the vehicle is driven for the day or after it has been shut off for at least three hours.

3   Remove the cap from the expansion tank. Clean the cap thoroughly, inside and out, with clean water. Also clean the filler neck on the expansion tank. The presence of rust or corrosion in the filler neck means the coolant should be changed (see Section 25). The coolant inside the radiator should be relatively clean and transparent. If it's rust colored, drain the system and refill it with new coolant.

Note: Diesel engine models are equipped with a secondary expansion tank mounted to the cooling fan shroud. The secondary system is used for the Charge Air Cooler (CAC) system on the turbocharger.

4   Carefully check the radiator hoses and the smaller diameter heater hoses (see illustration). Inspect each coolant hose along its entire length, replacing any hose which is cracked, swollen or deteriorated. Cracks will show up better if the hose is squeezed. Pay close attention to hose clamps that secure the hoses to cooling system components. Hose clamps can pinch and puncture hoses, resulting in coolant leaks. Some hoses are hidden from view so sometimes you'll have to trace a coolant leak.

5   Make sure that all hose connections are tight. A leak in the cooling system will usually show up as white or rust colored deposits on

the area adjoining the leak. If wire-type clamps are used on the hoses, it may be a good idea to replace them with screw-type clamps.

6   Clean the front of the radiator and air conditioning condenser with compressed air, if available, or a soft brush. Remove all bugs, leaves, etc., embedded in the radiator fins. Be extremely careful not to damage the cooling fins or cut your fingers on them.

7   If the coolant level has been dropping consistently and no leaks are detectable, have the radiator cap and cooling system pressure checked at a service station.

18   Fuel system check (every 15,000 miles or 12 months)

Warning: Gasoline and diesel fuels are flammable, so take extra precautions when you work on any part of the fuel system. Don't smoke or allow open flames or bare light bulbs near the work area, and don't work in a garage where a gas-type appliance (such as a water heater or clothes dryer) is present. Since fuel is carcinogenic, wear latex gloves when there's a possibility of being exposed to fuel, and, if you spill any fuel on your skin, rinse it off immediately with soap and water. Mop up any spills immediately and do not store fuel-soaked rags where they could ignite. When you perform any kind of work on the fuel system, wear safety glasses and have a Class B type fire extinguisher on hand. The fuel system is under constant pressure, so, before any lines are disconnected, the fuel system pressure must be relieved (see Chapter 4A).

1   If you smell gasoline or diesel fuel while driving or after the vehicle has been sitting in the sun, inspect the fuel system immediately.

2   Remove the fuel filler cap and inspect it for damage and corrosion. The gasket should have an unbroken sealing imprint. If the gasket is damaged or corroded, install a new cap.

3   Inspect the fuel feed and return lines for cracks. Make sure that the connections between the fuel lines and the fuel injection system and between the fuel lines and the in-line fuel filter are tight.

Warning: Your vehicle is fuel injected, so you must relieve the fuel system pressure before servicing fuel system components. The fuel system pressure relief procedure is outlined in Chapter 4A.

4   Since some components of the fuel system - the fuel tank and part of the fuel feed and return lines, for example - are underneath the vehicle, they can be inspected more easily with the vehicle raised on a hoist. If that's not possible, raise the vehicle and support it on jackstands.

5   With the vehicle raised and safely supported, inspect the gas tank and filler neck for punctures, cracks and other damage. The connection between the filler neck and the tank is particularly critical. Sometimes a rubber filler neck will leak because of loose clamps or deteriorated rubber. Inspect all fuel tank mounting brackets and straps to be sure that the tank is

19.6 Loosen the fuel filter bracket bolt

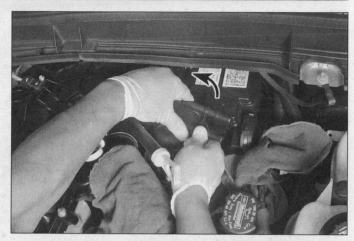

19.9a Rotate the filter counterclockwise until it is unlocked from the bracket . . .

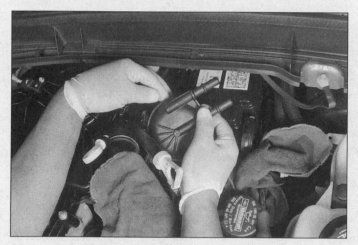

19.9b . . . then lift the filter up and out

19.12 Location of the primary fuel filter is in the diesel fuel conditioning module (DFCM)

securely attached to the vehicle.

Warning: *Do not, under any circumstances, try to repair a fuel tank (except rubber components). A welding torch or any open flame can easily cause fuel vapors inside the tank to explode.*

6   Carefully check all rubber hoses and metal lines leading away from the fuel tank. Check for loose connections, deteriorated hoses, crimped lines and other damage. Repair or replace damaged sections as necessary (see Chapter 4A).

19   Fuel filter replacement (every 15,000 miles or 12 months)

Warning: *Gasoline and diesel fuels are flammable, so take extra precautions when you work on any part of the fuel system. Don't smoke or allow open flames or bare light bulbs near the work area, and don't work in a garage where a gas-type appliance (such as a water heater or clothes dryer) is present. Since fuel is carcinogenic, wear latex gloves when there's a possibility of being exposed to fuel,*

and, if you spill any fuel on your skin, rinse it off immediately with soap and water. Mop up any spills immediately and do not store fuel-soaked rags where they could ignite. When you perform any kind of work on the fuel system, wear safety glasses and have a Class B type fire extinguisher on hand.

Gasoline engine

1   The fuel filter is located on the inlet of the fuel pump and is not serviceable separately from the fuel pump module (see Chapter 4A).

Diesel engines

2   Diesel engine models use two fuel filters, one located on the top of the engine and the under the vehicle on the frame rail that is part of the water separator or Diesel Fuel Conditioner Module (DFCM).

3   Relieve the fuel system pressure (see Chapter 4B).

4   Disconnect the battery cable(s) from the negative battery terminal(s).

Secondary (engine mounted) filter

5   Remove the engine cover (if equipped).

6   Loosen the fuel filter bracket bolt (see illustration).

7   Place a few rags or newspaper below the fuel filter and filter connections.

8   Disconnect the fuel filter quick-connect fittings and remove the lines from the filter (see Chapter 4B).

9   Rotate the filter housing to release it from the bracket, then lift the filter out of the housing (see illustrations).

10   Installation is the reverse of removal, start the engine and check for fuel leaks.

Water separator/Diesel Fuel Conditioner Module (DFCM) filter (primary filter)

11   This filter is part of the diesel fuel conditioning module (DFCM) that is bolted to the frame of the vehicle. For additional information on the (DFCM), refer to Chapter 4B.

12   Raise the vehicle and support it securely on jackstands. Locate the fuel conditioning module just behind the transfer case (if equipped) on the left frame rail (see illustration).

13   Place a drain pan and rags under the fuel conditioning module.

19.14 Disconnect the water-in-fuel sensor electrical connector

19.15 Loosen the fuel filter cover three full turns

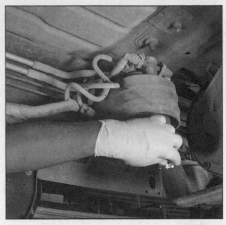

19.16 Loosen the drain plug and allow the fuel to drain into the pan

19.17 Lower the fuel filter cover (1) then remove the filter (2) from the cover

19.18 Remove the filter cover O-ring

19.20 Push the filter into the cover until the filter has snapped into place

14   Disconnect the water-in-fuel sensor electrical connector (see illustration).

15   Place a large socket and ratchet on the fuel filter cover (see illustration) then loosen the fuel filter cover and unscrew it, counterclockwise at least three full turns.

Note: *There is a check valve in the (DFCM), that will stop the fuel from siphoning out of the tank. It works when the filter cover is loosened three full turns or removed completely.*

16   Loosen the drain plug until the O-ring is starting to show and allow the fuel to drain (see illustration).

17   Lower the fuel filter cover down and remove the filter from the cover (see illustration).

18   Remove and replace the filter cover O-ring (see illustration).

19   Install a new filter cover O-ring onto the housing.

20   Install the fuel filter into the fuel filter cover (see illustration) until the filter snaps into the locking tabs at the base of the cover.

21   Install the fuel filter cover assembly onto the DFCM and rotate the cover clockwise until

it contacts the mechanical stops. DO NOT use power tools to tighten the cover or the DFCM or filter cover may be damaged.

22   Installation is the reverse of removal.

23   Start the engine and check for fuel leaks before driving the vehicle.

20   Steering and suspension check (every 15,000 miles or 12 months)

Note: *The steering linkage and suspension components should be checked periodically. Worn or damaged suspension and steering linkage components can result in excessive and abnormal tire wear, poor ride quality and vehicle handling and reduced fuel economy. For detailed illustrations of the steering and suspension components, refer to Chapter 10.*

Shock absorber check

1   Park the vehicle on level ground, turn the engine off and set the parking brake. Check the tire pressures.

2   Push down at one corner of the vehicle,

then release it while noting the movement of the body. It should stop moving and come to rest in a level position within one or two bounces.

3   If the vehicle continues to move up-and-down or if it fails to return to its original position, a worn or weak shock absorber is probably the reason.

4   Repeat the above check at each of the three remaining corners of the vehicle.

5   Raise the vehicle and support it securely on jackstands.

6   Check the shock absorbers for evidence of fluid leakage (see illustration). A light film of fluid is no cause for concern. Make sure that any fluid noted is from the shocks and not from some other source. If leakage is noted, replace the shocks as a set.

7   Check the shocks to be sure that they are securely mounted and undamaged. Check the upper mounts for damage and wear. If damage or wear is noted, replace the shocks as a set (front or rear).

8   If the shocks must be replaced, refer to Chapter 10 for the procedure.

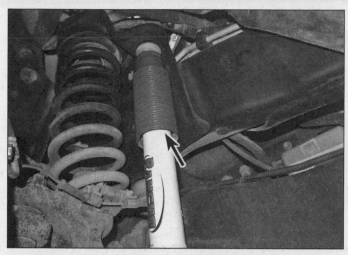

20.6 Check the shocks for leakage at the indicated area

20.9a Inspect the steering and suspension components for torn grease seals (arrows)

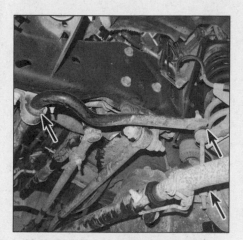

20.9b Check the stabilizer bar and link bushings for deterioration at the front and the rear of the vehicle

20.11 With the steering wheel locked and the vehicle raised, grasp the front tire as shown and try to move it back-and-forth - if any play is noted, check the tie-rod ends for looseness

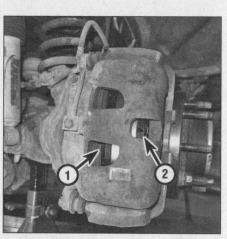

21.6 You will find an inspection hole(s) like this in each caliper - placing a ruler across the hole should enable you to determine the thickness of remaining pad material for the inner (1) and outer pads (2)

Steering and suspension check

9   Visually inspect the steering and suspension components for damage and distortion. Look for damaged seals, boots and bushings and leaks of any kind (see illustrations).
10   Clean the lower end of the steering knuckle. Have an assistant grasp the lower edge of the tire and move the wheel in-and-out while you look for movement at the steering knuckle-to-control arm balljoint. If there is 1/32-inch or more movement the suspension balljoint(s) must be replaced.
11   Grasp each front tire at the front and rear edges, push in at the front, pull out at the rear and feel for play in the steering system components. If any freeplay is noted, check the tie-rod ends for looseness (see illustration).
12   Additional steering and suspension system information and illustrations can be found in Chapter 10.

21   Brake check (every 15,000 miles or 12 months)

Warning: *The dust created by the brake system is harmful to your health. Never blow it out with compressed air and don't inhale any of it. An approved filtering mask should be worn when working on the brakes. Do not, under any circumstances, use petroleum-based solvents to clean brake parts. Use brake system cleaner only! Try to use non-asbestos replacement parts whenever possible.*
Note: *For detailed photographs of the brake system, refer to Chapter 9.*
1   In addition to the specified intervals, the brakes should be inspected every time the wheels are removed or whenever a defect is suspected.
2   Any of the following symptoms could indicate a potential brake system defect: The vehicle pulls to one side when the brake pedal is depressed; the brakes make squealing or dragging noises when applied; brake pedal travel is excessive; the pedal pulsates; brake fluid leaks, usually onto the inside of the tire or wheel.
3   Loosen the wheel lug nuts.
4   Raise the vehicle and place it securely on jackstands.
5   Remove the wheels (see *Jacking and towing* at the front of this book, or your owner's manual, if necessary).

Disc brakes

6   There are two pads (an outer and an inner) in each caliper. The pads are visible through inspection holes in each caliper (see illustration).
7   Check the pad thickness by looking at each end of the caliper and through the inspection hole in the caliper body. If the lining

21.11 Check along the brake hoses and at each fitting for deterioration and cracks

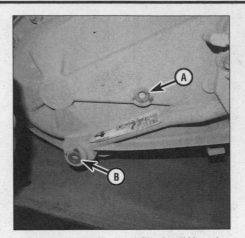

22.1 The transfer case fill plug (A) and drain plug (B) are located on the rear of the transfer case

23.2a Some check/fill plugs (arrow) are located at the front of the differential housing

material is less than the thickness listed in this Chapter's Specifications, replace the pads.

Note: *Keep in mind that the lining material is riveted or bonded to a metal backing plate and the metal portion is not included in this measurement.*

8   If it is difficult to determine the exact thickness of the remaining pad material by the above method, or if you are at all concerned about the condition of the pads, remove the caliper(s), then remove the pads from the calipers for further inspection (refer to Chapter 9).

9   Once the pads are removed from the calipers, clean them with brake cleaner and re-measure them with a ruler or a vernier caliper.

10   Measure the disc thickness with a micrometer to make sure that it still has service life remaining. If any disc is thinner than the specified minimum thickness, replace it (see Chapter 9). Even if the disc has service life remaining, check its condition. Look for scoring, gouging and burned spots. If these conditions exist, remove the disc and have it resurfaced (see Chapter 9).

11   Before installing the wheels, check all brake lines and hoses for damage, wear, deformation, cracks, corrosion, leakage, bends and twists, particularly in the vicinity of the rubber hoses at the calipers (see illustration). Check the clamps for tightness and the connections for leakage. Make sure that all hoses and lines are clear of sharp edges, moving parts and the exhaust system. If any of the above conditions are noted, repair, reroute or replace the lines and/or fittings as necessary (see Chapter 9).

Brake booster check

12   Sit in the driver's seat and perform the following sequence of tests.

13   With the brake fully depressed, start the engine - the pedal should move down a little when the engine starts.

14   With the engine running, depress the brake pedal several times - the travel distance should not change.

15   Depress the brake, stop the engine and hold the pedal in for about 30 seconds - the pedal should neither sink nor rise.

16   Restart the engine, run it for about a minute and turn it off. Then firmly depress the brake several times - the pedal travel should decrease with each application.

17   If your brakes do not operate as described, the brake booster has failed. Refer to Chapter 9 for the replacement procedure.

Parking brake

18   One method of checking the parking brake is to park the vehicle on a steep hill with the parking brake set and the transmission in Neutral (stay in the vehicle for this check!). If the parking brake cannot prevent the vehicle from rolling, it's in need of adjustment (see Chapter 9).

---

22   Transfer case lubricant level check (4WD models) (every 15,000 miles or 12 months)

---

1   The lubricant level is checked by removing a plug from the side of the case (see illustration). If the vehicle is raised to gain access to the plug, be sure to support it safely on jackstands - DO NOT crawl under the vehicle when it's supported only by a jack!

2   With the engine and transfer case cold, remove the plug. If lubricant immediately starts leaking out, thread the plug back into the case - the level is correct. If it doesn't, completely remove the plug and reach inside the hole with your little finger. The level should be even with the bottom of the plug hole.

3   If more lubricant is needed, use a syringe or small pump to add it through the opening.

4   Thread the plug back into the case and tighten it securely. Drive the vehicle, then check for leaks around the plug.

23.2b On other models the check/fill plug is on the differential cover - use a 3/8-inch drive ratchet or breaker bar and an extension to remove it

---

23   Differential lubricant level check (every 15,000 miles or 12 months)

---

1   The differential has a check/fill plug which must be removed to check the lubricant level. If the vehicle is raised to gain access to the plug, be sure to support it safely on jackstands - DO NOT crawl under the vehicle when it's supported only by the jack!

2   Remove the check/fill plug from the differential (see illustrations).

3   Use your little finger as a dipstick to make sure the lubricant level is even with the bottom of the plug hole. If not, use a syringe to add the recommended lubricant until it just starts to run out of the opening. On some models a tag is located in the area of the plug which gives information regarding lubricant type, particularly on models equipped with a Traction Lok differential.

4   Install the plug and tighten it securely.

24.6 Check the diesel engine air restriction gauge and change the filter element if necessary

24.7a Release the four filter cover clips . . .

24.7b . . . then pull the housing toward the front of the vehicle body to disengage the lock tabs

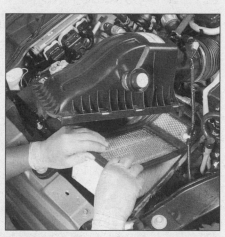

24.8 Remove the air filter element from the air filter housing

24.12 Reset the air restriction gauge by pressing the button on the end of the gauge

24 Air filter check and replacement (every 30,000 miles or 24 months)

Gasoline engines

1   The air filter is located inside a housing at the left (driver's) side of the engine compartment. To remove the air filter, disconnect the mass air flow electrical connector, release the clamp that secures the two halves of the air cleaner housing together, then separate the cover halves and remove the air filter element. On models with the MAF sensor located in the air intake tube, disconnect the sensor electrical connector before changing the air filter.

2   Inspect the outer surface of the filter element. If it is dirty, replace it. If it is only moderately dusty, it can be reused by blowing it clean from the back to the front surface with compressed air. Because it is a pleated paper type filter, it cannot be washed or oiled. If it cannot be cleaned satisfactorily with compressed air, discard and replace it. While the cover is off, be careful not to drop anything down into the housing.

Caution: *Never drive the vehicle with the air cleaner removed. Excessive engine wear could result and backfiring could even cause a fire under the hood.*

3   Wipe out the inside of the air cleaner housing.

4   Place the new filter into the air cleaner housing, making sure it seats properly.

5   Installation of the housing is the reverse of removal.

Diesel engines

6   Since an uninterrupted air supply is crucial to the operation of a diesel engine, these models have an air filter restriction gauge mounted on the intake side of the air cleaner housing to monitor the condition of the air filter element. The air filter restriction gauge should be checked periodically to determine if dirt in the element is restricting the air flow. Regardless of mileage, replace the filter with a new one if the gauge is at the "Change Filter" mark (see illustration).

7   To replace the filter, release the filter cover clips (see illustration) then pull the housing toward the front of the vehicle body to disengage the lock tabs (see illustration).

8   Lift the cover up and remove the air filter (see illustration).

9   Wipe out the inside of the air cleaner housing.

10   Place the new filter into the air cleaner housing, making sure it seats properly.

11   Installation of the housing is the reverse of removal.

12   After replacing the filter element, reset the air restriction gauge by pressing the button on the end (see illustration).

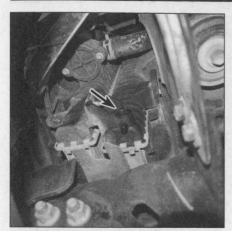

25.4 The radiator drain fitting is located at the lower corner of the radiator

25.5 The block drain plugs (arrow) are generally located about one to two inches above the oil pan - there is one on each side of the engine block

26.5 Remove the drain plug and allow the fluid to drain

25   Cooling system servicing (draining, flushing and refilling) (every 30,000 miles or 24 months)

Warning: *Do not allow antifreeze to come in contact with your skin or painted surfaces of the vehicle. Rinse off spills immediately with plenty of water. Antifreeze is highly toxic if ingested. Never leave antifreeze lying around in an open container or in puddles on the floor; children and pets are attracted by its sweet smell and may drink it. Check with local authorities about disposing of used antifreeze. Many communities have collection centers which will see that antifreeze is disposed of safely.*

1   Periodically, the cooling system should be drained, flushed and refilled to replenish the antifreeze mixture and prevent formation of rust and corrosion, which can impair the performance of the cooling system and cause engine damage. When the cooling system is serviced, all hoses and the expansion tank cap should be checked and replaced if necessary.

Draining

2   Apply the parking brake and block the wheels. If the vehicle has just been driven, wait several hours to allow the engine to cool down before beginning this procedure.

3   Once the engine is completely cool, remove the expansion tank cap.

4   Move a large container under the radiator drain to catch the coolant. Attach a 3/8-inch diameter hose to the drain fitting to direct the coolant into the container, then open the drain fitting (a pair of pliers may be required to turn it) (see illustration).

5   After the coolant stops flowing out of the radiator, move the container under the engine block drain plugs and allow the coolant in the block to drain (see illustration).

6   While the coolant is draining, check the condition of the radiator hoses, heater hoses and clamps (refer to Section 17 if necessary). Replace any damaged clamps or hoses.

7   Reinstall the block drain plugs and tighten them securely.

Flushing

8   Close the radiator drain fitting, then fill the cooling system with clean water, following the Refilling procedure (see Step 14).

9   Start the engine and allow it to reach normal operating temperature, then rev up the engine a few times.

10   Turn the engine off and allow it to cool completely, then drain the system as described earlier.

11   Repeat Steps 8 through10 until the water being drained is free of contaminants.

12   In severe cases of contamination or clogging of the radiator, remove the radiator (see Chapter 3) and have a radiator repair facility clean and repair it if necessary.

13   Many deposits can be removed by the chemical action of a cleaner available at auto parts stores. Follow the procedure outlined in the manufacturer's instructions.

Note: *When the coolant is regularly drained and the system refilled with the correct antifreeze/water mixture, there should be no need to use chemical cleaners or descalers.*

Refilling

14   Close and tighten the radiator drain.

15   Place the heater temperature control in the maximum heat position.

16   Slowly add new coolant (a 50/50 mixture of water and antifreeze) to the expansion tank until the level is between the MIN and MAX marks.

17   Leave the expansion tank cap off and run the engine in a well-ventilated area until the thermostat opens (coolant will begin flowing through the radiator and the upper radiator hose will become hot).

18   Turn the engine off and let it cool. Add more coolant mixture to bring the level between the MIN and MAX marks on the expansion tank.

19   Squeeze the upper radiator hose to expel air, then add more coolant mixture if necessary. Replace the expansion tank cap.

20   Start the engine, allow it to reach normal operating temperature and check for leaks. Also, set the heater and blower controls to the maximum setting and check to see that the heater output from the air ducts is warm. This is a good indication that all air has been purged from the cooling system.

26   Automatic transmission fluid and filter change (every 30,000 miles or 24 months)

1   At the specified intervals, the transmission fluid should be drained and replaced. Since the fluid will remain hot long after driving, perform this procedure only after the engine has cooled down completely.

2   Before beginning work, purchase the transmission fluid specified in *Recommended lubricants and fluids* at the front of this Chapter, a new torque converter drain plug, a new filter and, if necessary, a new gasket. Never reuse the old filter!

Note: *The gasket is reusable as long as it isn't damaged.*

3   Other tools necessary for this job include jackstands to support the vehicle in a raised position, a drain pan capable of holding at least ten quarts, newspapers and clean rags.

4   Raise the vehicle and support it securely on jackstands.

5   With the drain pan in place, remove the plug and allow the fluid to run into the pan (see illustration).

Note: *If your fluid pan isn't equipped with a drain plug, loosen all of the pan bolts, then remove the front and side bolts. Carefully pry the rear of the pan down and allow the fluid to drain. Now remove the rear pan bolts and detach the pan, being careful not to spill the remaining fluid.*

26.8 Make sure the pan is completely clean and dried

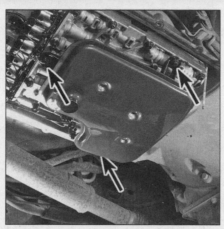

26.9a Remove the filter mounting bolts . . .

26.9b . . . then pull the filter straight down to remove it

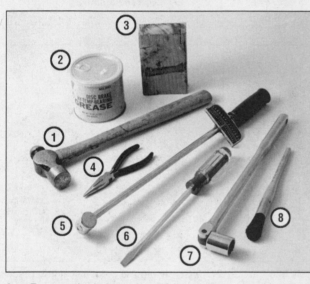

27.1 Tools and materials needed for front wheel bearing maintenance

1   **Hammer** - *A common hammer will do just fine*
2   **Grease** - *High-temperature grease that is formulated specially for front wheel bearings should be used*
3   **Wood block** - *If you have a scrap piece of 2x4, it can be used to drive the new seal into the hub*
4   **Needle-nose pliers** - *Used to straighten and remove the cotter pin in the spindle*
5   **Torque wrench** - *Used for pre-loading the bearing before adjustment*
6   **Screwdriver** - *Used to remove the seal from the hub (a long screwdriver is preferred)*
7   **Socket/breaker bar** - *Needed to loosen the nut on the spindle if it's extremely tight*
8   **Brush** - *Together with some clean solvent, this will be used to remove old grease from the hub and spindle*

6   Remove the bolts and carefully pry the transmission pan loose with a screwdriver. Don't damage the pan or transmission gasket surfaces or leaks could develop.

7   Carefully clean the gasket surface of the transmission to remove all traces of the old gasket and sealant.

8   Clean the transmission pan with solvent and dry it thoroughly (see illustration).

9   Remove the filter mounting bolts and the old filter from the transmission (see illustrations). If the filter seal did not come out with the filter, remove it from the transmission being careful not to gouge the seal bore in any way.

10   Install a new seal and filter.

11   Make sure the gasket surface on the transmission pan is clean, then install a new gasket (see illustration). Put the pan in place against the transmission and install the bolts. Working around the pan, tighten each bolt a little at a time until the final torque figure listed in this Chapter's Specifications is reached. Don't overtighten the bolts!

12   Lower the vehicle and add 6-1/2 quarts of automatic transmission fluid through the filler tube (see Section 7).

13   With the transmission in Park and the parking brake set, run the engine at a fast idle, but don't race it.

14   Move the gear selector through each range and back to Park. Check the fluid level. Add fluid if needed to reach the correct level.
Caution: *Add fluid 1/2-pint at a time to avoid overfilling the transmission.*

15   Check under the vehicle for leaks during the first few trips.

---

27   Front wheel bearing check, repack and adjustment (2WD models) (every 30,000 miles or 24 months)

Check and repack

1   In most cases the front wheel bearings will not need servicing until the brake pads are changed. However, the bearings should be checked whenever the front of the vehicle is raised for any reason. Several items, including a torque wrench and special grease, are required for this procedure (see illustration).

2   With the vehicle securely supported on jackstands, spin each wheel and check for noise, rolling resistance and freeplay.

3   Grasp the top of each tire with one hand and the bottom with the other (see illustration). Move the wheel in-and-out on the spindle. If there's any noticeable movement, the bearings should be checked and then repacked with grease or replaced if necessary.

27.3 To check the wheel bearings, move the tire in and out as shown above - if there's any noticeable freeplay, the bearings should be checked and then repacked with grease (or replaced if necessary)

27.6 Dislodge the dust cap by working around the outer circumference with a screwdriver or a hammer and chisel

27.7 Remove the cotter pin and discard it - use a new one when the disc assembly is reinstalled

27.10 After the washer and outer wheel bearing have been dislodged, pull the disc off the spindle

27.11 Use a seal puller or large screwdriver to remove the inner grease seal - note the seal installed position

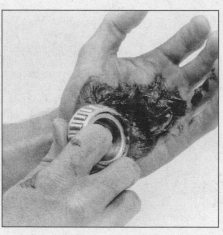

27.15 Work the grease completely into the bearing rollers

27.19 Gently tap the grease seal into place

4   Remove the wheel.

5   Remove the brake caliper (see Chapter 9) and hang it out of the way on a piece of wire.

6   Dislodge the dust cap from the hub/disc assembly using a screwdriver or hammer and chisel (see illustration).

7   Straighten the bent ends of the cotter pin, then pull the cotter pin out of the nut lock (see illustration). Discard the cotter pin and use a new one during reassembly.

8   Remove the nut lock, nut and washer from the end of the spindle.

9   Pull the hub/disc assembly out slightly, then push it back into its original position. This should force the outer bearing off the spindle enough so it can be removed.

10   Pull the disc assembly off the spindle (see illustration).

11   Use a screwdriver or a seal puller tool to pry the seal out of the rear of the hub (see illustration). Note how the seal is installed.

12   Remove the inner wheel bearing from the hub.

13   Use solvent to remove all traces of the old grease from the bearings, hub and spindle. A small brush may prove helpful; however make sure no bristles from the brush embed themselves inside the bearing rollers. Allow the parts to air dry.

14   Carefully inspect the bearings for cracks, heat discoloration, worn rollers, etc. Check the bearing races inside the hub for wear and damage. If the bearing races are defective, the hubs should be taken to a machine shop with the facilities to remove the old races and press new ones in. Note that the bearings and races come as matched sets and old bearings should never be installed on new races.

15   Use high-temperature front wheel bearing grease to pack the bearings. Work the grease completely into the bearings, forcing it between the rollers, cone and cage from the larger diameter side (see illustration).

16   Apply a thin coat of grease to the spindle at the outer bearing seat, inner bearing seat, shoulder and seal seat.

17   Put a small quantity of grease inboard of

each bearing race inside the hub.

18   Place the grease-packed inner bearing into the rear of the hub and put a little more grease outboard of the bearing.

19   Place a new seal over the inner bearing and tap the seal evenly into place until it's flush with the hub (see illustration).

20   Carefully place the hub assembly onto the spindle and push the grease-packed outer bearing into position.

Adjustment

21   Install the washer and spindle nut. Tighten the nut only slightly (no more than 12 ft-lbs of torque).

22   Spin the hub in a forward direction while tightening the spindle nut to the Step 1 torque torque listed in this Chapter's Specifications.

23   Loosen the spindle nut 1/2-turn, then tighten the nut to the Step 3 torque listed in this Chapter's Specifications. Install the nut lock and a new cotter pin through the hole in the spindle and the slots in the nut lock. If the

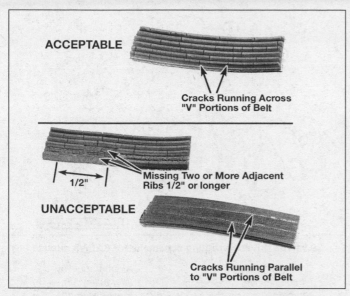

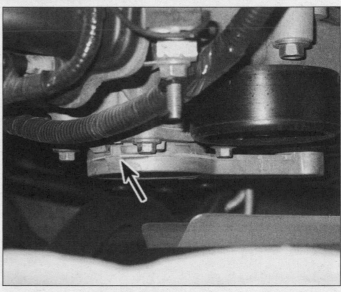

29.4 Small cracks in the underside of a V-ribbed belt are acceptable - lengthwise cracks, or missing pieces that cause the belt to make noise, are cause for replacement

29.5 Belt wear indicator marks are located on the side of the tensioner body - when the belt reaches the maximum wear mark it must be replaced

slots in the nut lock don't line up with the hole in the spindle, reposition the nut lock until they do.

24   Check that the hub/disc assembly spins freely with no noticeable freeplay. If freeplay exists, repeat Steps 22 and 23 until proper adjustment is obtained.

25   Bend the ends of the cotter pin until they're flat against the nut. Cut off any extra length which could interfere with the dust cap.

26   Install the dust cap, lightly tapping it into place with a hammer.

27   Install the brake caliper mount, pads and caliper in the reverse order of removal (see Chapter 9).

28   Install the wheel on the hub and tighten the lug nuts.

29   Lower the vehicle and tighten the lug nuts to the torque listed in this Chapter's Specifications.

28   Brake fluid change (every 30,000 miles or 24 months)

Warning: *Brake fluid can harm your eyes and damage painted surfaces, so use extreme caution when handling or pouring it. Do not use brake fluid that has been standing open or is more than one year old. Brake fluid absorbs moisture from the air. Excess moisture can cause a dangerous loss of braking effectiveness.*

1   At the specified intervals, the brake fluid should be drained and replaced. Since the brake fluid may drip or splash when pouring it, place plenty of rags around the master cylinder to protect any surrounding painted surfaces.

2   Before beginning work, purchase the specified brake fluid (see Recommended lubricants and fluids at the beginning of this Chapter).

3   Remove the cap from the master cylinder reservoir.

4   Using a hand suction pump or similar device, withdraw the fluid from the master cylinder reservoir.

5   Add new fluid to the master cylinder until it rises to the base of the filler neck.

6   Bleed the brake system as described in Chapter 9 at all four brakes until new and uncontaminated fluid is expelled from the bleeder screw. Be sure to maintain the fluid level in the master cylinder as you perform the bleeding process. If you allow the master cylinder to run dry, air will enter the system.

7   Refill the master cylinder with fluid and check the operation of the brakes. The pedal should feel solid when depressed, with no sponginess.

Warning: *Do not operate the vehicle if you are in doubt about the effectiveness of the brake system.*

29   Drivebelt check (every 3000 miles or 3 months) and replacement (every 60,000 miles or 48 months)

1   The drivebelt is located at the front of the engine and plays an important role in the overall operation of the vehicle and its components. Due to its function and material make-up, the drivebelt is prone to failure after a period of time and should be inspected and adjusted periodically to prevent major engine damage.

2   The vehicles covered by this manual are equipped with a single self-adjusting serpentine drivebelt, which is used to drive all of the accessory components such as the alternator, power steering pump, water pump and air conditioning compressor.

Inspection

3   With the engine off, open the hood and locate the drivebelt at the front of the engine. Using your fingers (and a flashlight, if necessary), move along the belts checking for cracks and separation of the belt plies. Also check for fraying and glazing, which gives the belt a shiny appearance. Both sides of each belt should be inspected, which means you will have to twist the belt to check the underside.

4   Check the ribs on the underside of the belt. They should all be the same depth, with none of the surface uneven (see illustration).

5   The tension of the belt is automatically adjusted by the belt tensioner and does not require any adjustments. Drivebelt wear can be checked visually by inspecting the wear indicator marks located on the side of the tensioner body. Locate the belt tensioner at the front of the engine on the right (passenger's) side, adjacent to the lower crankshaft pulley, then find the tensioner operating marks (see illustration). If the indicator mark is outside the operating range, the belt should be replaced.

Replacement

6   If you're working on a diesel model, remove the air intake duct and resonator (see Chapter 4B). Also disconnect the electrical connector from the fan clutch and remove the bracket bolt (see Chapter 3).

29.7 Rotate the tensioner arm to relieve belt tension

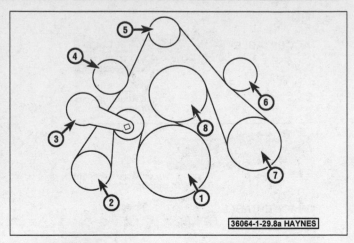

29.9a The drivebelt routing schematic for 6.2L V8 models

1	Crankshaft pulley	5	Alternator pulley
2	Air conditioning compressor	6	Idler pulley
3	Tensioner and pulley	7	Power steering pump pulley
4	Idler pulley	8	Water pump pulley

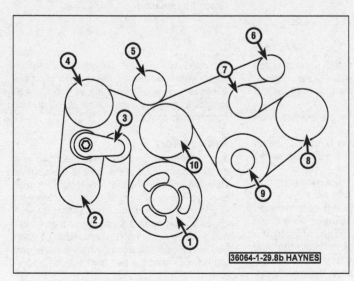

29.9b The drivebelt routing schematic for 6.7L diesel models (with single alternator)

1	Crankshaft pulley	6	Alternator
2	Air conditioning compressor	7	Idler pulley
3	Tensioner and pulley	8	Power steering pump
4	Secondary water pump	9	Water pump
5	Idler pulley	10	Cooling fan drive

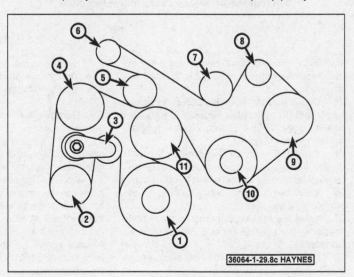

29.9c The drivebelt routing schematic for 6.7L diesel models (with dual alternators)

1	Crankshaft pulley	6	Alternator
2	Air conditioning compressor	7	Idler pulley
3	Tensioner and pulley	8	Alternator
4	Secondary water pump	9	Power steering pump
5	Idler pulley	10	Water pump
		11	Cooling fan drive

7   To replace the belt, rotate the tensioner to relieve the tension on the belt (see illustration). All models have a square hole in the tensioner arm that will accept a 3/8-inch drive breaker bar.

8   Remove the belt from the auxiliary components and carefully release the tensioner.

9   Route the new belt over the various pulleys, again rotating the tensioner to allow the belt to be installed, then release the belt tensioner. Make sure the belt fits properly into the pulley grooves - it must be completely engaged (see illustrations).

Note: *Most models have a drivebelt routing decal on the upper radiator panel to help during drivebelt installation.*

30 Positive Crankcase Ventilation (PCV) valve check (every 60,000 miles or 48 months)

Note: *To maintain efficient operation of the PCV system, clean the hoses and check the PCV valve at the intervals recommended in the* maintenance schedule. For additional information on the PCV system, refer to Chapter 6A.

Note: *This Section applies to models with gasoline engines only.*

1   The PCV valve is located in the left-side valve cover (see illustration).

Note: *When the PCV valve is removed, the retaining tabs on the valve are damaged and a new PCV valve must be installed.*

2   Start the engine and allow it to idle, then disconnect the PCV valve from the valve cover and feel for vacuum at the end of the hose. If vacuum is felt then the hose to the

30.1 PCV valve location

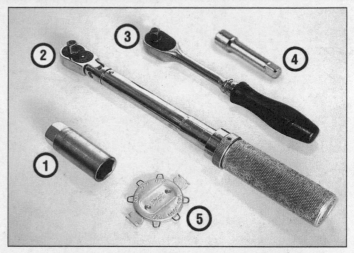

31.2 Tools required for changing spark plugs

1   **Spark plug socket** - This will have special padding inside to protect the spark plug's porcelain insulator
2   **Torque wrench** - Although not mandatory, using this tool is the best way to ensure the plugs are tightened properly
3   **Ratchet** - Standard hand tool to fit the spark plug socket
4   **Extension** - Depending on model and accessories, you may need special extensions and universal joints to reach one or more of the plugs
5   **Spark plug gap gauge** - This gauge for checking the gap comes in a variety of styles. Make sure the gap for your engine is included

PCV valve is not clogged. If a problem is still suspected at the valve replace the valve (see Chapter 6A).

3   If no vacuum is felt at the hose, check for a plugged or cracked hose between the PCV valve and the intake plenum.

4   If the valve is clogged, the hose is also probably plugged. Remove the hose between the valve and the intake manifold and clean it with solvent.

5   After cleaning the hose, inspect it for damage, wear and deterioration. Make sure it fits snugly on the fittings.

6   If necessary, install a new PCV valve. On 6.2L models rotate the PCV valve counterclockwise and remove the valve from the cover. To install the PCV valve insert the valve into the cover and rotate the valve until the retaining tabs click into place. If a new PCV valve is needed on the 6.8L models the valve cover must be replaced (see Chapter 2A).

31   Spark plug check and replacement (every 60,000 miles or 48 months)

1   There are two sets of spark plugs; one set located on the top of the engine (upper plugs) and another on the sides of the engine (lower plugs).

2   In most cases, the tools necessary for spark plug replacement include a spark plug socket which fits onto a ratchet (spark plug sockets are padded inside to prevent damage to the porcelain insulators on the new plugs), various extensions and a gap gauge to check and adjust the gaps on the new plugs (see illustration). A torque wrench should be used to tighten the new plugs.

3   The best approach when replacing the spark plugs is to purchase the new ones in advance, adjust them to the proper gap and replace the plugs one at a time. When buying the new spark plugs, be sure to obtain the correct plug type for your particular engine. This

31.5a Spark plug manufacturers recommend using a wire-type gauge when checking the gap - if the wire does not slide between the electrodes with a slight drag, adjustment is required

31.5b To change the gap, bend the side electrode only, as indicated by the arrows, and be very careful not to crack or chip the porcelain insulator surrounding the center electrode

information can be found in the Specifications Section at the beginning of this Chapter or in the owner's manual.

4   Allow the engine to cool completely before attempting to remove any of the plugs. These engines are equipped with aluminum cylinder heads, which can be damaged if the spark plugs are removed when the engine is hot. While you are waiting for the engine to cool, check the new plugs for defects and adjust the gaps.

5   The gap is checked by inserting the proper thickness gauge between the elec-

trodes at the tip of the plug (see illustration). The gap between the electrodes should be the same as the one specified on the Emissions Control Information label. The wire should just slide between the electrodes with a slight amount of drag. If the gap is incorrect, use the adjuster on the gauge body to bend the curved side electrode slightly until the specified gap is obtained (see illustration). If the side electrode is not exactly over the center electrode, bend it with the adjuster until it is. Check for cracks in the porcelain insulator (if any are found, the plug should not be used).

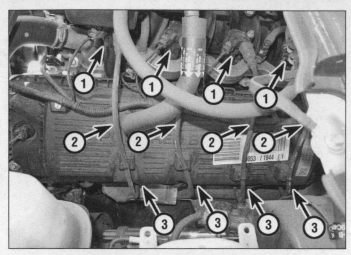

31.6 Spark plug system details

1   Ignition coil and upper   2   Spark plug wire
    spark plug   3   Lower spark plug

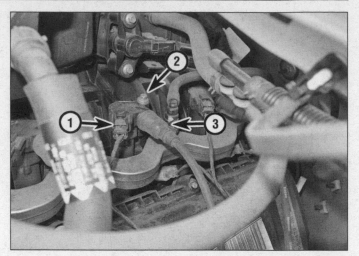

31.7 The ignition coils must be removed to access the upper spark plugs - disconnect the electrical connector (1) and remove the coil retaining screw (2), then remove the spark plug wire (3) - pull straight up and out to remove the coil

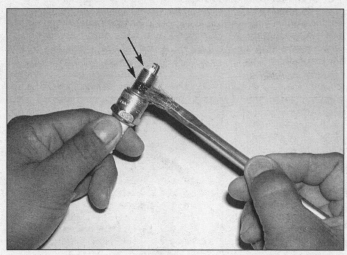

31.12a Apply a thin film of anti-seize compound to the spark plug threads to prevent damage to the cylinder head

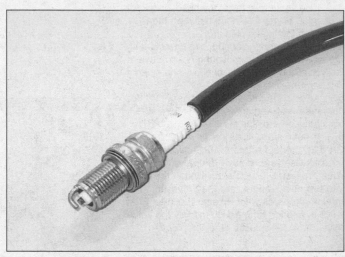

31.12b A length of rubber hose will save time and prevent damaged threads when installing the spark plugs

6   These models are equipped with individual ignition coils mounted on the upper spark plugs which must be removed to access the upper spark plugs, and with ignition wires that are mounted to the sides of the ignition coils and go to the lower spark plugs (see illustration).

7   To remove the upper spark plugs, the individual ignition coils must be removed first to access the spark plugs (see illustration).

8   To remove the lower spark plugs, carefully twist the spark plug wire boot off of the spark plug.

9   If compressed air is available, use it to blow any dirt or foreign material away from the spark plug to eliminate the possibility of debris falling into the cylinder as the spark plug is removed.

10   Place the spark plug socket over the plug and remove it from the engine by turning it in a counterclockwise direction.

11   Compare each old spark plug to those shown on the inside of the back cover to get an indication of the general running condition of the engine.

12   Apply a small amount of anti-seize compound to the spark plug threads (see illustration). Install one of the new plugs into the hole until you can no longer turn it with your fingers, then tighten it with a torque wrench (if available) or the ratchet. It is a good idea to slip a short length of rubber hose over the end of the plug to use as a tool to thread it into place (see illustration). The hose will grip the plug well enough to turn it, but will start to slip if the plug begins to cross-thread in the hole - this will prevent damaged threads and the accompanying repair costs.

13   Before pushing the ignition coil onto the end of the plug, inspect it following the procedures outlined in Section 32.

14   Before installing the spark wires, inspect them following the procedures outlined in Section 35.

15   Repeat the procedure for the remaining spark plugs.

32   Ignition coil check and replacement (every 60,000 miles or 48 months)

1   Clean the coils with a dampened cloth and dry them thoroughly.

2   Inspect each coil for cracks, damage and carbon tracking. Make sure the coil fits securely onto the upper spark plug and the spark plug wire fits securely on each coil. If damage exists, replace the coil (see Section 31).

35.6 Use a spark plug boot pulling tool to remove each end of a spark plug wire from the lower spark plugs (typical V8 shown) - never pull on the wire itself

33 Differential lubricant change (every 100,000 miles or 60 months)

Drain

1   This procedure should be performed after the vehicle has been driven so the lubricant will be warm and therefore flow out of the differential more easily.
2   Raise the vehicle and support it securely on jackstands.
3   The easiest way to drain the differential(s) is to remove the lubricant through the filler plug hole with a suction pump. If the differential cover gasket is leaking, it will be necessary to remove the cover to drain the lubricant (which will also allow you to inspect the differential).

Changing the lubricant with a suction pump

4   Remove the filler plug from the differential (see Section 23).
5   Insert the flexible hose.
6   Work the hose down to the bottom of the differential housing and pump the lubricant out.

Changing rear differential lubricant by removing the cover

7   Move a drain pan, rags, newspapers and wrenches under the vehicle.
8   On models equipped a rear cover, remove the bolts on the lower half of the cover. Loosen the bolts on the upper half and use them to loosely retain the cover. Allow the oil to drain into the pan, then completely remove the cover.
9   Using a lint-free rag, clean the inside of the cover and the accessible areas of the differential housing. As this is done, check for chipped gears and metal particles in the lubricant, indicating that the differential should be

more thoroughly inspected and/or repaired.
10   Thoroughly clean the gasket mating surfaces of the differential housing and the cover plate. Use a gasket scraper or putty knife to remove all traces of the old gasket.
11   Apply a thin layer of RTV sealant to the cover flange, then press a new gasket into position on the cover. Make sure the bolt holes align properly.

Refill

12   Use a hand pump, syringe or funnel to fill the differential housing with the specified lubricant until it's level with the bottom of the filler plug hole.
13   Install the fill plug and tighten it securely.

34 Transfer case lubricant change (4WD models) (every 100,000 miles or 60 months)

1   Drive the vehicle for at least 15 minutes to warm the lubricant in the case. Perform this warm-up procedure with 4WD engaged, if possible.
2   Raise the vehicle and support it securely on jackstands.
3   Remove the drain plug from the lower part of the case and allow the old lubricant to drain completely (see illustration 22.1).
4   After the lubricant has drained completely, reinstall the plug and tighten it securely.
5   Remove the filler plug from the case.
6   Fill the case with the specified lubricant until it is level with the lower edge of the filler hole.
7   Install the filler plug and tighten it securely.
8   Drive the vehicle for a short distance, then check the drain and fill plugs for leakage.

35 Spark plug wire check and replacement (every 100,000 miles or 60 months)

1   The spark plug wires should be checked at the recommended intervals and whenever new spark plugs are installed in the engine. V8 engines have eight ignition coil-on-plugs mounted on the upper spark plugs with eight spark plug wires that go from the ignition coils to the lower spark plugs.
2   Begin this procedure by making a visual check of the spark plug wires while the engine is running. In a darkened garage (make sure there is adequate ventilation) start the engine and observe each plug wire. Be careful not to come into contact with any moving engine parts. If there is a break in the wire, you will see arcing or a small spark at the damaged area. If arcing is noticed, make a note to obtain new wires.
3   Disconnect the plug wire from one lower spark plug (with the engine off). To do this, grab the rubber boot, twist slightly and pull the wire free. Do not pull on the wire itself, only on the rubber boot. A boot-pulling tool is helpful.
4   Check inside the boot for corrosion, which will look like a white crusty powder. Push the wire and boot back onto the end of the spark plug. It should be a tight fit on the plug. If it isn't, remove the wire and use a pair of pliers to carefully crimp the metal connector inside the boot until it fits securely on the end of the spark plug.
5   Using a clean rag, wipe the entire length of the wire to remove any built-up dirt and grease. Once the wire is clean, check for holes, burned areas, cracks and other damage. Don't bend the wire excessively or the conductor inside might break.
6   Disconnect the wire from the ignition coil assembly by using a slight twisting motion while pulling the wire straight out of the coil. Pull only on the rubber boot during removal (see illustration). Check for corrosion and a tight fit in the same manner as the spark plug end. Reattach the wire to the ignition coil.
7   Check the remaining spark plug wires one at a time, making sure they are securely fastened at both ends when the check is complete.
8   If new spark plug wires are required, purchase a new set for your specific engine model. Wire sets are available pre-cut, with the rubber boots already installed. Remove and replace the wires one at a time. The wire routing is extremely important, so be sure to note exactly how each wire is situated before removing it. Release the ignition wire loom clamps to exchange the wires, then snap the clamps back in place on the new wires.

Notes

Chapter 2 Part A
Gasoline engine

Contents

Specifications

General

Displacement	6.2 liters (379 cubic inches)
Bore and stroke	4.015 X 3.74 inches
Cylinder numbers (front to rear)	
Right side	1-2-3-4
Left (driver's) side	5-6-7-8
Firing order	1-5-4-8-6-3-7-2

Camshaft

Lobe lift	
Exhaust	0.309 inch
Intake	0.316 inch
Allowable lobe lift loss	0.005 inch
Endplay	0.0033 to 0.0061 inch
Journal diameter	1.125 to 1.126 inches
Bearing inside diameter	1.128 to 1.129 inches
Journal-to-bearing (oil) clearance	
Standard	0.001 to 0.003 inch
Service limit	0.005 inch
Runout	0.004 inch

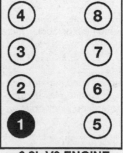

6.2L V8 ENGINE
1-5-4-8-6-3-7-2
36084-01-00.00b HAYNES

Cylinder numbering and firing order

Torque specifications

Ft-lbs (unless otherwise indicated)

Note: One foot-pound (ft-lb) of torque is equivalent to 12 inch-pounds (in-lbs) of torque. Torque values below approximately 15 foot-pounds are expressed in inch-pounds, because most foot-pound torque wrenches are not accurate at these smaller values.

Camshaft cap bolts	
Step 1	53 in-lbs
Step 2	Tighten an additional 45 degrees
Camshaft phaser and sprocket bolts*	
Step 1	80 in-lbs
Step 2	Tighten an additional 90 degrees
Crankshaft pulley bolt*	
Step 1	129
Step 2	Tighten an additional 90 degrees
Crankshaft rear oil seal retainer-to-block bolts	
Step 1	89 in-lbs
Step 2	Tighten an additional 45 degrees
Cylinder head bolts, in sequence (see illustrations 13.19a and 13.19b)*	
Step 1, (M12 bolts)	18
Step 2, (M12 bolts)	55
Step 3, (M12 bolts)	Loosen 180 degrees
Step 4, (M12 bolts)	18
Step 5, (M12 bolts)	37
Step 6, (M8 bolt)	89 in-lbs
Step 7, (M12 bolts)	Tighten an additional 90 degrees
Step 8, (M12 bolts)	Tighten an additional 90 degrees
Step 9, (M8 bolt)	177 in-lbs
Step 10, (M8 bolt)	Tighten an additional 45 degrees
Drivebelt idler and tensioner bolts	18
Engine mounts	
Engine mount-to-crossmember nuts*	148
Engine mount bracket-to-engine block bolts	
Step 1	22
Step 2	Tighten an additional 60 degrees
Engine mount through-bolt	258
Exhaust manifold-to-cylinder head nuts*	
Step 1	18
Step 2	24
Driveplate bolts*	
Step 1, bolts 1 through 4	177 in-lbs
Step 2, bolts 1 through 4	26
Step 3, bolts 1 through 4	Tighten additional 60 degrees
Step 4, bolts 5 through 8	177 in-lbs
Step 5, bolts 5 through 8	26
Step 6, bolts 5 through 8	Tighten an additional 60 degrees
Intake manifold-to-cylinder head bolts, in sequence (see illustration 11.15)*	
Step 1, intake manifold bolts	89 in-lbs
Step 2, intake manifold bolts	Tighten an additional 45 degrees
Step 3, fuel rail bolts	89 in-lbs
Step 4, fuel rail bolts	Tighten an additional 90 degrees
Oil filter adapter bolts	
Step 1	177 in-lbs
Step 2	Tighten an additional 180 degrees
Oil pan-to-engine block bolts, in sequence (see illustration 14.22) *	
Step 1	18 in-lbs
Step 2	89 in-lbs
Step 3	Tighten an additional 45 degrees
Oil pump screen and pick-up tube-to-oil pump bolts	89 in-lbs
Oil pump screen and pick-up tube-to-spacer nut	18
Oil pump-to-engine block bolts	
Step 1	18 in-lbs
Step 2	89 in-lbs
Step 3	177 in-lbs
Step 4, upper bolts only	Tighten an additional 45 degrees
Step 5, lower stud bolts only	Tighten an additional 60 degrees
Rocker arm shaft bolts	
Step1	89 in-lbs
Step 2	177 in-lbs
Step 3	Tighten an additional 60 degrees

Torque specifications (continued) Ft-lbs (unless otherwise indicated)

Note: *One foot-pound (ft-lb) of torque is equivalent to 12 inch-pounds (in-lbs) of torque. Torque values below approximately 15 foot-pounds are expressed in inch-pounds, because most foot-pound torque wrenches are not accurate at these smaller values.*

Timing chain cover bolts, in sequence (see illustration 6.56)*
    Step 1 .... 89 in-lbs
    Step 2 .... 177 in-lbs
    Step 3 .... Tighten an additional 45 degrees
Timing chain cover-to oil pan bolts*
    Step 1 .... 89 in-lbs
    Step 2 .... Tighten an additional 45 degrees
Timing chain guides
    Step 1 .... 89 in-lbs
    Step 2 .... Tighten an additional 45 degrees
Timing chain tensioners
    Step 1 .... 89 in-lbs
    Step 2 .... Tighten an additional 45 degrees
Valve cover bolts* .... 89 in-lbs

* Replace fasteners with new ones whenever they are removed.

1   General Information

1   This Part of Chapter 2 is devoted to in-vehicle repair procedures for the Single Overhead Cams (SOHC) 6.2L V8 engine. The engine has a modular design: a chain-driven single overhead camshaft for each cylinder head, aluminum heads and an iron block.

2   The V8 model use two valves per cylinder and is equipped with a variable camshaft timing system operated by the PCM.

3   The following repair procedures were written as if the engine were still in the vehicle. If the engine has been removed from the vehicle and mounted on a stand, many of the steps outlined in this Part of Chapter 2 will not apply. The Specifications included in this Part of Chapter 2 apply only to the procedures contained in this Part. Information about engine removal and installation, and overhaul is also in Chapter 2C.

2   Repair operations possible with the engine in the vehicle

1   Many major repair operations can be accomplished without removing the engine from the vehicle.

2   If possible, clean the engine compartment and the exterior of the engine with some type of pressure washer before any work is started. It will make the job easier and help keep dirt out of the internal areas of the engine.

3   If vacuum, exhaust, oil or coolant leaks develop, indicating a need for gasket or seal replacement, the repairs can generally be made with the engine in the vehicle. The intake and exhaust manifold gaskets, timing cover gasket, oil pan gasket, crankshaft oil seals and cylinder head gaskets are all accessible with the engine in place.

4   Exterior engine components, such as the intake and exhaust manifolds, the oil pan, the water pump, the starter motor, the alternator and the fuel system components can be removed for repair with the engine in place.

5   Since the cylinder heads can be removed without pulling the engine, valve component servicing can also be accomplished with the engine in the vehicle. Replacement of the timing chain and sprockets and oil pump is also possible with the engine in the vehicle.

6   In extreme cases caused by a lack of necessary equipment, repair or replacement of piston rings, pistons, connecting rods and rod bearings is also possible with the engine in the vehicle. However, this practice is not recommended because of the cleaning and preparation work that must be done to the components involved.

3   Top Dead Center (TDC) for number one piston - locating

1   Top Dead Center (TDC) is the highest point in the cylinder that each piston reaches as it travels up-and-down when the crankshaft turns. Each piston reaches TDC on the compression stroke and again on the exhaust stroke, but TDC generally refers to piston position on the compression stroke.

2   Positioning the number one piston at TDC is an essential part of certain procedures such as timing chain or camshaft replacement.

3   Remove the lower spark plugs (see Chapter 1) and install a compression gauge in the number one cylinder (see the cylinder identification diagrams at the beginning of this Chapter).

4   Turn the crankshaft clockwise until you see compression building up on the gauge - you are on the compression stroke for that cylinder. If you did not see compression build up, continue with one more complete revolution to achieve TDC for the number one cylinder.

5   Remove the compression gauge. Through the number one cylinder spark plug hole insert a wooden dowel or plastic rod and slowly push it down until it reaches the top surface of the piston crown.

Caution: *Don't insert a metal or sharp object into the spark plug hole as the piston crown may be damaged.*

6   With the dowel or rod in place on top of the piston crown, slowly rotate the crankshaft clockwise until the dowel or rod is pushed upward, stops, and then starts to move back down. At this point, rotate the crankshaft slightly counterclockwise until the dowel or rod has reached its uppermost travel. At this point the number one piston is at the TDC position.

7   After the number one piston has been positioned at TDC on the compression stroke, TDC for the next cylinder in the firing order can be located by turning the crankshaft another 90-degrees (1/4 turn) (refer to the firing order in this Chapter's Specifications).

4   Valve covers - removal and installation

Left valve cover

1   Remove the air intake duct (see Chapter 4A).

2   Disconnect the quick-connect fitting for the Positive Crankcse Ventilation (PCV) tube and detach the PCV tube from the PCV valve.

3   Disconnect the left-bank spark plug wires from the ignition coils, then detach the wire retainers from the valve cover.

4   Disconnect the left-side Variable Camshaft Timing (VCT) solenoid electrical connector, then disengage the harness retainers and move the harness out of the way.

5   Disconnect any interfering wiring and hoses such as the brake booster hose.

6   Unscrew the valve cover bolts (see illustration 4.9) and remove the valve cover.
Note: *The valve cover bolts are captive; they stay with the valve cover.*

7   Remove the old valve cover gasket and any gasket material that's stuck to the head.

8   Apply RTV sealant to the mating joints between the timing chain cover and the cylinder head, then install the valve cover and gasket.
Note: *Install the cover within five minutes of applying the RTV sealant.*

9   Tighten the valve cover bolts in sequence (see illustration) gradually and evenly, to the torque listed in this Chapter's Specifications.

10   The remainder of installation is the reverse of removal.

Right valve cover

11   Disconnect the quick-connect fitting for the crankcase ventilation tube and remove the tube.

12   Disconnect the right-bank spark plug wires, from the ignition coils, then detach the wire retainers from the valve cover.

13   Disconnect the right-side Variable Camshaft Timing (VCT) solenoid electrical connector, then disengage the harness retainers and move the harness out of the way.

14   Remove the engine oil dipstick tube mounting bolt and move the tube out of the way.

15   Remove the transmission fluid dipstick

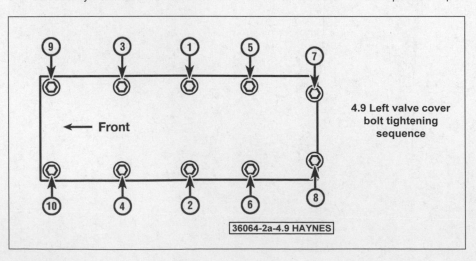

4.9 Left valve cover bolt tightening sequence

← Front

36064-2a-4.9 HAYNES

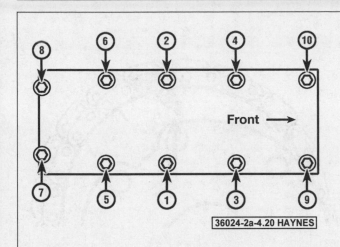

4.19 Left valve cover bolt tightening sequence

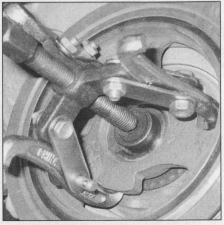

5.5 Remove the crankshaft pulley with a three-jaw puller

tube bolt and move the tube out of the way.

16 Unscrew the valve cover bolts (see illustration 4.19) and remove the valve cover.

Note: *The valve cover bolts are captive; they stay with the valve cover.*

17 Remove the old valve cover gasket and any gasket material that's stuck to the head.

18 Apply RTV sealant to the mating joints between the timing chain cover and the cylinder head, then install the valve cover and gasket.

Note: *Install the cover within five minutes of applying the RTV sealant.*

19 Tighten the valve cover bolts in sequence (see illustration) gradually and evenly, to the torque listed in this Chapter's Specifications.

20 The remainder of installation is the reverse of removal.

5 Crankshaft pulley - removal and installation

Removal

1 Remove the engine cooling fan and fan shroud (see Chapter 3).

2 Remove the accessory drivebelt (see Chapter 1).

3 Remove the flywheel/driveplate inspection cover (see Chapter 7A) and have an assistant wedge a large prybar or screwdriver into the starter ring gear teeth to prevent the crankshaft from turning.

4 Remove the large center bolt from the crankshaft pulley. Obtain a new bolt for installation.

5 Using a suitable puller, draw the crankshaft pulley off of the crankshaft (see illustration). Leave the Woodruff key in place in the crankshaft keyway.

Caution: *Don't use a puller with jaws that grip the outer edge of the crankshaft pulley. The puller jaws must bear on the crankshaft pulley hub only. Also, be sure to use proper size adapter on the end of the crankshaft, or the threads in the crankshaft could be damaged.*

Installation

6 Lubricate the oil seal contact surface of the crankshaft pulley hub (see illustration) with multi-purpose grease or clean engine oil. Apply RTV sealant to the keyway in the crankshaft pulley before installation.

Caution: *If the crankshaft pulley is not fully seated within four minutes, the RTV sealant must be removed and the area cleaned. New RTV sealant must be applied or oil will leak through the crankshaft key groove.*

7 Install the crankshaft pulley on the end of the crankshaft. The keyway in the crankshaft pulley must be aligned with the Woodruff key in the crankshaft. If the crankshaft pulley cannot be seated by hand, slip the large washer over the bolt, install the bolt and tighten it to pull the crankshaft pulley into place. Now loosen the bolt one full turn, then tighten the bolt to the torque listed in this Chapter's Specifications.

8 The remaining installation steps are the reverse of removal.

9 Check the oil level. Run the engine and check for oil leaks.

6 Timing chain cover, timing chains, tensioners and sprockets - removal, inspection and installation

Removal

Caution: *The timing system is complex. Severe engine damage will occur if you make any mistakes. Do not attempt this procedure unless you are highly experienced with this type of repair. If you are at all unsure of your abilities, consult an expert. Double-check all your work and be sure everything is correct before you attempt to start the engine.*

Caution: *Once the timing chain(s) have been removed, DO NOT ROTATE the crankshaft or the camshafts, or the valves and/or pistons will be damaged. Special tools are necessary to prevent the camshafts from moving when*

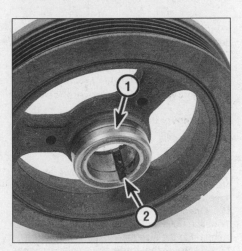

5.6 Inspect the crankshaft pulley for signs of damage or excessive wear

1   *Oil seal surface*
2   *Woodruff keyway*

the timing chain is removed. Read through the entire procedure and obtain the necessary tools before proceeding.

Note: *Because these engines are an interference design, there will be damage to the valves and/or pistons if a chain has broken, and this will require removal of the cylinder head(s).*

1 Remove the lower spark plugs (see Chapter 1) and position the number one cylinder at TDC on the compression stroke for cylinder no. 1 (see Section 3).

2 Disconnect the cable from the negative battery terminal (see Chapter 5).

3 Drain the engine oil and remove the oil filter (see Chapter 1), then drain the cooling system (see Chapter 1).

4 Remove the engine cooling fan (see Chapter 3).

5 Loosen the water pump pulley bolts, then remove the drivebelt (see Chapter 1) and the water pump pulley (see Chapter 3).

6 Remove the left and right Variable Camshaft Timing (VTC) solenoids from the cylin-

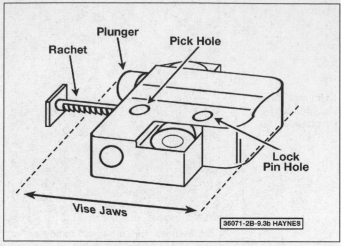

6.28 Timing chain tensioner details

6.32 Align the colored link with the groove in the VCT unit (sprocket)

der heads (see Chapter 6A).

7   Remove the crankshaft pulley and front oil seal (see Section 5). Always replace the crankshaft pulley bolt with a new one.

8   Remove both valve covers (see Section 4).

9   Remove the nut from the transmission fluid cooler line bracket and remove the bracket from the lower part of the timing chain cover.

10   Remove the nut from the power steering pressure hose bracket and move the pressure hose off to the side (see Chapter 10).

11   Drain the power steering fluid, remove the mounting bolts and position the power steering pump off to the side (see Chapter 10).

12   Remove the three drivebelt tensioner mounting bolts and tensioner, then the drivebelt idler pulley bolts and pulleys from the timing chain cover.

13   Remove the four front oil pan bolts (see Section 14).

Note: *Two of the fasteners are usually studs, which also should be removed.*

14   Remove the timing chain cover-to-engine block bolts. Note the locations of studs and different-length bolts so they can be reinstalled in their original locations.

15   Separate the timing chain cover from the engine block. If it's stuck, tap it gently with a soft-face hammer to break the gasket bond.

Caution: *DO NOT use excessive force or you may crack the cover. If the cover is difficult to remove, make sure all of the bolts and studs have been removed.*

16   Rotate the engine until the intake valve for cylinder Number 1 is open. Working on the left (driver's side) cylinder head, remove the ten intake rocker shaft mounting bolts and the shaft. Number the rocker arms (if they are not marked already), so they can be reinstalled in the same locations.

17   Rotate the engine clockwise until the exhaust valve for cylinder Number 1 is open. Working on the left (driver's side) cylinder head, remove the ten exhaust rocker shaft

mounting bolts and the shaft. Number the rocker arms (if they are not marked already), so they can be reinstalled in the same locations.

18   Rotate the engine clockwise until the intake valve for cylinder Number 1 is closed. Working on the right (passenger's side) cylinder head, remove the ten intake rocker shaft mounting bolts and the shaft. Number the rocker arms (if they are not marked already), so they can be reinstalled in the same locations.

19   Rotate the engine clockwise until the exhaust valve for cylinder Number 1 is closed. Working on the right (passenger's side) cylinder head, remove the ten exhaust rocker shaft mounting bolts and the shaft. Number the rocker arms (if they are not marked already), so they can be reinstalled in the same locations.

20   Rotate the engine clockwise until the crankshaft keyway is at the 11 o'clock position.

21   Push in on the right (passenger's side) primary chain tensioner arm to compress the plunger in the tensioner (one at each side of the engine). Keep it depressed until you can insert a large paper clip or drill bit in the hole in the tensioner (see illustration 9.24). Repeat for the tensioner on the other side.

22   Turn the crankshaft just a little, if necessary, to provide some slack in the timing chain.

23   Remove the right-side tensioner, then the tensioner arm and the stationary chain guide.

24   Remove the right-side primary timing chain.

25   Repeat Steps 21 through 23 on the left-side timing chain tensioner, guide and chain.

Inspection

26   Inspect the individual sprocket teeth and keyways for wear and damage.

27   Check the chain for cracked plates, and

pitted or worn rollers. Check the wear surface of the chain guides for wear and damage. Replace any worn or defective parts with new ones.

Caution: *If excessive plastic material is missing from the chain guides, the oil pan should be removed and cleaned of all debris (see Section 14). Check the oil pump pick-up tube screen, too.*

28   Check the timing chain tensioners:

a)   *Check the condition of the plunger. Depress the plunger to make sure it moves freely.*

b)   *Remove the plastic oil filter in the center hole of the VCT assemblies. Install new filters with the wide end out.*

c)   *The tensioners can be kept in the retracted position with a drill bit or paper clip.*

d)   *Compress the tensioner in a vise and place the drill bit or paper clip into the lock pin hole to hold it in the compressed position (see illustration).*

Caution: *Compress only the round plunger on the tensioner and not the ratchet mechanism.*

Installation

Caution: *Before starting the engine, carefully rotate the crankshaft by hand through at least two full revolutions (use a socket and breaker bar on the crankshaft pulley center bolt). If you feel any resistance, STOP! There is something wrong - most likely, valves are contacting the pistons. You must find the problem before proceeding.*

29   Install the timing chain stationary guides for both sides, and tighten the bolts to the torque listed in this Chapter's Specifications.

30   Install the crankshaft sprocket on the crankshaft with the timing marks facing forward.

31   Install the camshaft VCT units (sprockets), if removed.

32   Install the left-side timing chain, making sure a colored link in the timing chain align with the marks on the VCT unit (see illustration).

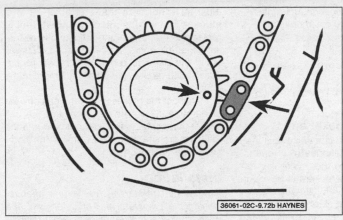

6.33 Align the colored link with the mark on the crankshaft sprocket

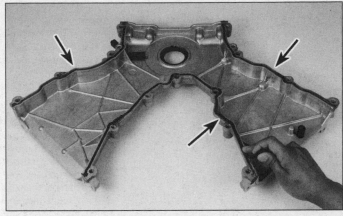

6.53 Install three new gaskets into the grooves in the back of the timing chain cover

33   Engage the crankshaft sprocket with the chain, aligning the colored link on the chain with the mark on the sprocket, then install the sprocket onto the crankshaft (see illustration). **Note:** *Make sure the crankshaft keyway is in the 11 o'clock position.*

34   Install the secondary chain guide, tensioner arm and tensioner.

35   Remove the pin holding the tensioner in the retracted position.

36   Engage the right-side timing chain with the crankshaft sprocket, aligning a colored link on the chain with the mark on the sprocket.

37   Loop the chain over the right-side VCT unit, aligning the other colored link with the mark on the VCT unit.

38   Install the secondary chain guide, tensioner arm and tensioner.

39   Remove the pin holding the tensioner in the retracted position.

40   Double-check to make sure all of the timing marks and colored links are in proper alignment.

41   Lubricate the camshafts and valve stems with camshaft installation lubricant or clean engine oil.

42   Turn the crankshaft clockwise so the intake cam lobe for cylinder number one is pointing toward the roller of the intake rocker arm (when it is installed) - so as to ensure the number one cylinder intake valve will be fully opened when the rocker shaft is installed.

43   Working on the left (driver's side) cylinder head, install the intake rocker shaft (see Section).

44   Turn the crankshaft clockwise so the exhaust cam lobe for cylinder number one is pointing toward the roller of the exhaust rocker arm (when it is installed) - so as to ensure the number one cylinder exhaust valve will be fully opened when the rocker shaft is installed.

45   Working on the left (driver's side) cylinder head, install the exhaust rocker shaft (see Section 7).

46   Turn the crankshaft clockwise so the intake cam lobe for cylinder number one is pointing away from the roller of the intake rocker arm (when it is installed) - so as to ensure the number one cylinder intake valve

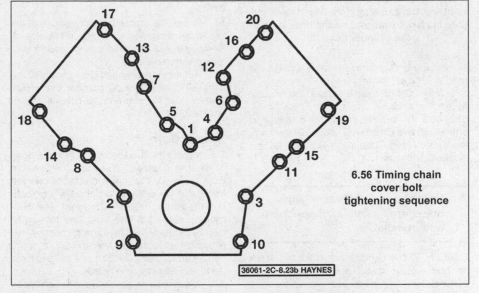

6.56 Timing chain cover bolt tightening sequence

will be fully closed when the rocker shaft is installed.

47   Working on the right-side cylinder head, install the intake rocker shaft (see Section 7).

48   Turn the crankshaft clockwise so the exhaust cam lobe for cylinder number one is pointing away from the roller of the exhaust rocker arm (when it is installed) - so as to ensure the number one cylinder exhaust valve will be fully closed when the rocker shaft is installed.

49   Install the right-side exhaust rocker shaft (see Section 7).

50   Slowly rotate the crankshaft in the normal direction of rotation (clockwise) at least two revolutions and again bring the engine to TDC. If you feel any resistance, stop and find out why.

51   Clean the mating surfaces of the timing chain cover, engine block and cylinder heads to remove all traces of old gasket material, oil and dirt.

Warning: *Be careful when cleaning any of the aluminum components. Use of a metal scraper could cause scratches or gouges that could lead to an oil leak later.*

52   Install a new crankshaft front oil seal (see Section 17).

53   Install new gaskets into the grooves of the timing chain cover (see illustration).

54   Apply a 1/8-inch bead of RTV sealant to the junctions of the oil pan-to-engine block and the cylinder head-to-engine block. Apply a small dab of RTV where the timing chain cover and engine block meet, and where the timing chain cover and cylinder heads meet at the valve cover surfaces.

55   Lubricate the timing chains and the lip of the crankshaft front oil seal with clean engine oil.

56   Install the timing chain cover on the engine within five minutes of applying the RTV sealant. Position the bottom/front edge of the timing chain cover flush with the front edge of the oil pan and tilt the top of the cover into place against the engine. Do not press the cover straight in against the engine or the sealant may be scraped off the front of the oil pan and cause a leak. Tighten the timing chain cover-to-engine block bolts in the recommended sequence (see illustration), to the torque listed in this Chapter's Specifications.

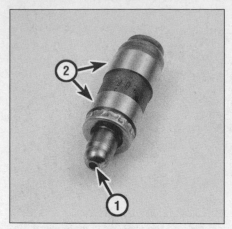

7.7 Inspect the lash adjuster for signs of excessive wear or damage, such as pitting, scoring or signs of overheating (bluing or discoloration), where the tip contacts the camshaft follower (1) and the side surfaces that contact the lifter bore in the cylinder head (2)

Tighten the timing chain cover-to-oil pan bolts (and studs if applicable).

57   Install the remaining parts in the reverse order of removal.

58   Add the proper type and quantity of engine oil (see Chapter 1), also refill the cooling system (see Chapter 1). Run the engine and check for leaks.

---

7   Rocker arms and valve lash adjusters - removal, inspection and installation

---

Caution: *These engines are difficult to work on and require special tools for many procedures. On any procedure involving timing chain, camshaft or cylinder head removal, the steps must be read carefully and disassembly must proceed using the special tools, otherwise damage to the engine could result.*

Removal

1   Remove the valve cover(s) (see Section 4).

2   Remove the lower eight spark plugs (see Chapter 1).

Left (driver's side) rocker arms

3   Rotate the crankshaft clockwise until the intake valve on cylinder Number 6 is open. Remove the 10 bolts and the intake rocker shaft assembly. Keep the rockers in position and number them if they aren't already.

4   Rotate the crankshaft clockwise until the exhaust valve on cylinder Number 6 is open. Remove the 10 bolts and the exhaust rocker shaft assembly.

Right (passenger's side) rocker arms

5   Rotate the crankshaft clockwise until the intake valve on cylinder Number 1 is closed. Remove the 10 bolts and the right-side intake rocker shaft assembly.

6   Rotate the crankshaft clockwise until the exhaust valve on cylinder Number 1 is closed. Remove the 10 bolts and the exhaust rocker shaft assembly.

Inspection

7   Inspect each adjuster carefully for signs of wear or damage. The areas of possible wear are the ball tip that contacts the cam follower and the sides of the adjuster that contact the bore in the cylinder head (see illustration). Since the lash adjusters frequently become clogged as mileage increases, we recommend replacing them if you're concerned about their condition or if the engine is exhibiting valve tapping noises.

8   A thin wire or paper clip can be placed in the oil hole to move the plunger and make sure it's not stuck. Note: The lash adjuster must have no more than 1/16-inch of total plunger travel. It's recommended that if replacement of any of the adjusters is necessary, that the entire set be replaced. This will avoid the need to repeat the repair procedure as the others require replacement in the future.

9   Inspect the rocker arms for signs of wear or damage. The areas of wear are the ball socket that contacts the lash adjuster and the roller where the follower contacts the camshaft (see illustration).

Installation

Note: *When re-starting the engine after replacing the adjusters, the adjusters will normally make some tapping noises, until all the air is bled from them. After the engine is warmed-up, raise the speed from idle to 3,000 rpm for one minute. Stop the engine and let it cool down. All of the noise should be gone when it is restarted.*

10   Clean, then lubricate the rocker arms and shafts with clean engine oil before installation.

Left (driver's side) rocker arms

11   Rotate the engine clockwise to position the left-side camshaft's intake lobe for cylinder Number 6 so that the intake valve would be opened fully if the rocker shafts were tightened down. Install the left-side intake rocker arms and rocker shaft. Install all the bolts hand tight, then tighten them, in sequence, to the torque listed in this Chapter's Specifications (see illustration).

12   Rotate the engine clockwise to position the left-side camshaft's exhaust lobe for cylinder Number 6 so that the exhaust valve would be opened fully if the rocker shafts were tightened down. Install the left-side intake rocker arms and rocker shaft. Install all the bolts hand tight, then tighten them, in sequence, to the torque listed in this Chapter's Specifications (see illustration 7.11).

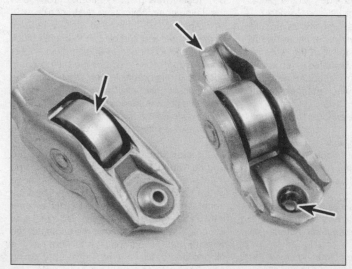

7.9 Check the roller surface of the rocker arms and the areas where the valve stem and lash adjuster contact the rocker

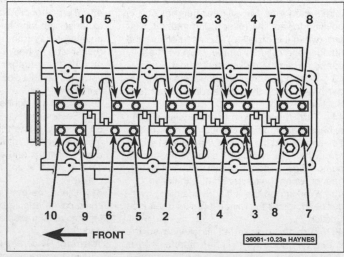

7.11 Left cylinder head intake (top) and exhaust (bottom) rocker shaft bolt tightening sequence

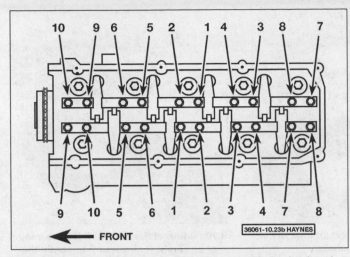

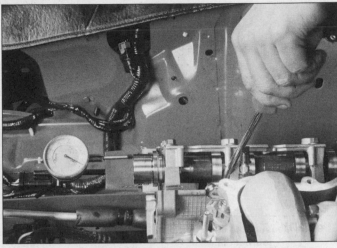

7.13 Right cylinder head intake (bottom) and exhaust (top) rocker shaft bolt tightening sequence

9.3 Camshaft endplay can be checked by setting up a dial indicator off the front of the camshaft and prying the camshaft gently forward and back

Right (passenger's side) rocker arms

13   Rotate the engine clockwise to position the right-side camshaft's intake lobe for cylinder Number 1 so that the intake valve would be closed if the rocker shafts were tightened down. Install the right-side intake rocker arms and rocker shaft. Install all the bolts hand tight, then tighten them, in sequence, to the torque listed in this Chapter's Specifications (see illustration).

14   Rotate the engine clockwise to position the right-side camshaft's exhaust lobe for cylinder Number 1 so that the exhaust valve would be closed if the rocker shafts were tightened down. Install the right-side exhaust rocker arms and rocker shaft. Install all the bolts hand tight, then tighten them, in sequence, to the torque listed in this Chapter's Specifications (see illustration 7.13).

15   The remainder of installation is the reverse of removal.

8   Variable Camshaft Timing (VCT) system - general information

1   The Variable Camshaft Timing (VCT) system consists of a VCT solenoid (or a pair of VCT solenoids on DOHC engines) mounted on the front of each cylinder head over the camshaft(s), a Camshaft Position (CMP) sensor with a trigger wheel and a variable camshaft timing sprocket mounted on each camshaft. When oil flow is sent to the timing sprockets by way of the VCT solenoids, the phasing of the camshafts is altered, thereby advancing or retarding the actuation of the valves. The Variable Camshaft Timing (VCT) system oil control solenoid replacement procedure is in Chapter 6A. The VCT units, sometimes called camshaft phasers, are removed and installed in Section 9 of this Chapter.

2   The Variable Camshaft Timing (VCT)

system should be diagnosed by a dealer service department or other qualified automotive repair facility. The VCT system is controlled and monitored by the Powertrain Control Module (PCM), thereby requiring a specialized scan tool for diagnostics.

9   Camshaft(s) - removal, inspection and installation

Caution: *These engines are difficult to work on and require special tools for many procedures. On any procedure involving timing chain, camshaft or cylinder head removal, the steps must be read carefully and disassembly must proceed using the special tools, otherwise damage to the engine could result.*

Removal

1   Remove the valve covers (see Section 4), and the timing chain cover (see Section 6).

2   Remove the timing chains (see Section 6) and the rocker arms (see Section 7).

3   Measure the thrust clearance (endplay)

of the camshaft(s) with a dial indicator (see illustration). If the clearance is greater than the value listed in this Chapter's Specifications, replace the camshaft and/or the cylinder head.

4   Remove the VCT assembly from the front end of each camshaft.

5   Hold the flats of the camshaft with a wrench while using a ratchet and socket to remove the VCT mounting bolts (see Section 6).

6   Loosen and remove the camshaft bearing cap bolts 1/4-turn at a time in the reverse of the tightening sequence (see illustration 9.15).

7   Lift the camshafts out of their bearing saddles and set the cams aside.

8   Keep the camshafts clean and mark them with paint to indicate left or right and intake or exhaust.

Inspection

9   Visually examine the cam lobes and bearing journals for score marks, pitting, galling and evidence of overheating (blue, discolored areas). Look for flaking of the hardened surface of each lobe (see illustration).

9.9 Areas to look for excessive wear or damage on the camshafts are the bearing surfaces and the camshaft lobes. Also inspect the bearing surfaces of the camshaft bearing caps for signs of excessive wear, damage or overheating

9.10a Measure the camshaft bearing journal diameter

9.10b Measure the camshaft lobe at its greatest dimension

9.10c Subtract the camshaft lobe diameter at its smallest dimension to obtain the lobe lift specification

10   Using a micrometer, measure the diameter of each camshaft journal and the lift of each camshaft lobe (see illustrations). Compare your measurements with the Specifications listed at the front of this Chapter, and if the diameter of any one of these is less than specified, replace the camshaft.

11   Check the oil clearance for each camshaft journal as follows:

a) Clean the bearing surfaces and the camshaft journals with lacquer thinner or acetone.

b) Carefully lay the camshaft(s) in place in the cylinder head. Don't install the rocker arms or lash adjusters and don't use any lubrication.

c) Lay a strip of Plastigage on each journal (see illustration).

d) Install the camshaft bearing caps.

e) Tighten the cap bolts, a little at a time, to the torque listed in this Chapter's Specifications. **Note:** Don't turn the camshaft while the Plastigage is in place.

f) Remove the bolts and detach the caps.

g) Compare the width of the crushed Plastigage (at its widest point) to the scale on the Plastigage envelope (see illustration).

9.11a Lay a strip of Plastigage on each of the camshaft journals

h) If the clearance is greater than specified, and the diameter of any journal is less than specified, replace the camshaft. If the journal diameters are within specifications but the oil clearance is too great, the cylinder head is worn and must be replaced.

9.11b Compare the width of the crushed Plastigage to the scale on the envelope to determine the oil clearance

12   Scrape off the Plastigage with your fingernail or the edge of a credit card - don't scratch or nick the journals or bearing surfaces.

Installation

13   If the lash adjusters and/or camshaft followers have been removed, install them in their original locations (see Section 7).

14   Apply moly-base grease or camshaft installation lube to the camshaft lobes and bearing journals, then install the camshaft(s).

15   Install the camshaft caps in the correct locations, and loosely install all the bolts. Align the camshaft sprockets before tightening the cap bolts (see Section 7). Follow the correct bolt-tightening sequence (see illustration), and tighten the bolts to the torque listed in this Chapter's Specifications.

16   Install the timing chain(s) (see Section 6).

17   The remainder of installation is the reverse of removal.

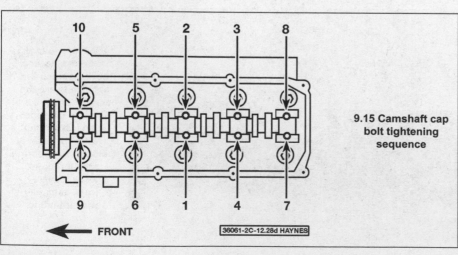

9.15 Camshaft cap bolt tightening sequence

FRONT

36061-2C-12.28d HAYNES

10.6 This is the air hose adapter that threads into the spark plug hole - they're commonly available from auto parts stores

10.14 Installation of a valve spring compressor more commonly available at local automotive parts stores (the camshaft must be removed for the use of this type of valve spring compressor)

10.15 Once the spring is compressed, remove the keepers with needle-nose pliers or a magnet, as shown here

10 Valve springs, retainers and seals - removal and installation

1   Broken valve springs and/or defective valve stem seals can be replaced without removing the cylinder heads. There is a method described in Section 10, using a tool recommended by the manufacturer, which accomplishes the removal of the valve springs and seals without the removal of the camshafts. The alternative method uses a more commonly available tool, but will require the removal of the camshaft (see Section 9) in order to remove the valve spring. Either method will achieve the same results, but it is much easier using the manufacturer's recommended procedure, if the correct tools can be located.

2   In either repair procedure, a compressed air source is normally required to perform this operation, so read through this Section carefully and rent or buy the tools before beginning the job.

Removal

3   Remove the valve cover (see Section 4).

4   Remove the lower spark plug from the cylinder with the defective component (see Chapter 1). If all of the valve stem seals are being replaced, remove all of the lower spark plugs.

5   Turn the crankshaft until the piston in the affected cylinder is at Top Dead Center (TDC) on the compression stroke (see Section 3). If you're replacing all of the valve stem seals, begin with cylinder number one and work on the valves for one cylinder at a time. Move from cylinder-to-cylinder following the firing order sequence (see this Chapter's Specifications).

6   Thread an air hose adapter into the spark plug hole (see illustration) and connect an air hose from a compressed air source to it. Most auto parts stores can supply the air

hose adapter.

Note: *Many cylinder compression gauges utilize a screw-in fitting that may work with your air hose quick-disconnect fitting.*

7   Apply compressed air to the cylinder.

Warning: *The piston may be forced down by compressed air, causing the crankshaft to turn suddenly. If the wrench used when positioning the number one piston at TDC is still attached to the bolt in the crankshaft nose, it could cause damage or injury when the crankshaft moves.*

8   The valves should be held in place by the air pressure.

9   Stuff shop rags into the cylinder head holes above and below the valves to prevent parts and tools from falling into the engine.

Method using special tools

10   Install a valve spring spacer, available at most auto parts stores.

Caution: *If the spacer isn't in place between one of the spring coils, the spring can be compressed too far, possibly damaging the valve seal.*

11   Compress the spring and remove the rocker arm (see Section 7). Rocker arms and hydraulic lash adjusters MUST be reinstalled with the same camshaft lobe that they were removed from. Label and store all components to avoid confusion during reassembly.

12   Keeping the spring compressed, remove the keepers with small needle-nose pliers or a magnet (see illustration 10.15). Remove the spring retainer and valve spring. Remove the valve stem seal (see illustration 10.17). If air pressure fails to hold the valve in the closed position during this operation, the valve face and/or seat is probably damaged. If so, the cylinder head will have to be removed for additional repair operations.

Alternative procedure

13   Remove the camshaft(s) (see Section 9).

14   Install the commonly available clamp-

10.17 Use pliers to firmly grasp the old seal and pull it off the valve guide

type valve spring compressor (see illustration).

15   Compress the spring and remove the keepers with small needle-nose pliers or a magnet (see illustration).

16   Remove the spring retainer and valve spring.

17   Remove the stem seal (see illustration). If air pressure fails to hold the valve in the closed position during this operation, the valve face and/or seat is probably damaged. If so, the cylinder head will have to be removed for additional repair operations.

Installation

18   Wrap a rubber band or tape around the top of the valve stem so the valve won't fall into the cylinder, then release the air pressure.

19   Inspect the valve stem for damage. Rotate the valve in the guide and check the end for eccentric movement, which would indicate that the valve is bent.

10.22a The valve stem seals combine a seal with the valve spring seat

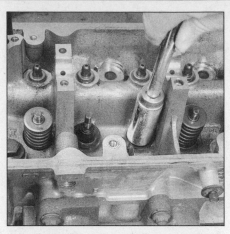

10.22b There is special valve seal installation tool available, but a deep socket that fits over the seal can be used to gently tap the seal into place

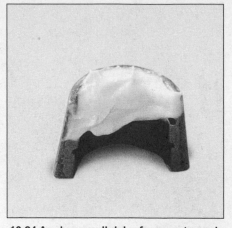

10.24 Apply a small dab of grease to each keeper as shown here before installation - it'll hold them in place on the valve stem as the spring is released

20   Move the valve up-and-down in the guide and make sure it doesn't bind. If the valve stem binds, either the valve is bent or the guide is damaged. In either case, the cylinder head will have to be removed for repair.

21   Reapply air pressure to the cylinder to retain the valve in the closed position, then remove the tape or rubber band from the valve stem.

22   Lubricate the valve stem with engine oil and install a new seal (see illustration). There is a special tool for the installation of the valve seal. If the tool isn't available, a socket that will fit over the seal and is deep enough to make contact with the seat (see illustration) can be used to carefully tap the new seal into place.

Caution: *The valve seal used on these engines is a combination seal and spring seat. Never place a valve spring directly against the aluminum cylinder head (without a seal/spring seat) - the hardened spring would damage the cylinder head.*

23   Make sure the garter spring on the seal is still in place, and install the spring in position over the valve.

24   Install the valve spring retainer. Compress the valve spring and carefully position the keepers in the groove. Apply a small dab of grease to the inside of each keeper to hold it in place if necessary (see illustration).

25   Remove the pressure from the spring tool and make sure the keepers are seated.

26   Disconnect the air hose and remove the adapter from the spark plug hole.

27   If the camshaft(s) were removed, reinstall them at this time (see Section 9).

28   Install the spark plug(s), spark plug wires and ignition coil(s) (see Chapter 1).

29   The remaining installation steps are the reverse of removal.

30   Start and run the engine, then check for oil leaks and unusual sounds coming from the valve cover area.

11   Intake manifold - removal and installation

Removal

1   Relieve the fuel system pressure (see Chapter 4A).

2   Disconnect the cable from the negative terminal of the battery (see Chapter 5).

3   Remove the throttle body (see Chapter 4A).

4   Detach the electrical harness where it attaches to the valve covers or intake manifold, then disconnect the electrical connectors at the ignition coils.

5   Remove the ignition coils (see Chapter 5).

6   Remove the alternator and the alternator support bracket (see Chapter 5).

7   Disconnect the EVAP tube (quick-connect) and electrical connector, and the PCV hoses.

8   Release the brake booster vacuum hose from its clip on the manifold and set it aside.

9   Remove the intake manifold bolts and remove the intake manifold with the fuel rail still attached (there are 12 intake mounting bolts and four fuel rail bolts).

10   The manifold may be stuck to the cylinder heads and force may be required to break the gasket seal(s). A prybar can be used to pry up the manifold, but make sure all fasteners have been removed first!

Caution: *Don't pry between the engine block and manifold or the cylinder heads, or damage to the gasket sealing surface may occur, leading to vacuum and oil leaks. Pry only at a manifold protrusion.*

11   Remove the intake manifold gaskets and clean all traces of gasket or sealant material from the sealing surfaces of the cylinder heads and intake manifold.

Caution: *The mating surfaces of the cylinder heads, engine block and intake manifold must be perfectly clean. Do not use metal scrapers,*

wire brushes, power abrasive discs or other abrasive means to clean sealing surfaces; these tools can scratch and gouge gasket surfaces, which creates leak paths. Instead, use gasket removal solvents, which are available at most auto parts stores, and may be helpful when removing old gasket material that's stuck to the cylinder heads and intake manifold. Since the cylinder heads are aluminum and the intake manifold is aluminum or plastic, aggressive scraping can cause damage! Follow the directions printed on the container, and use only a plastic-tipped scraper, not a metal one. While the manifold is removed, cover the open engine areas with shop rags to keep debris out of the engine. Use a vacuum cleaner to remove any debris that falls into the intake ports in the cylinder heads.

Installation

12   If you are replacing the manifold, transfer all components to the new unit.

13   Install new O-ring type intake manifold gaskets into their grooves in the manifold.

14   Carefully set the intake manifold in place.

15   Install the intake manifold bolts and following the recommended tightening sequence (see illustration), tighten them to the torque listed in this Chapter's Specifications.

16   The remainder of installation is the reverse of removal.

12   Exhaust manifolds - removal and installation

1   Park the vehicle with the wheels pointing straight ahead and secure the steering wheel in place if you're going to remove the left exhaust manifold on a V8 model.

2   Loosen the wheel lug nuts on the side to be worked on. Raise the vehicle and place it securely on jackstands. Remove the left or right front wheel.

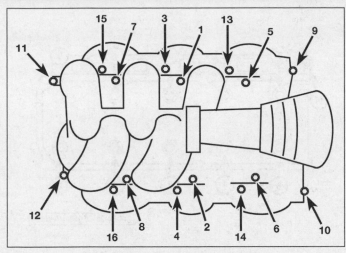

11.15 Intake manifold bolt tightening sequence

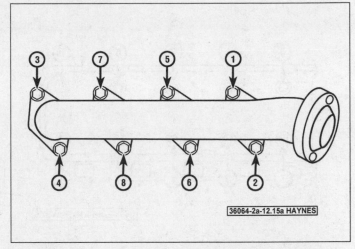

12.15a Left exhaust manifold nut tightening sequence (V8 engine)

3   Remove the left or right front inner fenderwell (see Chapter 11).
Warning: *Make sure that the steering shaft is not rotated with the steering gear removed or damage to the airbag clockspring assembly could occur. One method of preventing the steering shaft from rotating is to run the seat belt through the steering wheel and clip the seat belt buckles together.*
4   When removing the left exhaust manifold, remove the steering shaft-to-steering column bolt (see Chapter 10), then disconnect the shaft from the column. Always replace the shaft-to-column bolt with a new one.
5   If you're working on the right exhaust manifold, remove the starter (see Chapter 5).
6   If you're removing a right-side manifold, remove the starter (see Chapter 5).
7   Remove the transmission filler tube for access to the right-side manifold.
8   Working under the vehicle, apply penetrating oil to the exhaust pipe-to-exhaust manifold studs and nuts and to the exhaust manifold nuts and studs. They're usually corroded or rusty.
9   Remove the catalytic converter-to-exhaust manifold nuts.
10   Remove the heat shield bolts and remove the heat shield from the manifold.
11   Remove the eight exhaust manifold nuts and then remove the exhaust manifold and the old exhaust manifold gaskets.
12   Clean off all old gasket material from the cylinder head and from the exhaust manifold. The exhaust manifold and cylinder head mating surfaces must be clean before the exhaust manifolds are reinstalled. Use a gasket scraper to remove all carbon deposits.
13   Inspect the exhaust manifold for cracks. If a manifold is cracked, it must be replaced or repaired. Depending on the size and location of the crack, an automotive machine shop might be able to repair a cracked manifold by welding it. Sometimes, though, the welding might cost more than a new manifold. If a manifold has gotten hot enough to crack, it might very well be warped as well as

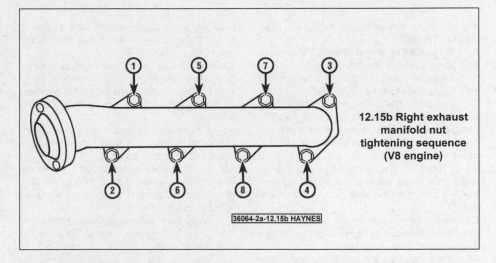

12.15b Right exhaust manifold nut tightening sequence (V8 engine)

cracked. Take the manifold to an automotive machine shop and get an estimate for any needed repairs, and then compare the estimate to the cost of a new manifold.
14   Make sure that the threads of the exhaust manifold studs are clean and undamaged. If they're damaged, clean them up with a thread chaser or replace them. To replace damaged exhaust manifold studs, "double-nut" them.
15   Install new gaskets, place the exhaust manifold in position and then, working in the specified sequence (see illustrations), tighten the exhaust manifold nuts to the torque listed in this Chapter's Specifications.
16   Installation is otherwise the reverse of removal.
17   Start the engine and check for exhaust leaks.

13   Cylinder heads - removal and installation

Warning: *Wait until the engine is completely cool before beginning this procedure.*
Caution: *These engines are difficult to work*

on and require special tools for many procedures. On any procedure involving timing chains, the steps must be read carefully and disassembly must proceed using the special tools, otherwise damage to the engine could result.
Caution: *The engine must be completely cool when the cylinder heads are removed. Failure to allow the engine to cool off could result in cylinder head warpage.*
Note: *The manufacturer recommends that the engine be removed from the vehicle to perform this procedure (see Chapter 2C).*

Removal

1   Relieve the fuel system pressure (see Chapter 4A).
2   Disconnect the cable from the negative battery terminal (see Chapter 5).
3   Drain the cooling system (see Chapter 1).
4   Remove the valve covers (see Section 4).
5   Remove the intake manifold (see Section 11).
6   Remove the engine from the vehicle (see Chapter 2C).

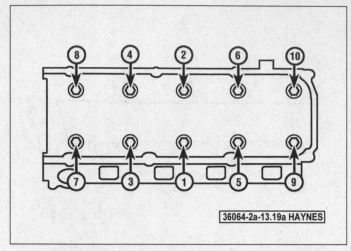

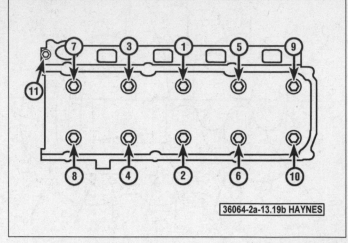

13.19a Driver's side cylinder head bolt tightening sequence   **13.19b Passenger's side cylinder head bolt tightening sequence**

7   Remove the exhaust manifolds (see Section 12).

8   Remove the timing chains, tensioners and sprockets (see Section 6).

9   Remove rocker arms (see Section 7) and the camshafts (see Section 9).

10   Following the reverse of the tightening sequence (see illustration 13.19a or 13.19b), use a breaker bar to remove the cylinder head bolts. Loosen the bolts in sequence 1/4-turn at a time.

11   Lift the cylinder head(s) off the engine. If resistance is felt, place a wood block against the end and strike the wood block with a hammer.

Caution: *The cylinder heads are aluminum; store them on wood blocks to prevent damage to the gasket sealing surfaces.*

12   If the head sticks, use a pry bar at the corners of the cylinder head-to-engine block mating surface to break the cylinder head gasket seal. Do not pry between the cylinder head and engine block in the gasket sealing area.

13   Remove the cylinder head gasket(s). Before removing, note which gasket goes on which side (they are different and cannot be interchanged).

Installation

Caution: *New cylinder head bolts must be used for reassembly. Failure to use new bolts may result in cylinder head gasket leakage and engine damage.*

14   The mating surfaces of the cylinder heads and engine block must be perfectly clean when the cylinder heads are installed. Use a gasket scraper to remove all traces of carbon and old gasket material, then clean the mating surfaces with silicone gasket remover and a plastic scraper. If there's oil on the mating surfaces when the cylinder heads are installed, the gaskets may not seal correctly and leaks may develop. When working on the engine block, cover the open areas of the engine with shop rags to keep debris out

during repair and reassembly. Use a vacuum cleaner to remove any debris that falls into the cylinders.

Caution: *Do not use abrasive wheels or metal scrapers on the heads or block surface; use a plastic scraper and chemical gasket remover, or the head gasket surfaces could have future leaks.*

15   Check the engine block and cylinder head mating surfaces for nicks, deep scratches and other damage.

16   Use a tap of the correct size to chase the threads in the cylinder head bolt holes. Dirt, corrosion, sealant and damaged threads will affect torque readings.

17   Make sure the new gaskets are installed on the correct cylinder banks. They are not interchangeable.

18   Position the new gasket(s) over the alignment dowels in the engine block.

19   Carefully position the cylinder heads on the engine block without disturbing the gaskets. Install the NEW cylinder head bolts (the cylinder head bolts are torque-to-yield design and they cannot be reused). Following the recommended sequence (see illustrations), tighten the cylinder head bolts, in stages, to the torque and angle of rotation listed in this Chapter's Specifications.

Note: *The method used for the cylinder head bolt tightening procedure is referred to as "torque-angle" or "torque-to-yield" method. Follow the procedure exactly. Tighten the bolts in the first step using a torque wrench, then use a breaker bar and a special torque-angle gauge (available at most auto parts stores) to tighten the bolts the required angle. If the adapter is not available, mark each bolt with a paint stripe to aid in the torque angle process.*

20   The remainder of installation is the reverse of removal.

21   Change the engine oil and filter and refill the cooling system (see Chapter 1), then start the engine and check carefully for oil and coolant leaks.

14   Oil pan - removal and installation

Removal

1   Disconnect the cable from the negative terminal of the battery (see Chapter 5).

2   Drain the engine oil and remove the oil filter (see Chapter 1).

3   Drain the engine coolant (see Chapter 1) and then remove the upper radiator hose (see Chapter 3).

4   Remove the air cleaner and the air intake duct (see Chapter 4A).

5   Remove the engine cooling fan and shroud (see Chapter 3).

6   Remove the alternator (see Chapter 5).

7   Remove the intake manifold (see Section 11).

8   Raise the vehicle and place it securely on jackstands.

9   Disconnect the exhaust "Y-pipe" from the exhaust manifold flanges and from the exhaust pipe flange in front of the catalytic converter and remove it (see Chapter 4A).

10   On automatic transmission models, detach any transmission cooler line brackets that are located near the oil pan and set the cooler lines aside.

11   Remove the flywheel/driveplate inspection cover from the transmission bellhousing.

12   Remove the driveshaft (see Chapter 8). On 4WD models, disconnect the rear end of the front driveshaft from the transfer case (see Chapter 8).

13   Remove the nuts from the front engine mounts (see Section 18) and remove the nuts from the transmission mount (see Chapter 7A).

14   Attach an engine hoist to the engine and carefully raise it up as high as possible.

Caution: *Raise the engine slowly while observing the location of the intake manifold. Do not allow the manifold to contact the vehicle or the manifold might be damaged.*

Note: *The transmission must also be raised for access.*

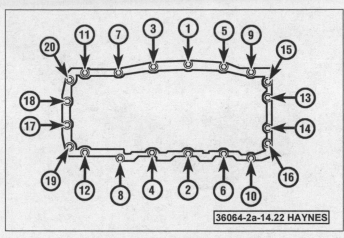

14.20 Apply a bead of RTV sealant at the junctions of the timing chain cover-to-engine block and the rear seal retainer-to-engine block (shown) before installing the oil pan

14.22 Oil pan bolt tightening sequence

15   After the engine is securely supported, unbolt the rear transmission mount from the crossmember (refer to Chapter 7A). Remove any sensors that could be damaged on automatic transmissions and plug the openings to prevent contamination. Position a block of wood on a jack and raise the rear of the transmission.

Caution: *Do not place the jack under any part of the transmission that could be dented or otherwise damaged.*

16   Remove the oil pan mounting bolts in the reverse order of the tightening sequence (see illustration 14.22). Carefully separate the oil pan from the engine block. Don't pry between the engine block and oil pan or damage to the sealing surfaces may result and oil leaks could develop. Instead, dislodge the oil pan with a large rubber mallet or a wood block and a hammer.

Note: *The oil pan gasket is reusable if care is taken in oil pan removal.*

17   You can't remove the oil pan yet because of interference from the oil pump pick-up tube. With the oil pan resting on the crossmember, remove the pick-up tube bolts and bracket nut and allow the pick-up tube to fall into the pan. Now you can remove the pan.

Installation

18   Use a gasket scraper or putty knife to remove all traces of old gasket material and sealant from the pan and engine block.

Caution: *Be careful not to gouge the oil pan or block, or oil leaks could develop later.*

19   Clean the mating surfaces with lacquer thinner or acetone. Make sure the bolt holes in the engine block are clean.

20   Apply a bead of RTV sealant to the four corner seams where the rear seal retainer meets the engine block, and the front cover meets the engine block (see illustration).

21   Install a new gasket on the oil pan flange. Place the pick-up tube in the pan.

22   Raise the pan into position on the crossmember and attach the pick-up tube. Tighten the bolts to the torque listed in this

Chapter's Specifications. Carefully position the oil pan against the engine block and install the bolts finger-tight. Make sure the gaskets haven't shifted, then tighten the bolts in the indicated sequence (see illustration) to the torque listed in this Chapter's Specifications.

23   The remaining steps are the reverse of removal.

Caution: *Don't forget to refill the engine with oil before starting it (see Chapter 1).*

24   Start the engine and check carefully for oil leaks at the oil pan. Drive the vehicle and check again.

---

15 Oil pump - removal and installation

Note: *The oil pump is available as a complete replacement unit only. No service parts or repair specifications are available from the manufacturer.*

Removal

1   Raise the vehicle and support it securely on jackstands.

2   Drain the engine oil (see Chapter 1).

3   Remove the oil pan as described in Sec-

15.3 Remove the two bolts retaining the pick-up tube to the oil pump

tion 14. Remove the the bolts that attach the oil pump pick-up tube to the oil pump (see illustration).

4   Remove the timing chain cover, the timing chains, the chain guides and the crankshaft timing chain sprockets (see Section 6).

5   Remove the four oil pump mounting bolts (see illustration) and separate the pump from the engine block.

15.5 Remove the four oil pump mounting bolts (arrows) and detach the oil pump from the engine block

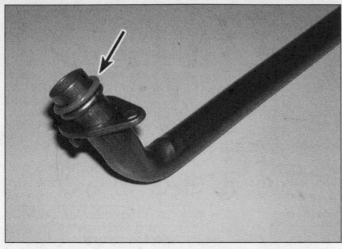

15.6 Before bolting the pick-up tube back into the oil pump, inspect the O-ring (arrow) and replace it if necessary

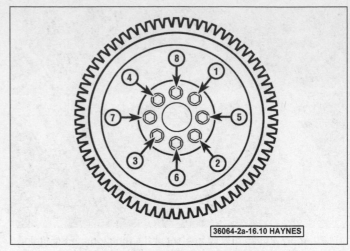

36064-2a-16.10 HAYNES

16.10 Flywheel bolt tightening sequence

Installation

6   Inspect the O-ring gasket on the pick-up tube (see illustration). If it's damaged, replace it.

7   Install the oil pump to the engine and tighten the bolts to the torque listed in this Chapter's Specifications.

Note: *Prime the oil pump prior to installation. Pour clean oil into the pick-up port and turn the pump by hand.*

8   The remainder of installation is the reverse of the removal procedure.

9   Fill the engine with the correct type and quantity of oil (see Chapter 1). Start the engine and check for leaks.

16 Driveplate - removal and installation

Removal

1   Disconnect the cable from the negative battery terminal.

2   Raise the vehicle and support it securely on jackstands.

3   Remove the transmission (see Chapter 7A).

4   Look for factory paint marks that indicate driveplate-to-crankshaft alignment. If they aren't there, scribe or paint marks on the driveplate and crankshaft to ensure correct alignment during reassembly. Rotate the driveplate until the two slotted holes are at the 6 o'clock position before removing it.

5   Remove the bolts that secure the driveplate to the crankshaft. Insert a screwdriver or punch through one of the holes in the driveplate to keep the crankshaft from turning while loosening the bolts.

6   Remove the driveplate from the crankshaft. Be sure to support it while removing the last bolt. After the driveplate is removed, there is a sheet metal cover plate located between the engine block and the driveplate. It's not necessary to remove the cover plate unless you're going to remove the crankshaft rear oil

seal retainer (see Section 17).

Warning: *The teeth on the driveplate may be sharp. Be sure to hold it with gloves or rags.*

Installation

7   Clean and inspect the mating surfaces of the driveplate and the crankshaft. And be sure to inspect the crankshaft rear oil seal. If the rear seal is leaking, replace it before reinstalling the driveplate (see Section 17).

8   If the reinforcement plate was removed, be sure to install it correctly positioned for the starter installation.

9   Check for cracked, broken or missing ring gear teeth. If any of these conditions are found, replace the flywheel/driveplate.

10   Install the driveplate, install the driveplate bolts and tighten the bolts to the torque listed in this Chapter's Specifications following the correct tightening sequence (see illustration). Position the two slotted holes in the driveplate at the 6 o'clock position before installing the bolts.

Caution: *The two slotted holes in the flex plate must be at the 6 o'clock position or the ignition pulse ring will not be in the correct position.*

11   The remainder of installation is the reverse of the removal procedure.

17 Crankshaft oil seals - replacement

Front seal

1   Remove the drivebelt (Chapter 1).

2   Remove the engine cooling fan and shroud assembly (Chapter 3).

3   Remove the crankshaft pulley (see Section 5).

4   Carefully remove the seal from the cover with a seal removal tool (see illustration). If a seal removal tool is not available, carefully use a screwdriver. If the timing cover is removed, use a chisel or small punch and hammer to drive the seal out of the cover from the back side. Support the cover as close to the seal bore as possible with two wood blocks. Be careful not to damage the cover or scratch the wall of the seal bore.

5   Check the seal bore and crankshaft, as well as the seal contact surface on the crank-

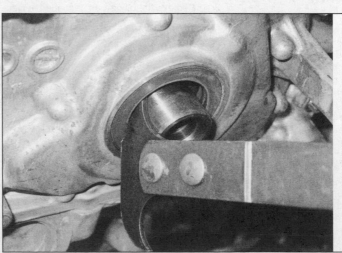

17.4 Using a special seal removal tool or screwdriver, remove the crankshaft front oil seal, being very careful not to scratch the crankshaft during seal removal

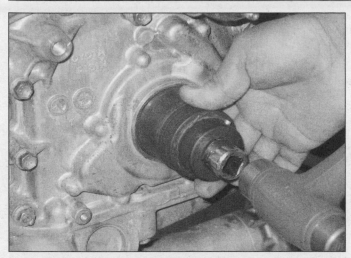

17.6 There is special tool for installing the front oil seal into the timing chain cover, but if the tool is unavailable a large socket or section of tubing (the same diameter as the seal) can be used to drive the seal into place

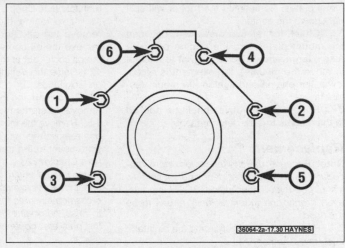

17.30 Rear seal housing tightening sequence

shaft pulley for nicks and burrs. Position the new seal in the bore with the open end of the seal facing IN. A small amount of engine oil applied to the outer edge of the new seal will make installation easier.

6   Drive the seal into the bore with a large socket and hammer until it's completely seated (see illustration). If the cover is removed, support the cover on wood blocks. Select a socket that's the same outside diameter as the seal (a section of pipe can be used if a socket isn't available).

7   Lubricate the lip of the seal with clean engine oil and install the crankshaft pulley on the end of the crankshaft. The keyway in the crankshaft pulley bore must be aligned with the Woodruff key in the crankshaft nose.

Note: *Before reinstalling the crankshaft pulley, apply RTV sealant to the Woodruff key slot in the pulley.*

Caution: *If the crankshaft pulley is not fully seated within four minutes, the RTV sealant must be removed and the area cleaned. New RTV sealant must be applied or oil will leak through the crankshaft key groove.*

8   If the crankshaft pulley can't be seated by hand, tap it into place with a soft-face hammer, or install the bolt and washer and tighten it to press the crankshaft pulley into place.

9   Tighten the crankshaft pulley bolt to the torque listed in this Chapter's Specifications.

10   Install the drivebelt.

11   Install the remaining parts removed for access to the seal.

12   Start the engine and check for leaks.

Rear seal and rear seal retainer

13   Disconnect the cable from the negative battery terminal. Raise the vehicle and support it securely on jackstands. Refer to Chapter 7A and remove the transmission.

14   Remove the flywheel/driveplate (see

Section 16), then remove the ignition pulse ring from the end of the crankshaft.

Rear seal

15   Use a seal removal tool or a screwdriver wrapped with tape to pry out the seal, being careful not to gouge or nick the housing.

16   Lubricate the crankshaft seal journal and the lip of the new seal with multi-purpose grease.

17   Install the seal with the seal lip toward the engine and the dust seal toward the transmission.

18   Tap the seal into place using a seal driver to make sure that it doesn't become tilted.

19   Install the seal so that its rear edge is flush with the seal retainer.

20   Install the flywheel/driveplate (see Section 16).

21   Install the transmission (see Chapter 7A).

22   The remainder of installation is the reverse of removal.

Rear seal retainer

23   Remove the bolts, detach the seal retainer (see illustration 17.30) and clean off all the old gasket and/or sealant material from both the engine block and the seal retainer.

24   Support the seal and retainer assembly on wood blocks and drive the old seal out from the back side with a punch and hammer.

25   Drive the new seal into the retainer with a wood block. Make sure to support the seal retainer when driving the new seal into the housing with a wood block (be careful not to cock the seal in the bore while installing).

26   Clean the crankshaft and seal bore with brake system cleaner. Check the seal contact surface on the crankshaft very carefully for scratches or nicks that could damage the new seal lip and cause oil leaks. If the crankshaft is damaged, the only alternative is a new or different crankshaft.

27   Lubricate the crankshaft seal journal and the lip of the new seal with engine oil.

28   Place a 1/16-inch wide bead of anaerobic sealant on either the engine block or the seal retainer.

29   Install the oil seal retainer by slowly and carefully pushing the seal onto the crankshaft. The seal lip is stiff, so work it onto the crankshaft with a smooth object such as the end of a socket extension as you push the retainer against the engine block.

30   Install and tighten the retainer bolts in sequence (see illustration), to the torque listed in this Chapter's Specifications.

31   Install the flywheel/driveplate (see Section 16).

32   Install the transmission (see Chapter 7A).

33   Check the oil level and add if necessary, run the engine and check for oil leaks.

18   Engine mounts - check and replacement

Check

1   Engine mounts seldom require attention, but broken or deteriorated mounts should be replaced immediately or the added strain placed on the driveline components may cause damage or wear.

2   During the check, the engine must be raised slightly to remove the weight from the mounts.

3   Raise the vehicle and support it securely on jackstands, then position a jack under the engine oil pan. Place a large wood block between the jack head and the oil pan, then carefully raise the engine just enough to take the weight off the mounts.

4   Inspect the mounts to see if the rubber is cracked, hardened or separated from the

metal plates. Sometimes the rubber will split right down the center.

5   Check for relative movement between the mount plates and the engine or frame (use a large screwdriver or pry bar to attempt to move the mounts). If movement is noted, lower the engine and tighten the mount fasteners.

6   Rubber preservative should be applied to the mounts to slow deterioration.

Replacement

Note: *Because the mount replacement procedure involves a great deal of disassembly, it's a good idea to replace BOTH mounts, even if only one mount actually needs to be replaced.*

7   Remove the air cleaner and the air intake duct (see Chapter 4A).

8   Remove the engine cooling fan and shroud (see Chapter 3).

9   Remove the alternator (see Chapter 5).

10   Remove the throttle body from the intake manifold (see Chapter 4A).

11   If replacing the driver's side mount, remove the oil filter (see Chapter 1) then remove the oil cooler center bolt and move the oil cooler out of the way.

12   Raise the vehicle and place it securely on jackstands.

13   If you're going to replace the right mount, remove the starter motor (see Chapter 5).

14   Remove the mount-to-crossmember nuts and remove the nuts that attach the transmission mount to the transmission crossmember (see Chapter 7A).

Note: *Even if you're only planning to replace one mount, you must remove the mount-to-crossmember nuts from both mounts in order to raise the engine. If the studs loosen when the nuts are loosened the studs must be replaced.*

15   Attach an engine hoist to the engine and carefully raise the engine until it's supported, i.e., the weight of the engine is no longer on the mounts.

16   Remove the four mount-to-engine block bolts.

17   Raise the engine enough to take the weight off of the mounts then remove the engine mount through bolt. Always replace the through bolt once it's been removed.

18   Raise the engine as high as necessary in order to pull out the mount(s). Basically, you need to raise the engine high enough so that the two studs protruding downward from each mount can be removed from their respective holes in the crossmember. Do NOT raise the engine so high that the intake manifold touches the cowl.

Caution: *Raise the engine slowly while observing the location of the intake manifold. Do not allow the intake manifold to contact the cowl, which might damage the manifold.*

19   Installation is the reverse of removal. Tighten the bolts and nuts to the torque listed in this Chapter's Specifications.

Chapter 2 Part B
Diesel engine

Contents

Specifications

General

Displacement	6.7L (409 cu. in.)
Cylinder numbering	
Left cylinder bank (front to rear)	5-6-7-8
Right cylinder bank (front to rear)	1-2-3-4
Firing order	1-3-7-2-6-5-4-8

Exhaust manifold

Allowable warpage	
Between exhaust ports	0.005 inch
Everywhere else	0.009 inch

DIESEL ENGINE
1-3-7-2-6-5-4-8

36064-01-00.01c HAYNES

Cylinder numbering diagram

Torque specifications   Ft-lbs (unless otherwise indicated)

Note: *One foot-pound (ft-lb) of torque is equivalent to 12 inch-pounds (in-lbs) of torque. Torque values below approximately 15 foot-pounds are expressed in inch-pounds, because most foot-pound torque wrenches are not accurate at these smaller values.*

Oil spray tube nuts	98 in-lbs
Crankshaft pulley fasteners	
Step 1	22
Step 2	Tighten an additional 90 degrees
Crankshaft rear oil seal retainer bolts	
Step 1	89 in-lbs
Step 2	35
Cylinder head bolts* (see illustrations 13.27a and 13.27b)	
Step 1, bolts 1 through 18	177 in-lbs
Step 2, bolts 1 through 18	36
Step 3, bolts 1 through 18	36
Step 4, bolts 1 through 18	36
Step 5, bolts 1 through 18	Tighten an additional 90 degrees
Step 6, bolts 1 through 18	Tighten an additional 90 degrees
Step 7, bolts 1 through 18	Tighten an additional 90 degrees
Step 8	
Right-side cylinder head, bolts 19 through 22	22
Left-side cylinder head, bolts 19 through 23	22
Engine front cover	18
Exhaust manifold nuts*	
2014 and earlier models	
Step 1	177 in-lbs
Step 2	22
Step 3	22
2015 and later models	
Step 1	159 in-lbs
Step 2	18
Step 3	18
Driveplate bolts	
Step 1	44 in-lbs
Step 2	74
Intake manifold bolts	
Upper intake manifold	
2014 and earlier models	89 in-lbs
2015 and later models	97 in-lbs
Lower intake manifold	18
Engine mounts	
Through-bolts	258
Mount-to-frame nuts	148
Oil pan	
Upper oil pan	
Step 1	18 in-lbs
Step 2	18
Lower oil pan	89 in-lbs
Rocker arm assembly bolts	
Step 1	Make sure the top of the pushrod is in contact with the socket on each rocker arm
Step 2	35 lb-in
Step 3	71 lb-in
Step 4	115 lb-in
Step 5	115 lb-in
Step 6	
2014 and earlier models	Repeat Step 5 if any bolt turned more than 5 degrees during Step 5
2015 and later models	Tighten an additional 80 degrees
Valve cover bolts	106 in-lbs

Replace nuts or bolts with new ones.

1 General Information

1 This part of Chapter 2 is devoted to in-vehicle repair procedures for the V8 diesel engines. For information regarding engine removal and installation and engine overhaul, see Chapter 2C of this Chapter. The specifications included in this part of Chapter 2 apply only to the procedures included here. You'll find more specifications in Chapter 2C.

2 Most of the repair procedures included in this Part are based on the assumption that the engine is still installed in the engine compartment. If you're going to refer to any of these procedures after the engine has been removed from the engine compartment, many of the steps included here will not apply.

2 Repair operations possible with the engine in the vehicle

1 Some major repairs are possible with the engine installed in the engine compartment.

2 Clean the engine compartment and the exterior of the engine with some type of pressure washer before doing any work. A clean engine makes the job easier and helps to keep dirt out of the internal areas of the engine.

3 Depending on the procedure, it might be a good idea to remove the hood to improve access to the engine (see Chapter 11).

4 Oil and coolant leaks usually indicate a need for a new gasket or seal. Most of these repairs can be made with the engine installed. The intake and exhaust manifold gaskets, engine front cover gasket and crankshaft oil seals are accessible without removing the engine.

5 The oil pump is installed in the front cover and cannot be serviced separately. If there is a problem with the oil pump, the engine front

cover must be replaced.

6 Some exterior engine components can be removed with the engine in place. They include:

a) The valve covers (see Section 3)
b) The rocker arms, lash adjusters and pushrods (see Section 4)
c) The intake manifold (see Section 5)
d) The exhaust manifolds (see Section 5)
e) The crankshaft pulley (see Section 6)
f) The crankshaft front oil seal (see Section 7)
g) The flywheel/driveplate (see Section 10)
h) The crankshaft rear oil seal (see Section 11)
i) The cylinder heads and tappets (see Section 13)
j) The oil pans (see Section 14)
k) The water pump (see Chapter 3)
l) The turbocharger and the fuel injection components (see Chapter 4B)
m) The starter motor and the alternator (see Chapter 5)
n) The transmission (see Chapter 7A)

7 However, the removal of other components - such as the camshaft - requires engine removal. The removal and installation procedures for those components are in Chapter 2C. But even the procedures in this Chapter sometimes require extensive removal of other components. So read through a procedure before deciding whether you want to tackle it. Pay particularly close attention to any fuel system components that have to be removed before certain components in this Chapter can be removed.

3 Valve covers - removal and installation

1 Disconnect the cables from the negative terminals of the batteries (see Chapter 5).

2 If you're removing the left valve cover, remove the crankcase vent oil separator (see Chapter 6B).

3 Remove the glow plugs, fuel injectors, fuel supply line and fuel rail (see Chapter 4B) from the side that you are working on.

4 Remove the wiring harness retainer bolts and unclip the wiring harness retainers from around the valve cover and move the harness out of the way.

5 Remove the valve cover bolts and stud bolts, then remove the valve cover. If you're removing the left valve cover, the engine oil dipstick tube will need to be held out of the way to remove the valve cover.

6 Remove the valve cover gasket from the valve cover.

7 Using a small screwdriver, remove the fuel injector-to-valve cover seal from the valve cover.

8 Remove the old valve cover gasket and any gasket material that's stuck to the head.

9 Install new fuel injector-to-valve cover seals; use a socket to seat the seals into the cover.

Caution: *Verify that the fuel injector-to-valve cover seal is squarely seated in the valve cover.*

10 Install the valve cover gasket into the recessed groove in the valve cover.

11 Install the valve cover to the cylinder head(s) and tighten the fasteners, in sequence (see illustrations) to the torque listed in (this Chapter's Specifications).

12 Install the fuel injectors (see Chapter 4B).

Warning: *The fuel injectors must be installed before the final tightening of the valve cover bolts or possible engine damage may occur.*

13 Tighten the valve cover fasteners again, in sequence (see illustrations 3.11a or 3.11b), to the torque listed in this Chapter's Specifications.

14 The remainder of installation is the reverse of removal.

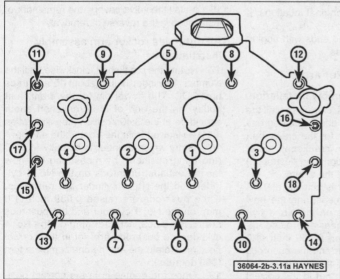

3.11a Left-side valve cover tightening sequence

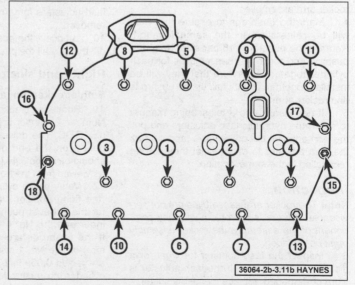

3.11b Right-side valve cover tightening sequence

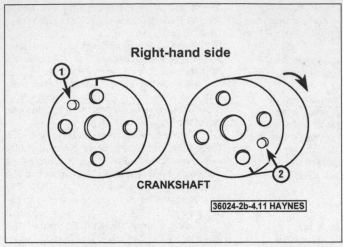

4.11 Starting from the No. 1 cylinder, rotate the crankshaft, clockwise until the Number 1 cylinder is just about 155-degrees past TDC.

1   *Dowel pin location at TDC No. 1 cylinder*
2   *Dowel pin at the 3:40 position (155-degrees past TDC)*

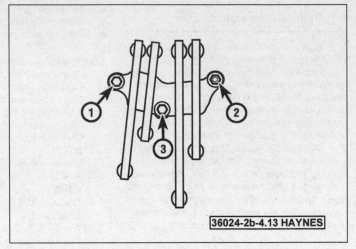

4.13 Rocker arm assembly tightening sequence - for single assembly

---

4   Rocker arms, lash adjusters and pushrods - removal, inspection and installation

Removal

1   Remove the valve covers (see Section 3).
2   Loosen the oil spray tube nuts evenly, a few turns at a time starting from the front and working toward the back. Then lift the tube up and off. Inspect the O-ring on the end of the tube and replace as necessary.
Note: *The oil spray tube nuts are captive and when loosened will stay on the tube.*
3   Mark the rocker arms to ensure that they will be reinstalled in the same location. Remove the rocker arm assembly bolts from each cylinder head and then remove the rocker arm assemblies.
4   Mark the pushrods to ensure that they will be reinstalled in the same location. Remove the pushrods, 16 per side. Keep the pushrod and rocker arm assemblies for each cylinder together to ensure that they will be installed together. Do not mix up the pushrods and rocker arms.
5   If necessary, use a telescoping magnet to reach the hydraulic lash adjusters and pull the lash adjuster out of the valve tappet. Mark the lash adjuster to ensure that they can be reinstalled in the same location.

Inspection

Note: *The rocker arm assemblies are not serviceable. If there is a problem with any component of the assembly, the rocker assembly must be replaced.*
6   Inspect the lash adjuster for wear, or a distorted or loose cap. If the lash adjuster is worn or damaged, replace the lash adjuster.
Note: *If there is damage to the lash adjuster,*

the valve tappet will have to be removed and inspected *(see Section 13).*
7   Clean the rocker assembly with solvent and dry the assembly. Look on the sides of the rocker arms and check the rocker arm ball for wear or abrasions. Inspect the pushrod cup on the arm and pivot pads on the arms for wear, cracks and any other damage. If any part of the rocker is worn or damaged, replace the assembly.
8   Inspect the pushrods for cracks and excessive wear at the ends. Roll each pushrod across a piece of plate glass to see if it is bent (if it wobbles, it is bent). Replace any damaged or worn pushrods.

Installation

9   Coat the lash adjusters with clean engine oil then use a telescoping magnet to place the lash adjusters into their original locations, if removed.
10   Lubricate the pushrod ends with engine oil, then install the pushrods.

Right-hand side rocker arms

Single rocker arm assembly installation

11   Rotate the crankshaft, clockwise until the Number 1 cylinder is just about 155-degrees past TDC. The dowel pin for the crankshaft pulley, on the end of the crankshaft snout should be at the 3:40 position (see illustration 4.11) as viewed from the front of the engine.
12   Verify which cylinder is actually at TDC (the firing position), by measuring the how far the exhaust pushrods on the No. 1 cylinder and the No. 4 cylinder are raised up. If the pushrods are raised 0.039 inch (1.0 mm) on the No. 1 cylinder and the pushrods are raised 0.059 inch (1.5 mm) on the No. 4 cylinder the pushrods are not in the correct position. Rotate the engine one complete turn (360 degrees).

13   Once the engine is in the correct position, install the rocker arm assembly as needed for the right-side cylinder head and tighten the assembly in sequence (see illustration) to the torque listed in this Chapter's Specifications.
14   Install the rocker arms and bolts and tighten them to the torque listed in this Chapter's Specifications.
Caution: *To prevent bent valves when installing the rocker arms, rotate the engine until the pushrods for each cylinder are at their lowest point.*
15   Install a new O-ring if necessary to the oil spray tube, then place the tube assembly onto the cylinder head. Start the captive nuts evenly, and a few turns starting from the left and working to the right tightening the nuts to the torque listed in this Chapter's Specifications.
16   Install the valve covers, the remainder of installation is the reverse of removal.

All right-side rocker arm assembly installation

17   Rotate the crankshaft, clockwise until the Number 1 cylinder is just about 155-degrees past TDC. The dowel pin for the crankshaft pulley, on the end of the crankshaft snout should be at the 3:40 position (see illustration 4.11) as viewed from the front of the engine.
18   Verify which cylinder is actually at TDC (the firing position), by measuring the how far the exhaust pushrods on the No. 1 cylinder and the No. 4 cylinder are raised up. If the pushrods are raised 0.039 inch (1.0 mm) on the No. 1 cylinder and the pushrods are raised 0.059 inch (1.5 mm) on the No. 4 cylinder the pushrods are not in the correct position. Rotate the engine one complete turn (360 degrees).
19   Once the engine is in the correct position, install the rocker arm assemblies for the right-side cylinder head and tighten the

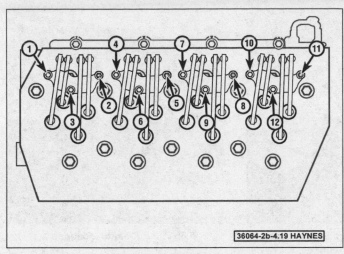

4.19 Rocker arm assembly tightening sequence - for all rocker assemblies on the right side

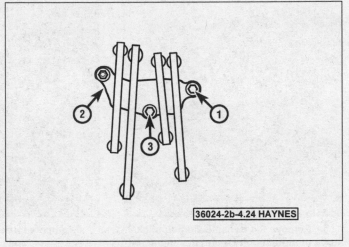

4.24 Rocker arm assembly tightening sequence - for single assembly

assembly in sequence (see illustration) to the torque listed in this Chapter's Specifications.

20   Install the rocker arms and bolts and tighten them to the torque listed in this Chapter's Specifications.

Caution: *To prevent bent valves when installing the rocker arms, rotate the engine until the pushrods for each cylinder are at their lowest point.*

21   Install the valve covers and the remainder of installation is the reverse of removal.

Left-hand side rocker arms

Single rocker arm assembly installation

22   Rotate the crankshaft, clockwise until the Number 1 cylinder is just about 155-degrees past TDC. The dowel pin for the crankshaft pulley, on the end of the crankshaft snout should be at the 3:40 position (see illustration 4.11) as viewed from the front of the engine.

23   Verify which cylinder is actually at TDC (the firing position), by measuring the how far the exhaust pushrods on the No. 6 cylinder and the No. 7 cylinder are raised up. If the pushrods are raised 0.039 inch (1.0 mm) on the No. 6 cylinder and the pushrods are raised 0.059 inch (1.5 mm) on the No. 7 cylinder the pushrods are not in the correct position. Rotate the engine one complete turn (360 degrees).

24   Once the engine is in the correct position, install the rocker arm assembly as needed for the left-side cylinder head and tighten the assembly in sequence (see illustration) to the torque listed in this Chapter's Specifications.

25   Install the rocker arms and bolts and tighten them to the torque listed in this Chapter's Specifications.

Caution: *To prevent bent valves when installing the rocker arms, rotate the engine until the pushrods for each cylinder are at their lowest point.*

26   Install a new O-ring if necessary to the oil spray tube then place the tube assembly onto the cylinder head. Start the captive nuts evenly, and a few turns starting from the left and work-

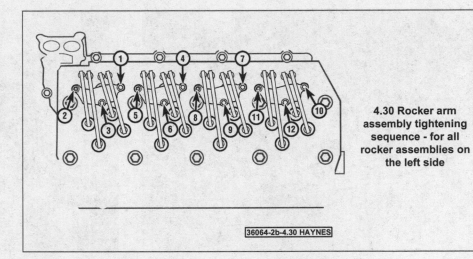

4.30 Rocker arm assembly tightening sequence - for all rocker assemblies on the left side

ing to the right tightening the nuts to the torque listed in this Chapter's Specifications.

27   Install the valve covers, the remainder of installation is the reverse of removal.

All left-side rocker arm assembly installation

28   Rotate the crankshaft, clockwise until the Number 1 cylinder is just about 155-degrees past TDC. The dowel pin for the crankshaft pulley, on the end of the crankshaft snout should be at the 3:40 position (see illustration 4.11) as viewed from the front of the engine.

29   Verify which cylinder is actually at TDC (the firing position), by measuring the how far the exhaust pushrods on the No. 6 cylinder and the No. 7 cylinder are raised up. If the pushrods are raised 0.039 inch (1.0 mm) on the No. 6 cylinder and the pushrods are raised 0.059 inch (1.5 mm) on the No. 7 cylinder the pushrods are not in the correct position. Rotate the engine one complete turn (360 degrees).

30   Once the engine is in the correct position, install the rocker arm assemblies for the right-side cylinder head and tighten the

assembly in sequence (see illustration) to the torque listed in this Chapter's Specifications.

31   Install the rocker arms and bolts and tighten them to the torque listed in this Chapter's Specifications.

Caution: *To prevent bent valves when installing the rocker arms, rotate the engine until the pushrods for each cylinder are at their lowest point.*

32   Install the valve covers and the remainder of installation is the reverse of removal.

5   Intake manifold – removal and installation

Warning: *Wait until the engine is completely cool before beginning this procedure.*

Removal

Upper intake manifold

1   Disconnect the cables from the negative battery terminals.

2   Disengage the coolant hose retainer

5.4 Remove the transmission dipstick tube mounting bolt (1) and the engine oil dipstick tube mounting bolt (2)

5.6a Upper intake manifold, left side bolt locations

5.6b . . . right side locations . . .

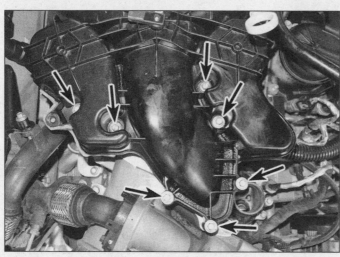

5.6c . . . and front locations

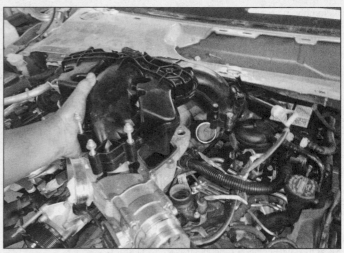

5.7 Lift the upper manifold up and off

from the upper right side of the manifold.

3   Disconnect the manifold absolute pressure (MAP) sensor, the engine oil pressure (EOP) sensor electrical connectors (see

5.11 EGR bypass outlet pipe mounting bolt locations

Chapter 6B) then disengage the wiring harness retainers and move the harness out of the way.

4   Remove the engine oil dipstick tube

mounting bolt and the transmission dipstick tube mounting bolt from the upper intake manifold (see illustration).

5   Remove the charge air cooler (CAC) outlet pipe (see Chapter 4B) and move the pipe out of the way.

6   Remove the upper intake manifold mounting bolts (see illustrations).

7   Lift the upper intake manifold straight up and off (see illustration), do not slide the upper intake manifold across the lower manifold.

Lower intake manifold

8   Remove the upper intake manifold (see Steps 1 through 7).

9   Remove the charge air cooler (CAC) outlet hose (see Chapter 4B).

10   Disconnect the electrical connectors to the throttle body and the EGR temperature (EGRT) sensor (see Chapter 6B).

11   Remove the EGR bypass outlet pipe mounting bolts (see illustration) and remove the pipe. Replace the gaskets for the pipe.

5.12 Loosen the turbocharger-to-lower intake hose clamps

5.13 Disconnect the crankcase ventilation hose quick-connect connector from the intake

5.14 Lower intake manifold mounting bolt locations

5.18 Install new gaskets into the manifold making sure they are fully seated

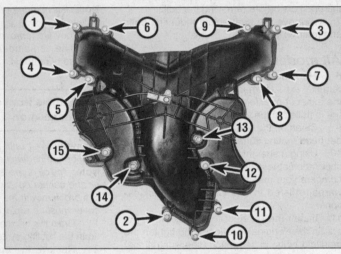

5.19 Upper intake manifold bolt tightening sequence

12   Remove the charge air cooler (CAC) inlet pipe from the turbocharger and loosen the inlet hose clamps to the turbocharger (see illustration).

13   Disconnect the crankcase ventilation hose from the lower intake manifold (see illustration).

14   Remove the lower intake manifold mounting bolts (see illustration) and remove the manifold.

Installation

15   Thoroughly clean all sealing surfaces prior to installation.

Lower intake manifold

16   Be sure to install the new intake manifold gaskets then insert the lower intake manifold into the turbocharger inlet hose. Install the bolts and tighten the bolts to the torque listed in this Chapter's Specifications.

17   Remaining installation is the reverse of

removal. Install the upper intake manifold.

Upper intake manifold

18   Install new gaskets into the upper intake manifold (see illustration).

19   Place the upper intake manifold on to the lower intake manifold (do not slide the manifold or the gaskets can get damaged) then install the bolts and tighten then bolts in sequence (see illustration) to the torque listed in this Chapter's Specifications.

20   The remainder of installation is the reverse of removal.

---

6   Exhaust manifolds - removal and installation

Warning: *Wait until the engine is completely cool before beginning this procedure.*

1   Remove the upper intake manifold (see Section 5).

2   Remove the left side turbocharger inlet pipe, outlet pipe and the turbocharger (see Chapter 4B).

Right-side manifold

3   Disconnect the electrical connector and vacuum hoses from the bypass solenoid (see Chapter 6B).

4   Remove the EGR valve and tube bolts then remove the tube and gaskets (see Chapter 6B).

5   Remove the fuel rail supply line bracket bolts, then remove the heat shield mounting bolts.

Note: *It will be necessary to push the fuel line toward the front of the engine to remove the heat shield.*

Left-side manifold

6   Remove the injector fuel line isolator-to-bracket bolts and separate the isolators from the lines and bracket, then remove the

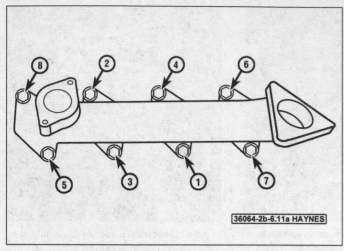

6.11a Right side manifold tightening sequence

6.11b Left side manifold tightening sequence

bracket bolt and bracket. Make sure to mark the brackets before removing the bolts.
7   Remove the heat shield mounting bolts and remove the shield.

All models

8   Mark the exhaust manifold spacers, then remove exhaust manifold nuts, spacers and the exhaust manifold. Make sure the spacers are installed in the same locations.
9   Clean off the exhaust manifold and cylinder head mating surfaces.
10   Using a straightedge and a feeler gauge, check the flatness of the exhaust manifold and compare your measurements to the allowable warpage listed in this Chapter's Specifications.
11   Install the exhaust manifold, the spacers (in their original locations) and tighten the manifold bolts, in sequence (see illustrations) to the torque listed in this Chapter's Specifications.
12   The remainder of installation is the reverse of removal.

7   Crankshaft pulley - removal and installation

1   Disconnect the cable from the negative battery terminal (see Chapter 5).
2   Remove the accessory drivebelt (refer to Chapter 1).
3   Remove the cooling fan and shroud (see Chapter 3).
4   Raise the vehicle and support it securely on jackstands.
5   Remove the crankshaft pulley bolts. A strap wrench can be used to hold the pulley (see illustration).
6   Remove the pulley. A crankshaft pulley puller is not required.
7   Install the dowel pin in the hole in the nose of the crankshaft, then place the pulley on the crankshaft snout and bolt it in place.
8   Tighten the new bolts to the torque listed in this Chapter's Specifications, in a criss-

cross pattern.
Note: *It is mandatory to replace the crankshaft pulley bolts with lightly-oiled new ones any time they are removed.*
9   The remainder of installation is the reverse of removal.

8   Engine front cover - removal and installation

Warning: *Wait until the engine is completely cool before beginning this procedure.*
Note: *The oil pump is installed in the front cover and cannot be serviced separately. If there is a problem with the oil pump the engine front cover must be replaced.*
1   Drain the engine oil (see Chapter 1), then drain the cooling system (see Chapter 1).
2   Remove the upper and lower oil pans (see Section 14).
3   Remove the cooling module (see Chapter 3).

7.5 Remove the pulley bolts. Use a strap wrench to hold the crankshaft pulley while you loosen the pulley bolts (it's a good idea to wrap a piece of old drivebelt around the pulley to prevent damage)

4   Remove the cooling fan shroud, cooling fan and cooling fan stator (see Chapter 3).
5   Remove the drivebelt (see Chapter 1), then remove the drivebelt tensioner mounting bolt and tensioner.
6   Remove the air conditioning compressor mounting fasteners and move the compressor out of the way (see Chapter 3) without disconnecting the lines to the compressor.
7   Remove the brake booster vacuum pump (see Chapter 9, Section 11).
8   Disconnect the electrical connector to the camshaft position sensor (see Chapter 6B). Remove the power steering pump hose retainers, then disengage the wiring harness connectors and secure the harness out of the way.
9   Remove the lower radiator hose retaining clip and detach the hose from the front cover.
10   Remove the crankshaft pulley (see Section 7).
11   Remove the crankshaft front seal (see Section 10).
12   Remove the engine front cover bolts and remove the front cover. Note the locations of studs and different-length bolts so they can be reinstalled in their original locations. Remove and discard the engine front cover gaskets. Remove the engine front cover dowel pins and store them in a plastic bag.
13   Clean off any gasket material from the front cover and from the cover mating surface on the engine block.
14   Rotate the crankshaft so that the flats for the oil pump dive are horizontal.
Note: *The dowel pin on the crankshaft should be at the 9 o'clock position.*
15   Using a new gasket, install the front cover aligning the gear rotor flats of the oil pump with the flats on the crankshaft, then hand tighten bolts 1 and 2 (see illustration 8.17) into the front cover.
16   The alignment of the bottom of the front cover to the oil pan rail is critical. Place a steel ruler or gauge block on the oil pan rail and the bottom of the engine cover, then measure between the bottom of the front cover and

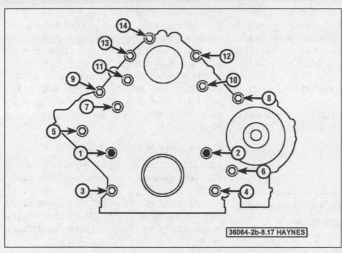

8.17 Front cover bolt tightening sequence

9.3 Remove the oil dipstick tube mounting bolt

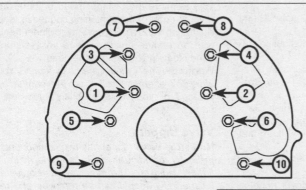

9.11 Rear engine cover tightening sequence

10.2 Use a seal removal tool to pry out the old crankshaft front seal

the ruler or gauge block with feeler gauges. Clearance should be 0.009 inch, never past. If the clearance is OK, tighten the bolts 1 and 2 enough to hold the cover in place then install the remaining bolts and tighten them finger-tight. If the clearance is not OK, loosen the bolts and move the front cover.

17 Tighten the cover bolts, in sequence (see illustration) to the torque listed in this Chapter's Specifications.

18 The remainder of installation is the reverse of removal.

9 Engine rear cover - removal and installation

1 Disconnect the cable from the negative battery(ies) terminal(s) (see Chapter 5).

2 Drain the engine oil (see Chapter 1).

3 Remove the oil dipstick tube mounting bolt (see illustration) from the engine compartment, then slowly twist and pull the tube out from the top of the engine rear cover.

4 Remove the transmission (see Chapter 7A).

5 Remove the upper and lower oil pans (see Section 14).

6 Remove the crankshaft rear seal (see Section 12).

7 Remove the turbocharger down pipe bracket bolt (see Chapter 4B).

8 Disconnect the crankshaft sensor (CKP) electrical connector (see Chapter 6B).

9 Remove the rear cover mounting bolts and cover.

10 Clean off any gasket material from the rear cover and from the cover mating surface on the engine block.

11 Install a new gasket to the cover and install the cover. Then tighten the bolts in sequence (see illustration), to the torque listed in this Chapter's Specifications.

12 Remaining installation is the reverse of removal.

10 Crankshaft front oil seal - replacement

1 Remove the crankshaft pulley (see Section 7).

2 Using a seal removal tool, carefully remove the old crank seal (see illustration).

3 Coat the new seal with multi-purpose grease.

4 Using a seal driver or a big socket, drive the new seal into place (see illustration).

5 Installation is otherwise the reverse of removal.

10.4 Use a big socket with an outside diameter slightly smaller than the outside diameter of the seal to drive the new seal into place - typical installation shown

11 Driveplate - removal and installation

Caution: *Do not use any type of magnetic screwdriver or other similar tool around the Crankshaft Position (CKP) sensor and the crankshaft sensor pulse ring. The magnet can damage the sensor or pulse ring if it comes in contact with either component.*

1   Remove the transmission (see Chapter 7A).

2   Remove the bolts (and discard) and remove the driveplate from the crankshaft flange.

3   Install the driveplate, install the new driveplate bolts and tighten the bolts to the torque listed in this Chapter's Specifications-following the correct tightening sequence (see illustration).

4   Install the transmission (see Chapter 7A).

12 Crankshaft rear oil seal - replacement

1   Remove the transmission (see Chapter 7A).

2   Remove the Driveplate (see Section 11).

3   Remove the crankshaft sensor pulse ring from the end of the crankshaft.

4   Remove the cylinder block opening plastic plug to access the crankshaft sensor, then disconnect the electrical connector to the sensor and remove it (see Chapter 6B).

5   Using a seal removal tool, carefully remove the old crank seal from the rear housing.

Caution: *Do not scratch the crankshaft or rear engine cover while removing the seal.*

6   Coat the outside of the seal and the crankshaft surface with engine oil only, do not use grease or petroleum jelly.

Warning: *Grease and petroleum jelly may cause oil to leak past the seal.*

7   Install the seal installation sleeve that comes with a new sleeve, over the end of the crankshaft.

8   Slide the seal onto the end of the crankshaft, then remove the seal installation tool. Use either installation tool #303-1514 or a seal driver and make sure that it doesn't become tilted. Press the seal into place until it is fully seated.

9   Install the crankshaft position (CKP) sensor (see Chapter 6B).

10   Install the crankshaft pulse ring onto the end of the crankshaft, then install the driveplate (see Section 11).

11   Installation is otherwise the reverse of removal.

13 Cylinder heads and valve tappets - removal and installations

Warning: *Wait until the engine is completely cool before beginning this procedure.*

Removal

1   Drain the engine oil (see Chapter 1) and the cooling system (see Chapter 1).

2   Remove the cooling fan and cooling stator (see Chapter 3).

3   Remove the drivebelt (see Chapter 1).

4   Remove the valve cover(s) (see Section 3).

5   Remove the rocker arms (see Section 4). If you're only removing one cylinder head, only remove the rocker arms from that side.

6   Remove the exhaust manifold(s) (see Section 6).

7   Remove the coolant crossover housing (Chapter 3).

Left-side cylinder head

8   Disconnect the charge air cooler (CAC) sensor electrical connector and disengage the wiring harness retainer.

9   Remove the CAC tube (see Chapter 4B).

10   Remove the power steering pump (see Chapter 10).

11   Disconnect the alternator electrical connectors (see Chapter 5), then remove alternator/power steering bracket bolts and remove the bracket and alternator as an assembly.

12   Remove the cylinder head bolts in the reverse order of the tightening sequence (see illustration 13.27b). Remove the cylinder head with an engine hoist, or have an assistant handy to help you remove the head.

Warning: *The cylinder head is heavy. You shouldn't attempt to lift it off the engine by yourself or you could seriously injure your back.*

Right-side cylinder head

13   On models equipped with dual alternators, remove the second alternator (see Chapter 5).

14   Disconnect the hose connectors to the secondary water pump, then remove the secondary pump bolts and pump (see Chapter 3).

15   Disconnect the electrical connector to the fuel volume control valve and position the harness out of the way.

16   Remove the cylinder head bolts in the reverse order of the tightening sequence (see illustration 13.27a). Remove the cylinder head with an engine hoist, or have an assistant handy to help you remove the head.

Warning: *The cylinder head is heavy. You shouldn't attempt to lift it off the engine by yourself or you could seriously injure your back.*

Valve tappets

Note: *The valve tappet has two components: the body including the roller and the hydraulic lash adjuster. The hydraulic lash adjuster for the valve tappet can be removed and replaced without removing the cylinder head(s) (see Section 4).*

17   Remove the cylinder head(s) (see Steps 1 through 16) as necessary.

18   Mark the tappet guide so it can be installed into its original locations, then unhook the valve tappet guide from the cylinder block and lift the guide/tappet assembly out of the engine block. Mark the valve tappets to the guide so that they can be installed in their original location.

Warning: *The valve tappets must be kept in the order in which they were removed and installed back in their original positions or possible engine damage can happen.*

19   Remove the valve tappet from the guide.

Caution: *Always keep the valve tappet upright or the hydraulic lash adjuster may fall out.*

20   Inspect the valve tappet roller for pitting or wear and replace as necessary.

Installation

Valve tappets

21   Coat the valve tappets with clean engine oil and insert the tappets into the tappet guides, if removed.

22   Install the tappet guide assemblies into the cylinder block, making sure the guide is seated properly into the cylinder block.

23   The remainder of installation is the reverse of removal.

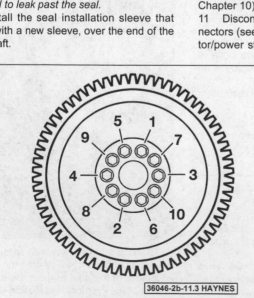

11.3 Driveplate bolt tightening sequence

36046-2b-11.3 HAYNES

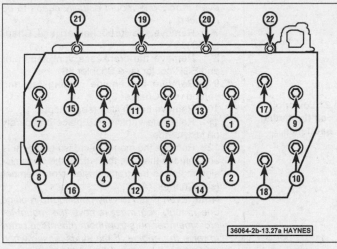

13.27a Right side cylinder head bolt tightening sequences

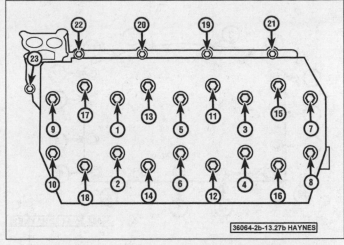

13.27b Left side cylinder head bolt tightening sequences

Cylinder heads

24 Remove and discard the old cylinder head gasket and cylinder head dowels. Thoroughly clean the mating surfaces of the cylinder head and engine block.

25 Chase the threads of the cylinder head bolt holes and then blow out the holes with compressed air.

26 Install new cylinder head dowels and a new gasket and then, using an engine hoist or at least one assistant, very carefully place the cylinder head in position. **Warning:** *The cylinder head is VERY heavy. Do NOT attempt to place it in position on the engine block by yourself or you could seriously injure your back.*
Caution: *Make SURE that you don't damage the cylinder head gasket when installing the head on the block. Even the slightest nick or scratch can cause the head gasket to leak very soon after installation.*

27 Lightly oil the NEW cylinder head bolts with clean engine oil, install the cylinder head bolts and then tighten, following the indicated sequences (see illustrations), to the torque listed in this Chapter's Specifications.

28 The remainder of installation is the reverse of removal.

14 Oil pan(s) - removal and installation

Removal

Lower oil pan

1 Raise the vehicle and support it securely on jackstands and block the rear wheels.
2 Drain the engine oil (see Chapter 1).
3 Remove the oil pan bolts.
4 Carefully cut the oil pan sealant all the way around the circumference of the pan, then carefully pry the pan away from the block and remove it.

Upper oil pan

5 Remove the dipstick form the dipstick tube.
6 Remove the transmission (see Chapter 7A).
7 Remove the flywheel/driveplate (see Section 11).
8 Remove the lower oil pan (see Steps 1 through 4).
9 Remove the oil filter and replace the oil filter (see Chapter 1).
10 Disconnect the electrical connector to

the engine oil temperature (EOT) sensor and engine oil pressure switch.
11 Remove the engine harness retainers along the perimeter of the upper oil pan and move the harness out of the way.
12 Remove the oil pan bolts.
13 On 2013 and earlier models and early built 2014 models, carefully pry the pan away from the block and remove it. Discard the gasket.
14 On 2014 late built models and 2015 and later models, carefully cut the oil pan sealant all the way around the circumference of the pan, then carefully pry the pan away from the block and remove it. Discard the three gaskets pressed into the pan.

Installation

Upper oil pan

15 Clean off all traces of old sealant from the oil pan flange and from the flange surface on the engine block.
16 On 2013 and earlier models and early built 2014 models, press the gasket into the groove along the pan, then place four dime size beads of RTV sealant at the cover joints on the pan.
Caution: *The pan must be installed within 60 minutes of applying the sealant.*
17 On 2014 late built models and 2015 and later models, press the three O-ring type gaskets into the pan then apply a 1/8-inch bead of RTV sealant along the flanges of the oil pan.
Caution: *The pan must be installed within 60 minutes of applying the sealant.*
18 Align the pan with the pan mating surface of the block, carefully press the pan against the block and install the oil pan bolts. After all the bolts are installed, tighten them gradually and evenly until they're snug. Then tighten the oil pan bolts, in sequence (see illustration) to the torque listed in this Chapter's Specifications.
19 Remaining installation is the reverse of removal.

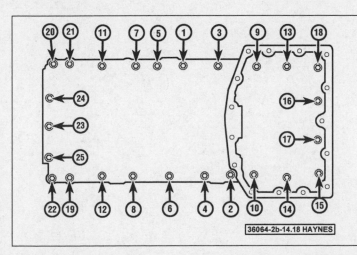

14.18 Upper oil pan bolt tightening sequence

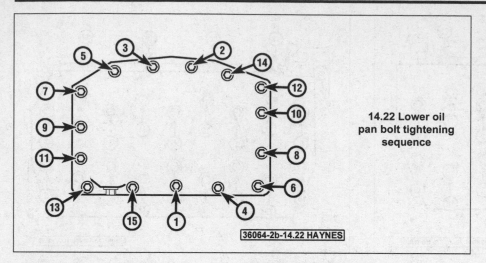

14.22 Lower oil pan bolt tightening sequence

36064-2b-14.22 HAYNES

Lower oil pan

20 Clean off all traces of old sealant from the oil pan flange and from the flange surface on the upper oil pan.

21 Apply a 1/8-inch bead of RTV sealant along the flanges of the oil pan.

22 Align the pan with the pan mating surface of the upper oil pan, carefully press the pan against the upper oil pan and install the oil pan bolts. After all the bolts are installed, tighten them gradually and evenly until they're snug. Then tighten the oil pan bolts, in sequence (see illustration) to the torque listed in this Chapter's Specifications.

23 Remaining installation is the reverse of removal.

15 Engine mounts - check and replacement

Check

1 Engine mounts seldom require attention, but broken or deteriorated mounts should be replaced immediately or the added strain placed on the driveline components may cause damage or wear.

2 During the check, the engine must be raised slightly to remove the weight from the mounts.

3 Raise the vehicle and support it securely on jackstands, then position a jack under the engine oil pan. Place a large wood block between the jack head and the oil pan, then carefully raise the engine just enough to take the weight off the mounts.

4 Inspect the mounts to see if the rubber is cracked, hardened or separated from the metal plates. Sometimes the rubber will split right down the center.

5 Check for relative movement between the mount plates and the engine or frame (use a large screwdriver or pry bar to attempt to move the mounts). If movement is noted, lower the engine and tighten the mount fasteners.

6 Rubber preservative should be applied to the mounts to slow deterioration.

Replacement

Note: *Because the mount replacement procedure involves a great deal of disassembly, it's a good idea to replace BOTH mounts,* even if only one mount actually needs to be replaced.

7 Remove the turbocharger (see Chapter 4B).

8 Remove the crankcase vent separator and EGR cooler (see Chapter 6B).

9 Remove the engine cooling fan and shroud (see Chapter 3).

10 Remove the oil filter (see Chapter 1).

11 Raise the vehicle and place it securely on jackstands.

12 Remove the mount-to-crossmember nuts and remove the nuts that attach the transmission mount to the transmission crossmember (see Chapter 7A).

Note: *Even if you're only planning to replace one mount, you must remove the mount-to-crossmember nuts from both mounts in order to raise the engine. If the studs loosen when the nuts are loosened the studs must be replaced.*

13 Attach an engine hoist to the engine and carefully raise the engine until it's supported, i.e., the weight of the engine is no longer on the mounts.

14 On the right-side mount, remove the four mount-to-frame bolts and on the left-side mount, remove the three mount-to-frame bolts.

15 Raise the engine enough to take the weight off of the mounts, then remove the engine mount through bolt. Always replace the through bolt once it's been removed.

16 Raise the engine as high as necessary in order to pull out the mount(s). Basically, you need to raise the engine high enough so that the two studs protruding downward from each mount can be removed from their respective holes in the crossmember. Do NOT raise the engine so high that the intake manifold touches the cowl.

Caution: *Raise the engine slowly making sure nothing is hitting the cowl and no hoses or wires are being streched.*

17 Installation is the reverse of removal. Tighten the fasteners to the torque listed in this Chapter's Specifications.

Chapter 2 Part C
General engine overhaul procedures

Contents

Specifications

General

Displacement
- Gasoline engine .... 379 cubic inches (6.2L)
- Diesel engine .... 409 cubic inches (6.7L)

Bore and stroke
- Gasoline engine .... 4.015 x 3.74 inches
- Diesel engine .... 3.897 x 4.251 inches

Oil pressure
- Gasoline engine
  - At idle .... 8 psi
  - At 2,000 rpm .... N/A
- Diesel engine
  - At 700 rpm .... 10 psi
  - At 1,200 rpm .... 20 psi
  - At 1,800 rpm .... 35 psi

Cylinder compression .... Lowest cylinder must be within 75 percent of highest cylinder

Note: *On diesel engines the oil pump is installed in the engine front cover and cannot be serviced separately. If there is a problem with the oil pump the engine front cover must be replaced (see Chapter 2B).*

Camshaft (diesel engine)

Lobe lift
  Intake .... 0.224 inch
  Exhaust .... 0.232 inch
Journal diameter... 2.361 to 2.362 inches
Bearing inside diameter.. 2.363 to 2.364 inches
Journal-to-bearing (oil) clearance... 0.0009 to 0.002 inch
Backlash
  Camshaft gear-to-high-pressure injection pump gear...................... 0.0007 to 0.007 inch
  Camshaft gear-to-crankshaft gear .... 0.0009 to 0.006 inch

Note: *For gasoline engine camshaft specifications, refer to Chapter 2A.*

Torque specifications   Ft-lbs (unless otherwise indicated)

Note: *One foot-pound (ft-lb) of torque is equivalent to 12 inch-pounds (in-lbs) of torque. Torque values below approximately 15 foot-pounds are expressed in inch-pounds, because most foot-pound torque wrenches are not accurate at these smaller values.*

Camshaft thrust plate bolts - diesel engine 18
Camshaft gear bolt - diesel engine
  Step 1 .... 72 in-lbs
  Step 2 .... Tighten an additional 60 degrees
Connecting rod bearing cap bolts*
  Gasoline engine
    Step 1 .... 177 in-lbs
    Step 2 32
    Step 3 .... Tighten an additional 135 degrees
  Diesel engine
    Step 1 33
    Step 2 .... Tighten an additional 90 degrees
Main bearing cap bolts*
  Gasoline engine (see illustration 11.19a)
    Vertical (outer) bolts (1 through 10)
      Step 1.. 26
      Step 2.. 37
      Step 3.. Tighten an additional 90 degrees
    Vertical (inner) bolts (11 through 20)
      Step 1.. 26
      Step 2.. 48
      Step 3.. Tighten an additional 90 degrees
    Side bolts (21 through 30)
      Step 1.. 177 in-lbs
      Step 2.. 26
      Step 3.. Tighten an additional 60 degrees
  Diesel engine (see illustration 11.19b)
    Step 1 (bolts 1 through 10) 59
    Step 2 (bolts 11 through 20) 118
    Step 3 (bolts 1 through 20) .... Tighten an additional 90 degrees
    Step 4 (bolts 21 through 30) 30
    Step 5 (bolts 21 through 30) .... Tighten an additional 90 degrees

* New bolts must be installed.

1.1 An engine block being bored. An engine rebuilder will use special machinery to recondition the cylinder bores

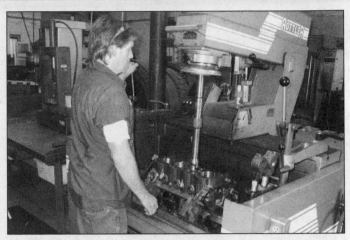

1.2 If the cylinders are bored, the machine shop will normally hone the engine on a machine like this

1   General information - engine overhaul

Included in this portion of Chapter 2 are general information and diagnostic testing procedures for determining the overall mechanical condition of your engine.

The information ranges from advice concerning preparation for an overhaul and the purchase of replacement parts and/or components to detailed, step-by-step procedures covering removal and installation.

The following Sections have been written to help you determine whether your engine needs to be overhauled and how to remove and install it once you've determined it needs to be rebuilt. For information concerning in-vehicle engine repair, see Chapter 2A or 2B.

The Specifications included in this Part are general in nature and include only those necessary for testing the oil pressure and checking the engine compression. Refer to Chapter 2A or 2B for additional engine Specifications.

It's not always easy to determine when, or if, an engine should be completely overhauled, because a number of factors must be considered.

High mileage is not necessarily an indication that an overhaul is needed, while low mileage doesn't preclude the need for an overhaul. Frequency of servicing is probably the most important consideration. An engine that's had regular and frequent oil and filter changes, as well as other required maintenance, will most likely give many thousands of miles of reliable service. Conversely, a neglected engine may require an overhaul very early in its service life.

Excessive oil consumption is an indication that piston rings, valve seals and/or valve guides are in need of attention. Make sure that oil leaks aren't responsible before deciding that the rings and/or guides are bad. Perform a cylinder compression check to determine the extent of the work required (see

Section 3). Also check the vacuum readings under various conditions (see Section 4).

Check the oil pressure with a gauge installed in place of the oil pressure sending unit and compare it to this Chapter's Specifications (see Section 2). If it's extremely low, the bearings and/or oil pump are probably worn out.

Loss of power, rough running, knocking or metallic engine noises, excessive valve train noise and high fuel consumption rates may also point to the need for an overhaul, especially if they're all present at the same time. If a complete tune-up doesn't remedy the situation, major mechanical work is the only solution.

An engine overhaul involves restoring the internal parts to the specifications of a new engine. During an overhaul, the piston rings are replaced and the cylinder walls are reconditioned (rebored and/or honed) **(see illustrations 1.1 and 1.2)**. If a rebore is done by an automotive machine shop, new oversize pistons will also be installed. The main bearings, connecting rod bearings and camshaft bearings are generally replaced with new

ones and, if necessary, the crankshaft may be reground to restore the journals **(see illustration 1.3)**. Generally, the valves are serviced as well, since they're usually in less-than-perfect condition at this point. While the engine is being overhauled, other components, such as the starter and alternator, can be rebuilt or replaced as well. The end result should be similar to a new engine that will give many trouble free miles.

Note: *Critical cooling system components such as the hoses, drivebelts, thermostat and water pump should be replaced with new parts when an engine is overhauled. The radiator should be checked carefully to ensure that it isn't clogged or leaking (see Chapter 3). If you purchase a rebuilt engine or short block, some rebuilders will not warranty their engines unless the radiator has been professionally flushed. Also, we don't recommend overhauling the oil pump - always install a new one when an engine is rebuilt.*

Overhauling the internal components on today's engines is a difficult and time-consuming task which requires a significant amount of specialty tools and is best left to a

1.3 A crankshaft having a main bearing journal ground

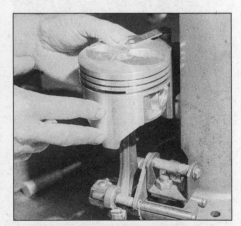

1.4 A machinist checks for a bent connecting rod, using specialized equipment

professional engine rebuilder **(see illustrations 1.4, 1.5 and 1.6)**. A competent engine rebuilder will handle the inspection of your old parts and offer advice concerning the reconditioning or replacement of the original engine, never purchase parts or have machine work done on other components until the block has been thoroughly inspected by a professional machine shop. As a general rule, time is the primary cost of an overhaul, especially since the vehicle may be tied up for a minimum of two weeks or more. Be aware that some engine builders only have the capability to rebuild the engine you bring them while other rebuilders have a large inventory of rebuilt exchange engines in stock. Also be aware that many machine shops could take as much as two weeks time to completely rebuild your engine depending on shop workload. Sometimes it makes more sense to simply exchange your engine for another engine that's already rebuilt to save time.

2   Oil pressure check

1   Low engine oil pressure can be a sign of an engine in need of rebuilding. A "low oil pressure" indicator (often called an "idiot light") is not a test of the oiling system. Such indicators only come on when the oil pressure is dangerously low. Even a factory oil pressure gauge in the instrument panel is only a relative indication, although much better for driver information than a warning light. A better test is with a mechanical (not electrical) oil pressure gauge.

2   Locate the oil pressure indicator sending unit on the engine block:

a) *On gasoline engines, the oil pressure sending unit is located on the left side of the engine block, on the oil filter adapter in front of the left engine mount (see illustration).*

b) *On diesel engines, the oil pressure sending unit is located on the front side of the oil filter housing (see illustration).*

1.5 A bore gauge being used to check the main bearing bore

3   Unscrew and remove the oil pressure sending unit and then screw in the hose for your oil pressure gauge. If necessary, install an adapter fitting. Use Teflon tape or thread sealant on the threads of the adapter and/or the fitting on the end of your gauge's hose.

4   Connect an accurate tachometer to the engine, according to the tachometer manufacturer's instructions.

5   Check the oil pressure with the engine running (normal operating temperature) at the specified engine speed, and compare it to this Chapter's Specifications. If it's extremely low, the bearings and/or oil pump are probably worn out.

3   Cylinder compression check

Gasoline engines

1   A compression check will tell you what mechanical condition the upper end of your engine (pistons, rings, valves, head gaskets) is in. Specifically, it can tell you if the compression is down due to leakage caused by

1.6 Uneven piston wear like this indicates a bent connecting rod

worn piston rings, defective valves and seats or a blown head gasket.

Note: *The engine must be at normal operating temperature and the battery must be fully charged for this check.*

2   Begin by cleaning the area around the spark plugs before you remove them (compressed air should be used, if available). The idea is to prevent dirt from getting into the cylinders as the compression check is being done.

3   Remove all of the spark plugs from the engine (see Chapter 1).

4   Disable the fuel pump circuit (see Chapter 4A).

5   Install a compression gauge in the spark plug hole (see illustration).

6   Have an assistant depress the accelerator pedal and crank the engine over at least seven compression strokes while you watch the gauge. The compression should build up quickly in a healthy engine. Low compression on the first stroke, followed by gradually increasing pressure on successive strokes, indicates worn piston rings. A low compression reading on the first stroke, which doesn't build

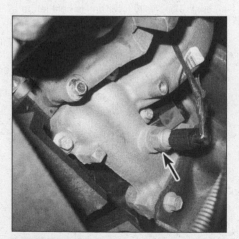

2.2a On gasoline engines, the oil pressure sending unit is mounted to the oil filter adapter on the left side of the engine block, in front of the left engine mount

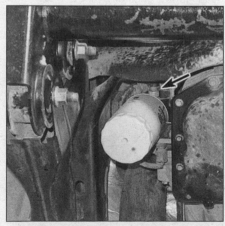

2.2b On diesel engines, the oil pressure sending unit is located on the front side of the oil filter housing

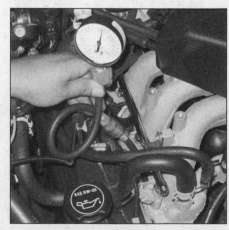

3.5 Use a compression gauge with a threaded fitting for the spark plug hole, not the type that requires hand pressure to maintain the seal

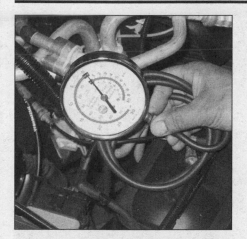

4.4 A simple vacuum gauge can be handy in diagnosing engine condition and performance

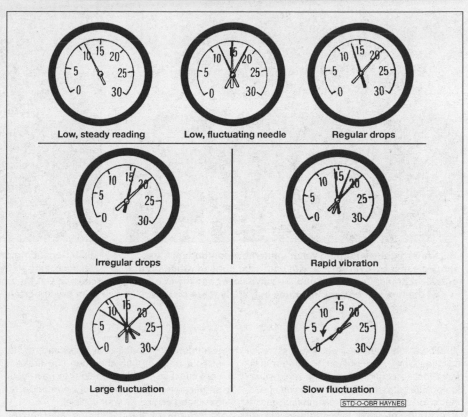

Low, steady reading   Low, fluctuating needle   Regular drops

Irregular drops   Rapid vibration

Large fluctuation   Slow fluctuation

STD-O-OBR HAYNES

4.6 Typical vacuum gauge readings

up during successive strokes, indicates leaking valves or a blown head gasket (a cracked head could also be the cause). Deposits on the undersides of the valve heads can also cause low compression. Record the highest gauge reading obtained.

7   Repeat the procedure for the remaining cylinders and compare the results to this Chapter's Specifications.

8   Add some engine oil (about three squirts from a plunger-type oil can) to each cylinder, through the spark plug hole, and repeat the test.

9   If the compression increases after the oil is added, the piston rings are definitely worn. If the compression doesn't increase significantly, the leakage is occurring at the valves or head gasket. Leakage past the valves may be caused by burned valve seats and/or faces or warped, cracked or bent valves.

10   If two adjacent cylinders have equally low compression, there's a strong possibility that the head gasket between them is blown. The appearance of coolant in the combustion chambers or the crankcase would verify this condition.

11   If one cylinder is slightly lower than the others, and the engine has a slightly rough idle, a worn lobe on the camshaft could be the cause.

12   If the compression is unusually high, the combustion chambers are probably coated with carbon deposits. If that's the case, the cylinder head(s) should be removed and decarbonized.

13   If compression is way down or varies greatly between cylinders, it would be a good idea to have a leak-down test performed by an automotive repair shop. This test will pinpoint exactly where the leakage is occurring and how severe it is.

Diesel engine

14   Because of the special tools needed to check compression on a diesel engine, this procedure is beyond the scope of the home mechanic. Have the compression checked by a dealer service department or other qualified

repair shop with the right tools.

4   Vacuum gauge diagnostic checks (gasoline engines)

1   A vacuum gauge provides inexpensive but valuable information about what is going on in the engine. You can check for worn rings or cylinder walls, leaking head or intake manifold gaskets, incorrect carburetor adjustments, restricted exhaust, stuck or burned valves, weak valve springs, improper ignition or valve timing and ignition problems.

2   Unfortunately, vacuum gauge readings are easy to misinterpret, so they should be used in conjunction with other tests to confirm the diagnosis.

3   Both the absolute readings and the rate of needle movement are important for accurate interpretation. Most gauges measure vacuum in inches of mercury (in-Hg). The following references to vacuum assume the diagnosis is being performed at sea level. As elevation increases (or atmospheric pressure decreases), the reading will decrease. For every 1,000 foot increase in elevation above approximately 2000 feet, the gauge readings will decrease about one inch of mercury.

4   Connect the vacuum gauge directly to the intake manifold vacuum, not to ported (throttle body) vacuum (see illustration). Be sure no hoses are left disconnected during the test or false readings will result.

5   Before you begin the test, allow the engine to warm up completely. Block the wheels and set the parking brake. With the transmission in Park, start the engine and allow it to run at normal idle speed.

Warning: *Keep your hands and the vacuum gauge clear of the fans.*

6   Read the vacuum gauge; an average, healthy engine should normally produce about 17 to 22 in-Hg with a fairly steady needle (see illustration). Refer to the following vacuum gauge readings and what they indicate about the engine's condition:

7   A low steady reading usually indicates a leaking gasket between the intake manifold and cylinder head(s) or throttle body, a leaky vacuum hose, late ignition timing or incorrect camshaft timing. Check ignition timing with a timing light and eliminate all other possible causes, utilizing the tests provided in this Chapter before you remove the timing chain cover to check the timing marks.

8   If the reading is three to eight inches below normal and it fluctuates at that low reading, suspect an intake manifold gasket leak at an intake port or a faulty fuel injector.

9   If the needle has regular drops of about two-to-four inches at a steady rate, the valves are probably leaking. Perform a compression check or leak-down test to confirm this.

10   An irregular drop or down-flick of the needle can be caused by a sticking valve or an ignition misfire. Perform a compression check or leak-down test and read the spark plugs.

11   A rapid vibration of about four in-Hg

6.3a After tightly wrapping water-vulnerable components, use a spray cleaner on everything, with particular concentration on the greasiest areas, usually around the valve cover and lower edges of the block. If one section dries out, apply more cleaner

6.3b Depending on how dirty the engine is, let the cleaner soak in according to the directions and then hose off the grime and cleaner. Get the rinse water down into every area you can get at; then dry important components with a hair dryer or paper towels

vibration at idle combined with exhaust smoke indicates worn valve guides. Perform a leak-down test to confirm this. If the rapid vibration occurs with an increase in engine speed, check for a leaking intake manifold gasket or head gasket, weak valve springs, burned valves or ignition misfire.

12   A slight fluctuation, say one inch up and down, may mean ignition problems. Check all the usual tune-up items and, if necessary, run the engine on an ignition analyzer.

13   If there is a large fluctuation, perform a compression or leak-down test to look for a weak or dead cylinder or a blown head gasket.

14   If the needle moves slowly through a wide range, check for a clogged PCV system, incorrect idle fuel mixture, carburetor/throttle body or intake manifold gasket leaks.

15   Check for a slow return after revving the engine by quickly snapping the throttle open until the engine reaches about 2,500 rpm and let it shut. Normally the reading should drop to near zero, rise above normal idle reading (about 5 in-Hg over) and then return to the previous idle reading. If the vacuum returns slowly and doesn't peak when the throttle is snapped shut, the rings may be worn. If there is a long delay, look for a restricted exhaust system (often the muffler or catalytic converter). An easy way to check this is to temporarily disconnect the exhaust ahead of the suspected part and redo the test.

5   Engine rebuilding alternatives

1   The do-it-yourselfer is faced with a number of options when purchasing a rebuilt engine. The major considerations are cost, warranty, parts availability and the time required for the rebuilder to complete the project. The decision to replace the engine block, piston/connecting rod assemblies and crank-

shaft depends on the final inspection results of your engine. Only then can you make a cost effective decision whether to have your engine overhauled or simply purchase an exchange engine for your vehicle.

2   Some of the rebuilding alternatives include:

3   **Individual parts** - If the inspection procedures reveal that the engine block and most engine components are in reusable condition, purchasing individual parts and having a rebuilder rebuild your engine may be the most economical alternative. The block, crankshaft and piston/connecting rod assemblies should all be inspected carefully by a machine shop first.

4   **Short block** - A short block consists of an engine block with a crankshaft and piston/connecting rod assemblies already installed. All new bearings are incorporated and all clearances will be correct. The existing camshafts, valve train components, cylinder head and external parts can be bolted to the short block with little or no machine shop work necessary.

5   **Long block** - A long block consists of a short block plus an oil pump, oil pan, cylinder head, valve cover, camshaft and valve train components, timing sprockets and chain or gears and timing cover. All components are installed with new bearings, seals and gaskets incorporated throughout. The installation of manifolds and external parts is all that's necessary.

6   **Low mileage used engines** - Some companies now offer low mileage used engines which is a very cost effective way to get your vehicle up and running again. These engines often come from vehicles which have been totaled in accidents or come from other countries which have a higher vehicle turnover rate. A low mileage used engine also usually has a warranty similar to the newly remanufactured engines.

7   Give careful thought to which alternative

is best for you and discuss the situation with local automotive machine shops, auto parts dealers and experienced rebuilders before ordering or purchasing replacement parts.

6   Engine removal - methods and precautions

1   If you've decided that an engine must be removed for overhaul or major repair work, several preliminary steps should be taken. Read all removal and installation procedures carefully prior to committing to this job. Some engines are removed by lowering to the floor and then raising the vehicle sufficiently to slide it out; this will require a vehicle hoist.

2   Locating a suitable place to work is extremely important. Adequate work space, along with storage space for the vehicle, will be needed. If a shop or garage isn't available, at the very least a flat, level, clean work surface made of concrete or asphalt is required.

3   Cleaning the engine compartment and engine before beginning the removal procedure will help keep tools clean and organized (see illustrations).

4   An engine hoist or A-frame will also be necessary. Make sure the equipment is rated in excess of the combined weight of the engine and transmission. Safety is of primary importance, considering the potential hazards involved in lifting the engine out of the vehicle.

5   If you're a novice at engine removal, get at least one helper. One person cannot easily do all the things you need to do to lift a big heavy engine out of the engine compartment. Also helpful is to seek advice and assistance from someone who's experienced in engine removal.

6   Plan the operation ahead of time. Arrange for or obtain all of the tools and

6.7a Get an engine hoist that's strong enough to easily lift your engine in and out of the engine compartment (the V10 and the diesel engine are particularly heavy, so make sure that you obtain a heavy-duty hoist if you're planning to lift either of these engines out of the engine compartment); an adapter, like the one shown here, can be used to change the angle of the engine as it's being removed or installed

6.7b Get an engine stand sturdy enough to firmly support the engine while you're working on it. Stay away from three-wheeled models: they have a tendency to tip over more easily, so get a four-wheeled unit. The V10 and the diesel engine are particularly heavy, so make sure that you obtain a heavy-duty stand capable of supporting one of these engines

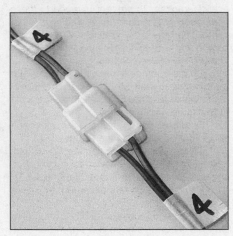

7.6 Label both ends of each wire and hose before disconnecting it - do the same for vacuum hoses

equipment you'll need prior to beginning the job (see illustrations 6.7a and 6.7b). Some of the equipment necessary to perform engine removal and installation safely and with relative ease are (in addition to an engine hoist) a heavy duty floor jack, complete sets of wrenches and sockets as described in the front of this manual, wooden blocks, plenty of rags and cleaning solvent for mopping up spilled oil, coolant and fuel. If the hoist must be rented, make sure that you arrange for it in advance and have everything disconnected and/or removed before bringing the hoist home. This will save you money and time.

7   Plan for the vehicle to be out of use for quite a while. A machine shop can do the work that is beyond the scope of the home mechanic. Machine shops often have a busy schedule, so before removing the engine, consult the shop for an estimate of how long it will take to rebuild or repair the components that may need work (see illustrations).

7   Engine - removal and installation

Warning: *DO NOT use a cheap engine hoist designed for lifting four-cylinder engines. Obtain a heavy-duty hoist designed for lifting heavy engines. And always be extremely careful when removing and installing the engine. Serious injury can result from careless actions.*

Warning: *Gasoline and diesel fuel are flammable, so take extra precautions when you work on any part of the fuel system. Don't smoke or allow open flames or bare light bulbs near the work area, and don't work in a garage where a gas-type appliance (such as a water heater or a clothes dryer) is present. Since fuel is carcinogenic, wear latex gloves when there's a possibility of being exposed to fuel, and, if you spill any fuel on your skin, rinse it off immediately with soap and water. Mop up any spills immediately and do not store fuel-soaked rags where they could ignite. The fuel*

system is under constant pressure, so, if any fuel lines are to be disconnected, the fuel pressure in the system must be relieved first (see Chapter 4A or Chapter 4B for more information). When you perform any kind of work on the fuel system, wear safety glasses and have a Class B type fire extinguisher on hand.

Warning: *The air conditioning system is under high pressure, and refrigerant is expensive. Have a dealer service department or an automotive air conditioning shop discharge the system before beginning this procedure.*

Caution: *On diesel engine models, the manufacturer states that removing the body/cab from the frame before removing the engine greatly eases the engine removal procedure. However, this would require a garage/barn with a high ceiling and equipment sturdy enough to safely lift the cab high enough to clear the engine (then the truck would be rolled out from underneath it). It is possible to remove the engine without removing the body/cab, and that is the procedure that is described in the Section. If you elect to remove the cab, refer to Chapter 11.*

Caution: *The engine oil should be changed any time the turbocharger is removed from the engine. The passages in the block are located underneath the turbocharger and are direct passages to the lubrication system. It's possible that coolant and or debris could directly enter the lubrication system which could cause engine damage.*

1   On vehicles with air conditioning, have the air conditioning system discharged.

2   Relieve the fuel system pressure (see Chapter 4A).

3   Remove the battery(ies), and on diesel models remove the left side battery tray (see Chapter 5).

4   Remove the hood (see Chapter 11) and cover the fenders and cowl. Special pads are available to protect the fenders, but an old bedspread or blanket will also work.

5   Remove the air cleaner assembly and the air intake duct (see Chapter 4A).

6   Label all vacuum lines, emissions system hoses, wiring harness electrical connectors and ground straps to ensure correct reinstallation, then disconnect them. Pieces of masking tape with numbers or letters written on them work well (see illustration). So does colored electrical tape. If there's any possibility of confusion, make a sketch of the engine compartment and clearly label the lines, hoses and wires. You can also use a camera to take photos of connectors, grounds, harness routing, etc.

7   Disconnect the electrical connectors to the PCM (see Chapter 6A). On diesel models, remove the PCM and bracket (see Chapter 6B).

8   Disconnect the fuel lines running from the engine to the chassis (see Chapter 4A). Plug or cap all open fittings/lines.

9   On gasoline models, disconnect the throttle cable from the throttle linkage on the throttle body and remove the throttle body (see Chapter 4A). Remove the fuel rail and the injectors (see Chapter 4A). Remove the intake manifold (see Chapter 2A). Also remove all fuel and/or emission control components that might be damaged during engine removal (see Chapter 4A or Chapter 6A).

10   On diesel models, remove the ducts between the charge air cooler and the turbocharger and between the turbocharger and the intake manifold, remove the turbocharger manifold, then remove the charge air cooler, turbocharger and the turbocharger pedestal (see Chapter 4B). Disconnect the fuel lines from the secondary fuel filter and injection pump return line (see Chapter 4B). Also disconnect the electrical connector from the high-pressure oil system sensor, then disconnect the high-pressure oil hose from each head (see Chapter 4B). Remove the glow plug relay and relay bracket (see Chapter 4B). Remove the fuel filter/water separator (see Chapter 4B).

Caution: *The fuel injection components are manufactured to very precise tolerances and fine clearances. To prevent damage to these components or system as a whole, it is essential that any area around a fuel injection component be absolutely clean when you're working with these components.*

11   On diesel models, once the turbocharger is removed, install special engine removal mounting plate to the cylinder block where the turbocharger was mounted.

12   On diesel models, remove the upper and lower intake manifolds (see Chapter 2B).

13   Clearly label and disconnect all coolant and heater hoses. Remove the cooling fan, shroud, radiator and fan clutch (see Chapter 3).

14   Remove the coolant expansion tank (see Chapter 3).

15   Remove the accessory drivebelt(s) (see Chapter 1) and remove the alternator (see Chapter 5).

16   On vehicles with air conditioning, remove the condenser and the compressor (see Chapter 3). Look carefully at the air conditioning system plumbing. If some section(s) of the plumbing - particularly any section consisting of rigid metal lines - looks like it's going to impede engine removal and installation, detach it from the engine and/or vehicle and set it aside. Secure it with wire if necessary to make sure that it won't be damaged by the engine when the engine is lifted out of the engine compartment. On diesel models, disconnect evaporator-to-compressor line and cover the openings to prevent contamination.

17   On diesel engines, remove the intercooler (see Chapter 4B). On some models equipped with an automatic transmission, there is an external oil cooler, which is located below the space occupied by an intercooler on diesel models. This auxiliary cooler, which is installed in addition to the regular oil cooler that's integrated into the radiator, must also be removed (see Chapter 7A).

18   Remove the grille (see Chapter 11) and mounting bracket, then the radiator crossmember, which is secured by two bolts on each end. Removing the grille and the upper radiator crossmember gives you more room when maneuvering the engine forward to clear the cowl, and it lowers the minimum height which the bottom of the engine must clear when it's hoisted out of the engine compartment.

19   Disconnect the oxygen sensors electrical connectors (see Chapter 6A).

20   On diesel models, remove the secondary fuel filter (see Chapter 1).

21   Remove the power steering pump from its mounting bracket (see Chapter 10) and secure it with wire so that it won't interfere with engine removal.

22   If you're going to be replacing the block, now is a good time to remove all large brackets such as the alternator, air conditioning compressor and power steering pump brackets.

23   Raise the vehicle and place it securely on jackstands.

Note: *On 4WD models, and on models with large tires, this step may not be necessary, because some models already have sufficient ground clearance to allow disconnection of the exhaust system, the engine mounts, transmission, etc., from underneath the vehicle. Raising these vehicles any higher might even make engine removal more difficult because it might position the vehicle too high to lift the engine out of the engine compartment with a hoist.*

24   Drain the engine oil and remove the oil filter (see Chapter 1). Detach the engine oil dipstick tube bracket and remove the engine oil dipstick tube.

25   On diesel models, remove the lower oil pan (see Chapter 2B).

26   Disconnect the engine oil cooler hoses, then remove the oil cooler mounting bolt and remove the oil cooler.

Note: *When the oil cooler is removed, inspect the cooler for metal debris in the cooler. If there is debris present there is a mechanical problem with the engine.*

27   Drain the cooling system (see Chapter 1).

28   Remove the clip that retains the lower radiator hose to the cylinder block.

29   Remove the part of the exhaust system that's routed underneath the engine, between the exhaust manifolds and the downstream catalytic converter (see Chapter 4A). It's not absolutely necessary to remove the exhaust manifolds in order to remove the engine, but removing the manifolds will shave a little weight off the engine.

30   On diesel models, remove the EGR outlet pipe bolts (Chapter 6B) and discard the gasket, then remove the EGR cooler.

31   Remove the starter motor (see Chapter 5).

32   Mark the torque converter to the driveplate, then unbolt the torque converter from the driveplate. Also remove the transmission-to-engine bolts accessible from under the vehicle.

Note: *On diesel models, always replace the torque converter nuts once they are removed.*

33   Remove and discard the transmission mount retaining nuts.

Caution: *If the transmission mount stud loosens when the nuts are being removed the stud must be replaced.*

34   Lower the vehicle. Locate the lifting brackets on the engine. Roll a heavy-duty hoist into position and then attach it to the lifting brackets with a couple pieces of heavy-duty chain. Take up the slack in the sling or chain, but don't lift the engine.

Warning: *DO NOT use a cheap hoist designed to lift four-cylinder engines. Obtain a heavy-duty lift designed for lifting heavy engines. And DO NOT place any part of your body under the engine when it's supported only by a hoist or other lifting device.*

35   Support the transmission with a jack (preferably one made for this purpose), then remove the remaining transmission-to-engine bolts and dipstick tube.

36   Remove the engine mount fasteners (see Chapter 2A). Recheck to be sure nothing is still connecting the engine to the transmission or vehicle. Disconnect anything still remaining. Raise the engine slightly and inspect it thoroughly once more to make sure that nothing is still attached, then slowly lift the engine out of the engine compartment. Check carefully to make sure nothing is hanging up.

Caution: *On diesel models, while removing the engine, the engine oil cooler will come very close to the frame crossmember. Make sure to slowly and carefully remove the engine, checking to make sure the engine oil cooler does not hit the frame crossmember as the engine is being removed.*

Note: *The engine mount through-bolts must be replaced with new ones.*

37   Remove the driveplate (see Chapter 2A or Chapter 2B) and then mount the engine on an engine stand.

Installation

38   Install the driveplate (see Chapter 2A or Chapter 2B).

39   Carefully lower the engine into the engine compartment and then reattach it to the engine mounts.

40   Guide the torque converter into the crankshaft following the procedure outlined in Chapter 7A. Install the transmission-to-engine bolts and tighten them securely.

Caution: *DO NOT use the bolts to force the transmission and engine together! Tighten the torque-converter-to-driveplate bolts to the torque listed in the (Chapter 7A).*

41   Reinstall the remaining components in the reverse order of removal.

42   Add coolant, oil and transmission fluid (see Chapter 1).

43   Run the engine and check for leaks and proper operation of all accessories, then install the hood and test drive the vehicle.

44   Have the air conditioning system recharged and leak tested, if it was discharged.

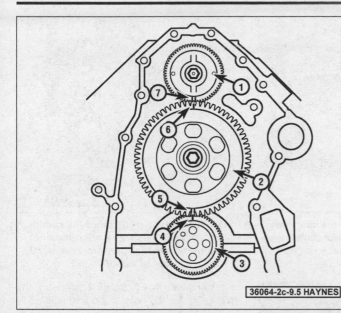

9.5 Timing gears and high-pressure fuel injection pump alignment details

1   *High-pressure fuel injection pump drive gear*
2   *Camshaft drive gear*
3   *Crankshaft drive gear*
4   *Crankshaft drive gear single timing mark*
5   *Camshaft gear double timing mark*
6   *Camshaft gear single timing mark*
7   *High-pressure fuel injection pump drive gear double timing mark*

36064-2c-9.5 HAYNES

8   Engine overhaul - disassembly sequence

1   It's much easier to remove the external components if the engine is mounted on a portable engine stand. A stand can often be rented quite cheaply from an equipment rental yard. Before the engine is mounted on a stand, the flywheel/driveplate should be removed from the engine.

2   If a stand isn't available, it's possible to remove the external engine components with it blocked up on the floor. Be extra careful not to tip or drop the engine when working without a stand.

3   If you're going to obtain a rebuilt engine, all external components must come off first, to be transferred to the replacement engine. These components include:

*Clutch and flywheel (models with
   manual transmission)*
*Driveplate (models with automatic
   transmission)*
Emissions-related components
Engine front cover (diesel engine)
Engine mounts and mount brackets
*Engine rear cover (spacer plate between
   flywheel/driveplate and engine block)*
Fuel injection components
Ignition coils
Intake/exhaust manifolds
Oil filter
Spark plug wires and spark plugs
Thermostat and housing assembly
Water pump

Note: *When removing the external components from the engine, pay close attention to details that may be helpful or important during installation. Note the installed position of gaskets, seals, spacers, pins, brackets, washers, bolts and other small items.*

4   If you're going to obtain a short block (assembled engine block, crankshaft, pistons and connecting rods), then remove the timing belt, cylinder head, oil pan, oil pump pick-up tube, oil pump and water pump from your engine so that you can turn in your old short block to the rebuilder as a core. See *Engine rebuilding alternatives* for additional information regarding the different possibilities to be considered.

9   Camshaft (diesel engines) - removal, inspection and installation

Removal

1   Remove the engine (see Section 7).

2   Remove the valve covers (see Chapter 2B).

3   Remove the rocker arms (see Chapter 2B).

4   Remove the cylinder heads and lifters (see Chapter 2B).

Note: *The crankshaft gear has a single timing mark, the camshaft gear has a single and double timing mark and the high-pressure fuel injection pump drive gear has a double timing mark.*

5   Rotate the engine clockwise, (viewed from the front) until the single mark on the crankshaft drive gear is aligned with the double mark on the camshaft gear, and the single mark on the camshaft drive gear is aligned with the double mark on the high-pressure fuel injection pump drive gear (see illustration).

Note: *The keyway on the high-pressure fuel injection pump drive gear must be at the 12 o'clock position.*

6   Place a dial indicator on the cylinder block, then measure and record the high-pressure fuel injection pump drive gear backlash. Remove the high-pressure fuel injection pump drive gear nut, then remove the high-pressure fuel injection pump fasteners (see Chapter 4B) and slide the pump toward the

rear of the block until the gear is flush with the block. Then carefully tap the pump shaft with a rubber mallet to free the drive gear.

7   Once the high-pressure fuel pump and drive gear are removed, place a dial indicator on the cylinder block, then measure and record the camshaft gear backlash, the crankshaft and camshaft endplay and compare them to the specifications listed in (this Chapter's Specifications).

8   If the backlash or endplay is not within specifications, the drive gear camshaft thrust plate or camshaft bearings may need to be replaced.

9   Remove the camshaft drive gear mounting bolt, then use two large screwdrivers, one on each side of the gear, to pry/walk the gear off of the camshaft.

Caution: *Once the camshaft gear has been removed, the camshaft can slide out of the opposite end of the cylinder block causing damage to the camshaft bearings.*

Note: *It may be necessary to remove the crankshaft (see Section 11) to allow the camshaft gear to be removed.*

10   Remove the two camshaft thrust plate retaining bolts and thrust plate.

11   Screw a long bolt into the end of the camshaft to act as a handle.

Caution: *If the camshaft is being replaced, the camshaft bearings must be installed and machined by a machine shop or serious engine damage may occur.*

12   Carefully pull the camshaft out. Support the cam near the block so the lobes don't nick or gouge the bearings as it's withdrawn.

Inspection

13   Wipe off the camshaft with a clean shop rag. Inspect the cam lobes for score marks, pitting, galling and evidence of overheating (blue, discolored areas). Look for flaking of the hardened surface of each lobe. If the camshaft shows any signs of excessive wear, replace it.

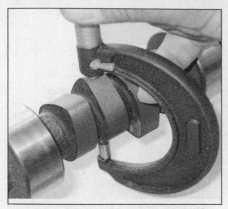

9.14a To calculate lobe lift after the camshaft has been removed from the engine, measure lobe height . . .

9.14b . . . then measure the camshaft base circle (diameter) and subtract the base circle from the lobe height

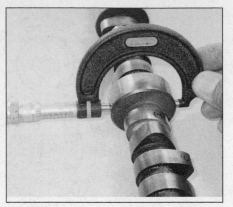

9.17 Measure the diameter of the camshaft bearing journals with a micrometer to determine whether they are excessively worn or out-of-round

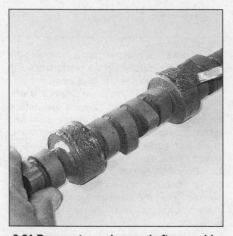

9.21 Be sure to apply camshaft assembly lube to the cam lobes and bearing journals before installing the camshaft

10.1 Before you try to remove the pistons, use a ridge reamer to remove the raised material (ridge) from the top of the cylinders

Camshaft lobe lift check

14   Measure the camshaft lobe height and the diameter of the circular part of the camshaft, or "base circle" (see illustrations). The difference between the two measurements is the lobe lift (lobe height - base circle = lobe lift). Record this figure for future reference and repeat the check on the remaining camshaft lobes.

15   After all of the lobe lifts have been measured, compare your measurements. If the lobe lift is significantly lower for some lobes than for others, replace the camshaft.

Bearing journals, lobes and bearings

16   Inspect the bearing journals for uneven wear, pitting and signs of seizure. If the journals are damaged, the bearing inserts in the block are probably damaged as well. Both the camshaft and bearings will have to be replaced.

Note: *Camshaft bearing replacement requires special tools and expertise that place it beyond the scope of the average home mechanic. The tools for bearing removal and installation are*

available at stores that carry automotive tools, possibly even found at a tool rental business. But this is not a job for beginners. If the bearings are bad, we recommend that you take the engine to an automotive machine shop to ensure that the job is done correctly.*

17   Measure the bearing journals with a micrometer to determine if they are excessively worn or out-of-round (see illustration).

18   Inspect the camshaft lobes for heat discoloration, score marks, chipped areas, pitting and uneven wear.

19   If the lobes are in good condition and if the lobe lift measurements recorded earlier are uniform, the camshaft can be reused.

20   Inspect the rocker arms and lash adjusters for wear (see Chapter 2B).

Installation

21   Lubricate the camshaft bearing journals and cam lobes with camshaft and lifter assembly lube (see illustration).

22   Slide the camshaft into the engine. Support the cam near the block and be careful not to scrape or nick the bearings.

23   Apply clean engine oil to both sides of

the camshaft thrust plate, then install the plate. Tighten the bolts to the torque listed in this Chapter's Specifications.

24   Install the camshaft drive gear, aligning the timing marks with the timing mark on the crankshaft gear (see illustration 9.5) then tighten the bolt to the torque listed in (this Chapter's Specifications).

25   Install the high-pressure fuel injection pump (see Chapter 4B), then the drive gear, aligning the timing marks on the pump gear with the timing mark on the camshaft gear (see illustration 9.5) and tighten the high-pressure injection pump drive gear nut to the torque listed in (this Chapter's Specifications).

26   Lubricate the lifters with clean engine oil and install them in the lifter retainers. Be sure to align the flats on the lifters with the flats in the lifter retainers. Install the retainer and lifters into the engine block as an assembly. If the original lifters are being installed, be sure to return them to their original locations. If a new camshaft is being installed, install new lifters as well.

27   The remainder of installation is otherwise the reverse of removal.

28   When you're done, install the engine (see Section 7).

29   Before starting and running the engine, change the oil and install a new oil filter (see Chapter 1).

10   Pistons and connecting rods - removal and installation

Removal

Note: *Prior to removing the piston/connecting rod assemblies, remove the cylinder heads and oil pan.*

1   Use your fingernail to feel if a ridge has formed at the upper limit of ring travel (about 1/4-inch down from the top of each cylinder). If carbon deposits or cylinder wear have produced ridges, they must be completely removed with a special tool (see illustration). Follow the manufacturer's instructions

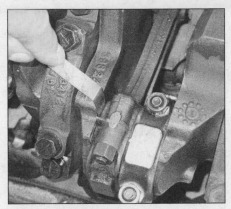

10.3 Checking the connecting rod endplay (side clearance)

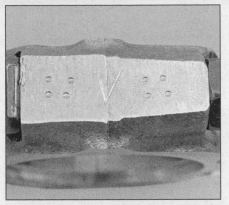

10.4 If the connecting rods and caps are not marked, use a center punch or numbered impression stamps to mark the caps to the rods by cylinder number (for example, this would be the No. 4 connecting rod)

10.12 Install the piston ring into the cylinder then push it down into position using a piston so the ring will be square in the cylinder

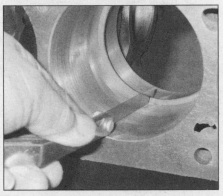

10.13 With the ring square in the cylinder, measure the ring end gap with a feeler gauge

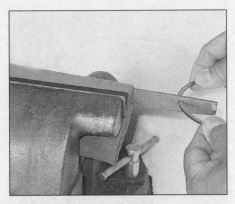

10.14 If the ring end gap is too small, clamp a file in a vise as shown and file the piston ring ends - be sure to remove all raised material

provided with the tool. Failure to remove the ridges before attempting to remove the piston/connecting rod assemblies may result in piston breakage.

2   After the cylinder ridges have been removed, turn the engine so the crankshaft is facing up.

3   Before the main bearing cap assembly and connecting rods are removed, check the connecting rod endplay with feeler gauges. Slide them between the first connecting rod and the crankshaft throw until the play is removed (see illustration). Repeat this procedure for each connecting rod. The endplay is equal to the thickness of the feeler gauge(s). Check with an automotive machine shop for the endplay service limit. If the play exceeds the service limit, new connecting rods will be required. If new rods (or a new crankshaft) are installed, the endplay may fall under the minimum listed in this Chapter's Specifications. If it does, the rods will have to be machined to restore it. If necessary, consult an automotive machine shop for advice.

4   Check the connecting rods and caps for identification marks (see illustration). If they aren't plainly marked, use a small center-punch to make the appropriate number of indentations on each rod and cap (1, 2, 3, etc., depending on the cylinder they're associated with).

5   Loosen each of the connecting rod cap bolts 1/2-turn at a time until they can be removed by hand. Remove the number one connecting rod cap and bearing insert. Don't drop the bearing insert out of the cap.

6   Remove the bearing insert and push the connecting rod/piston assembly out through the top of the engine. Use a wooden or plastic hammer handle to push on the upper bearing surface in the connecting rod. If resistance is felt, double-check to make sure that all of the ridge was removed from the cylinder.

7   Repeat the procedure for the remaining cylinders.

Caution: *Always replace the connecting rod cap bolts with new ones. The old bolts have stretched and cannot be reused.*

8   After removal, reassemble the connecting rod caps and bearing inserts in their respective connecting rods and install the cap bolts finger-tight. Leaving the old bearing inserts in place until reassembly will help prevent the connecting rod bearing surfaces from being accidentally nicked.

Caution: *The connecting rod caps are a "cracked type" design and must mate perfectly with the connecting rod ends. The bearing clearance will be excessive and may result in engine damage if not mated correctly.*

9   The pistons and connecting rods are now ready for inspection and overhaul at an automotive machine shop.

Piston ring installation

10   Before installing the new piston rings, the ring end gaps must be checked. It's assumed that the piston ring side clearance has been checked and verified correct.

11   Lay out the piston/connecting rod assemblies and the new ring sets so the ring sets will be matched with the same piston and cylinder during the end gap measurement and engine assembly.

12   Insert the top (number one) ring into the first cylinder and square it up with the cylinder

walls by pushing it in with the top of the piston (see illustration). The ring should be near the bottom of the cylinder, at the lower limit of ring travel.

13   To measure the end gap, slip feeler gauges between the ends of the ring until a gauge equal to the gap width is found (see illustration). The feeler gauge should slide between the ring ends with a slight amount of drag. Check with an automotive machine shop for the correct end gap for your engine. If the gap is larger or smaller than specified, double-check to make sure you have the correct rings before proceeding.

14   If the gap is too small, it must be enlarged or the ring ends may come in contact with each other during engine operation, which can cause serious damage to the engine. The end gap can be increased by filing the ring ends very carefully with a fine file. Mount the file in a vise equipped with soft jaws, slip the ring over the file with the ends contacting the file face and slowly move the ring to remove material from the ends. When performing this operation, file only by pushing the ring from the outside end of the file towards the vise (see illustration).

10.18a Installing the spacer/expander in the oil ring groove

10.18b DO NOT use a piston ring installation tool when installing the oil control side rails

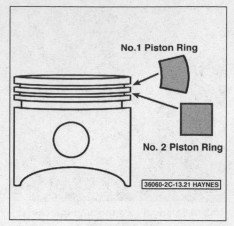

10.20 Top and middle piston ring details

10.21 Use a piston ring installation tool to install the second and top rings - be sure the mark on the piston ring(s) is facing toward the top of the piston (the appearance of the marks may vary)

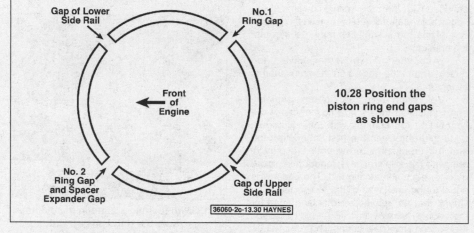

10.28 Position the piston ring end gaps as shown

15   Repeat the procedure for each ring that will be installed in the first cylinder and for each ring in the remaining cylinders.
16   Remember to keep rings, pistons and cylinders matched up.
17   Once the ring end gaps have been checked/corrected, the rings can be installed on the pistons.
18   The oil control ring (lowest one on the piston) is usually installed first. It's composed of three separate components. Slip the spacer/expander into the groove (see illustration). If an anti-rotation tang is used, make sure it's inserted into the drilled hole in the ring groove. Next, install the upper side rail in the same manner (see illustration). Don't use a piston ring installation tool on the oil ring side rails, as they may be damaged. Instead, place one end of the side rail into the groove between the spacer/expander and the ring land, hold it firmly in place and slide a finger around the piston while pushing the rail into the groove. Finally, install the lower side rail.
19   After the three oil ring components have

been installed, check to make sure that both the upper and lower side rails can be rotated smoothly inside the ring grooves.
20   The number two (middle) ring is installed next. It's usually stamped with a mark which must face up, toward the top of the piston. Do not mix up the top and middle rings, as they have different cross-sections (see illustration).
Note: *Always follow the instructions printed on the ring package or box - different manufacturers may require different approaches.*
21   Use a piston ring installation tool and make sure the identification mark is facing the top of the piston, then slip the ring into the middle groove on the piston (see illustration). Don't expand the ring any more than necessary to slide it over the piston.
22   Install the number one (top) ring in the same manner. Make sure the mark is facing up. Be careful not to confuse the number one and number two rings (see illustration 10.21).
23   Repeat the procedure for the remaining pistons and rings.

Installation

24   Before installing the piston/connecting rod assemblies, the cylinder walls must be perfectly clean, the top edge of each cylinder bore must be chamfered, and the crankshaft must be in place.

25   Remove the cap from the end of the number one connecting rod (refer to the marks made during removal). Remove the original bearing inserts and wipe the bearing surfaces of the connecting rod and cap with a clean, lint-free cloth. They must be kept spotlessly clean.

Connecting rod bearing oil clearance check

26   Clean the back side of the new upper bearing insert, then lay it in place in the connecting rod. Make sure the tab on the bearing fits into the recess in the rod. Don't hammer the bearing insert into place and be very careful not to nick or gouge the bearing face. Don't lubricate the bearing at this time.
Note: *On diesel engines, the upper rod bearing is black and the lower rod bearing (the one that fits in the cap) is bright.*
27   Clean the back side of the other bearing insert and install it in the rod cap. Again, make sure the tab on the bearing fits into the recess in the cap, and don't apply any lubricant. It's critically important that the mating surfaces of the bearing and connecting rod are perfectly clean and oil free when they're assembled.
28   Position the piston ring gaps at 90-degree intervals around the piston as shown (see illustration).

10.33 Use a plastic or wooden hammer handle to push the piston into the cylinder

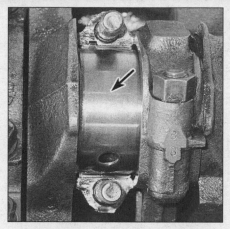

10.35 Place Plastigage on each connecting rod bearing journal parallel to the crankshaft centerline

10.39 Use the scale on the Plastigage package to determine the bearing oil clearance - be sure to measure the widest part of the Plastigage and use the correct scale; it comes with both standard and metric scales

29   Lubricate the piston and rings with clean engine oil and attach a piston ring compressor to the piston. Leave the skirt protruding about 1/4-inch to guide the piston into the cylinder. The rings must be compressed until they're flush with the piston.

30   Rotate the crankshaft until the number one connecting rod journal is at BDC (bottom dead center) and apply a liberal coat of engine oil to the cylinder walls.

31   With the weight designation mark, or arrow, on top of the piston facing the front of the engine, gently insert the piston/connecting rod assembly into the number one cylinder bore and rest the bottom edge of the ring compressor on the engine block.

32   Tap the top edge of the ring compressor to make sure it's contacting the block around its entire circumference.

33   Gently tap on the top of the piston with the end of a wooden or plastic hammer handle (see illustration) while guiding the end of the connecting rod into place on the crankshaft journal. The piston rings may try to pop out of the ring compressor just before entering the cylinder bore, so keep some downward pressure on the ring compressor. Work slowly, and if any resistance is felt as the piston enters the cylinder, stop immediately. Find out what's hanging up and fix it before proceeding. Do not, for any reason, force the piston into the cylinder - you might break a ring and/or the piston.

34   Once the piston/connecting rod assembly is installed, the connecting rod bearing oil clearance must be checked before the rod cap is permanently installed.

35   Cut a piece of the appropriate size Plastigage slightly shorter than the width of the connecting rod bearing and lay it in place on the number one connecting rod journal, parallel with the journal axis (see illustration).

36   Clean the connecting rod cap bearing face and install the rod cap. Make sure the mating mark on the cap is on the same side as the mark on the connecting rod (see illustration 10.4).

37   Install the old rod bolts at this time, and

tighten them to the torque listed in this Chapter's Specifications, working up to it in three steps.

Note: *Use a thin-wall socket to avoid erroneous torque readings that can result if the socket is wedged between the rod cap and the bolt. If the socket tends to wedge itself between the fastener and the cap, lift up on it slightly until it no longer contacts the cap. DO NOT rotate the crankshaft at any time during this operation.*

38   Remove the fasteners and detach the rod cap, being very careful not to disturb the Plastigage. Discard the cap bolts at this time as they cannot be reused.

Caution: *You MUST use new connecting rod bolts.*

39   Compare the width of the crushed Plastigage to the scale printed on the Plastigage envelope to obtain the oil clearance (see illustration). The connecting rod oil clearance is usually about 0.001 to 0.002 inch. Consult an automotive machine shop for the clearance specified for the rod bearings on your engine.

40   If the clearance is not as specified, the bearing inserts may be the wrong size (which means different ones will be required). Before deciding that different inserts are needed, make sure that no dirt or oil was between the bearing inserts and the connecting rod or cap when the clearance was measured. Also, recheck the journal diameter. If the Plastigage was wider at one end than the other, the journal may be tapered. If the clearance still exceeds the limit specified, the bearing will have to be replaced with an undersize bearing.

Caution: *When installing a new crankshaft always use a standard size bearing.*

Final installation

41   Carefully scrape all traces of the Plastigage material off the rod journal and/or bearing face. Be very careful not to scratch the bearing - use your fingernail or the edge of a plastic card.

42   Make sure the bearing faces are perfectly clean, then apply a uniform layer of clean moly-base grease or engine assembly

lube to both of them. You'll have to push the piston into the cylinder to expose the face of the bearing insert in the connecting rod.

Caution: *Always replace the connecting rod cap bolts with new ones. The old bolts have stretched and cannot be reused.*

43   Slide the connecting rod back into place on the journal, install the rod cap, install the new bolts and tighten them to the torque listed in this Chapter's Specifications. Again, work up to the torque in three steps.

Caution: *The connecting rod caps are a "cracked type" design and must mate perfectly with the connecting rod ends. The bearing clearance will be excessive and may result in engine damage if not mated correctly.*

44   Repeat the entire procedure for the remaining pistons/connecting rods.

45   The important points to remember are:

a) *Keep the back sides of the bearing inserts and the insides of the connecting rods and caps perfectly clean when assembling them.*

b) *Make sure you have the correct piston/ rod assembly for each cylinder.*

c) *The mark on the piston must face the front of the engine.*

d) *Lubricate the cylinder walls liberally with clean oil.*

e) *Lubricate the bearing faces when installing the rod caps after the oil clearance has been checked.*

46   After all the piston/connecting rod assemblies have been correctly installed, rotate the crankshaft a number of times by hand to check for any obvious binding.

47   As a final step, check the connecting rod endplay again (see Step 3). If it was correct before disassembly and the original crankshaft and rods were reinstalled, it should still be correct. If new rods or a new crankshaft were installed, the endplay may be inadequate. If so, the rods will have to be removed and taken to an automotive machine shop for resizing.

ENGINE BEARING ANALYSIS

Debris

Babbitt bearing embedded with debris from machinings

Microscopic detail of debris

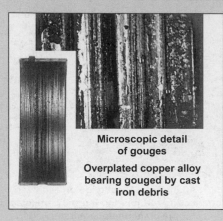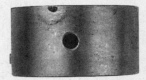

Microscopic detail of gouges

Overplated copper alloy bearing gouged by cast iron debris

Aluminum bearing embedded with glass beads

Microscopic detail of glass beads

Damaged lining caused by dirt left on the bearing back

Misassembly

Result of a lower half assembled as an upper - blocking the oil flow

Excessive oil clearance is indicated by a short contact arc

Polished and oil-stained backs are a result of a poor fit in the housing bore

Result of a wrong, reversed, or shifted cap

Overloading

Damage from excessive idling which resulted in an oil film unable to support the load imposed

Damaged upper connecting rod bearings caused by engine lugging; the lower main bearings (not shown) were similarly affected

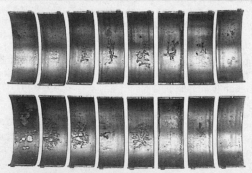

The damage shown in these upper and lower connecting rod bearings was caused by engine operation at a higher-than-rated speed under load

Misalignment

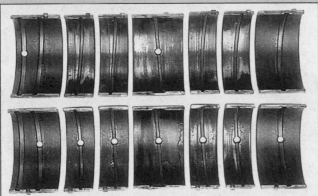

A warped crankshaft caused this pattern of severe wear in the center, diminishing toward the ends

A poorly finished crankshaft caused the equally spaced scoring shown

A tapered housing bore caused the damage along one edge of this pair

A bent connecting rod led to the damage in the "V" pattern

Lubrication

Result of dry start: The bearings on the left, farthest from the oil pump, show more damage

Result of a low oil supply or oil starvation

Severe wear as a result of inadequate oil clearance

Corrosion

Microscopic detail of corrosion

Corrosion is an acid attack on the bearing lining generally caused by inadequate maintenance, extremely hot or cold operation, or inferior oils or fuels

Microscopic detail of cavitation

Example of cavitation - a surface erosion caused by pressure changes in the oil film

Damage from excessive thrust or insufficient axial clearance

Bearing affected by oil dilution caused by excessive blow-by or a rich mixture

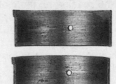

11.1 Checking crankshaft endplay with a dial indicator

11.3 Checking crankshaft endplay with feeler gauges at the thrust bearing journal

11   Crankshaft - removal and installation

Removal

Note: *The crankshaft can be removed only after the engine has been removed from the vehicle. It's assumed that the flywheel or driveplate, crankshaft pulley, timing chains (6.2L models), oil pan, oil pump body, oil filter, cylinder heads and piston/connecting rod assemblies have already been removed. The rear main oil seal retainer and ignition pulse ring on 6.2L models or rear cover on diesel models must be unbolted and separated from the block before proceeding with crankshaft removal.*

Note: *On diesel engine models, align the timing marks on the high-pressure fuel injection pump, camshaft gear and crankshaft gear before removing the crankshaft (see Section 9).*

1   Before the crankshaft is removed, measure the endplay. Mount a dial indicator with the indicator in line with the crankshaft and touching the end of the crankshaft as shown (see illustration).

2   Pry the crankshaft all the way to the rear and zero the dial indicator. Next, pry the crankshaft to the front as far as possible and check the reading on the dial indicator. The distance traveled is the endplay. A typical crankshaft endplay will fall between 0.003 and 0.010-inch. If it's greater than that, check the crankshaft thrust surfaces for wear after its removed. If no wear is evident, new main bearings should correct the endplay.

3   If a dial indicator isn't available, feeler gauges can be used. Gently pry the crankshaft all the way to the front of the engine. Slip feeler gauges between the crankshaft and the front face of the thrust bearing or washer to determine the clearance (see illustration).

4   Loosen the main bearing vertical bolts and side bolts 1/4-turn at a time each, until they can be removed by hand.

5   Gently tap the main bearing caps with a soft-face hammer. Pull the main bearing caps straight up and off the cylinder block. Try not to drop the bearing inserts if they come out with the caps.

6   Carefully lift the crankshaft out of the engine. Have an assistant available, since the crankshaft is quite heavy and awkward to handle. With the bearing inserts in place inside the engine block and main bearing caps, reinstall the main bearing caps onto the engine block and tighten the bolts finger tight. Make sure you install the main bearing caps with the arrows facing the front end of the engine.

Installation

7   Crankshaft installation is the first step in engine reassembly. It's assumed at this point that the engine block and crankshaft have been cleaned, inspected and repaired or reconditioned.

8   Position the engine block with the bottom facing up.

9   Remove the vertical and side mounting bolts then lift off the main bearing caps.

10   If they're still in place, remove the original bearing inserts from the block and from the main bearing caps. Wipe the bearing surfaces of the block and main bearing caps with a clean, lint-free cloth. They must be kept spotlessly clean. This is critical for determining the correct bearing oil clearance.

Main bearing oil clearance check

11   Without mixing them up, clean the back sides of the new upper main bearing inserts (with grooves and oil holes) and lay one in each main bearing saddle in the block. Each upper bearing has an oil groove and oil hole in it. All engines have a two-piece thrust washer installed on the number five bearing cap (V8 models), which is closest to the flywheel/driveplate, main bearing saddle in the block,

and a single thrust washer on the rear side of the last main bearing cap. The grooves in the thrust washers must face the crankshaft (the plain sides against the main bearing saddle or cap). Clean the back sides of the lower main bearing inserts (without grooves) and lay them in the caps. Make sure the tab on the bearing insert fits into the recess in the block or main bearing caps.

Caution: *The oil holes in the block must line up with the oil holes in the upper bearing inserts.*

Caution: *Do not hammer the bearing insert into place and don't nick or gouge the bearing faces. DO NOT apply any lubrication at this time.*

12   Clean the faces of the bearing inserts in the block and the crankshaft main bearing journals with a clean, lint-free cloth.

13   Check or clean the oil holes in the crankshaft, as any dirt here can go only one way - straight through the new bearings.

14   Once you're certain the crankshaft is clean, carefully lay it in position in the cylinder block.

15   Before the crankshaft can be permanently installed, the main bearing oil clearance must be checked.

16   Cut several strips of the appropriate size of Plastigage (they must be slightly shorter than the width of the main bearing journal).

17   Place one piece on each crankshaft main bearing journal, parallel with the journal axis as shown (see illustration).

18   Clean the faces of the bearing inserts in the main bearing caps. Hold the bearing inserts in place and install the assembly onto the crankshaft and cylinder block. DO NOT disturb the Plastigage. Make sure you install the main bearing caps with the arrows facing the front (timing chain end) of the engine.

19   Apply clean engine oil to all bolt threads prior to installation, then install all bolts finger-tight. Tighten the main bearing cap bolts in the sequence shown (see illustrations) progressing in mutiple steps, to the torque listed in this

11.17 Place the Plastigage onto the crankshaft bearing journal
as shown

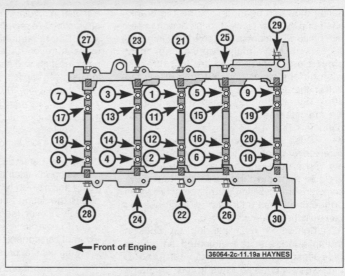

← Front of Engine

`36064-2c-11.19a HAYNES`

11.19a Main bearing cap bolt tightening sequence
(V8 gasoline engine)

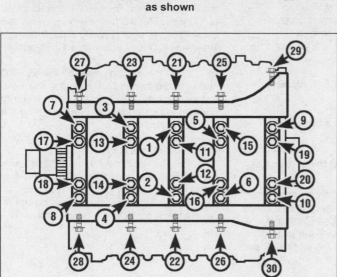

`36064-2c-11.19b HAYNES`

11.19b Main bearing cap bolt tightening sequence
(6.7L diesel engine)

11.21 Use the scale on the Plastigage package to determine the
bearing oil clearance - be sure to measure the widest part of the
Plastigage and use the correct scale; it comes with both standard
and metric scales

Chapter's Specifications. DO NOT rotate the crankshaft at any time during this operation.

20   Remove the bolts in the reverse order of the tightening sequence and carefully lift the main bearing caps straight up and off the block. Do not disturb the Plastigage or rotate the crankshaft. If a main bearing cap is difficult to remove, tap it gently from side-to-side with a soft-face hammer to loosen it.

21   Compare the width of the crushed Plastigage on each journal to the scale printed on the Plastigage envelope to determine the main bearing oil clearance (see illustration). A typical main bearing oil clearance should fall between 0.0015 to 0.0023-inch on gasoline engines or 0.0015 to 0.0045-inch on diesel engines. Check with an automotive machine shop for the oil clearance for your engine.

22   If the clearance is not acceptable, the bearing inserts may be the wrong size (which means different ones will be required). Before deciding if different inserts are needed, make sure that no dirt or oil was between the bearing inserts and the cap assembly or block when the clearance was measured. If the Plastigage was wider at one end than the other, the crankshaft journal may be tapered. If the clearance still exceeds the limit specified, the bearing insert(s) will have to be replaced with an undersize bearing insert(s).

Caution: *When installing a new crankshaft always install a standard bearing insert set.*

23   Carefully scrape all traces of the Plastigage material off the main bearing journals and/or the bearing insert faces. Be sure to remove all residue from the oil holes. Use your fingernail or the edge of a plastic card - don't nick or scratch the bearing faces.

Final installation

24   Carefully lift the crankshaft out of the cylinder block.

25   Clean the bearing insert faces in the cylinder block, then apply a thin, uniform layer of moly-base grease or engine assembly lube to each of the bearing surfaces. Be sure to coat the thrust faces of the thrust washers/bearings as well. See Step 11 for thrust washer/bearing locations.

26   Make sure the crankshaft journals are clean, then lay the crankshaft back in place in the cylinder block. On diesel engines, if the camshaft is still installed in the block, the timing marks must be aligned as the crankshaft is set in place.

27   Clean the bearing insert faces and then apply the same lubricant to them.

28   Hold the bearing inserts in place and

install the main bearing caps on the crankshaft and cylinder block. Tap the bearing caps into place with a brass or a soft-face hammer.

29 Using NEW main bearing cap bolts, apply clean engine oil to the bolt threads, wipe off any excess oil and then install the bolts finger-tight.

30 Tighten the main bearing cap bolts in the indicated sequence to 120 to 144 inch-pounds.

31 Push the crankshaft forward using a screwdriver or prybar to seat the thrust bearing. Once the crankshaft is pushed fully forward to seat the thrust bearing, leave the screwdriver in position so that pressure stays on the crankshaft until after all main bearing cap bolts have been tightened.

32 Tighten the main bearing cap bolts in multiple steps in the indicated sequence (see illustrations 11.19a or 11.19b) and to the torque and angle listed in this Chapter's Specifications.

33 Recheck crankshaft endplay with a feeler gauge or a dial indicator. The endplay should be correct if the crankshaft thrust faces aren't worn or damaged and if new bearings have been installed.

34 Rotate the crankshaft a number of times by hand to check for any obvious binding.

35 Install the new rear main oil seal (see Chapter 2A or Chapter 2B).

12 Engine overhaul - reassembly sequence

Note: *Diesel engines have several components in addition to the standard ones covered here. These include the high pressure oil system, the internal engine oil cooler, etc. Make sure to get additional information or as-* *sistance regarding these components before attempting to assemble a diesel engine.*

1 Before beginning engine reassembly, make sure you have all the necessary new parts, gaskets and seals as well as the following items on hand:

> *Common hand tools*
> *A 1/2-inch drive torque wrench*
> *New engine oil*
> *Gasket sealant*
> *Thread locking compound*

2 If you obtained a short block it will be necessary to install the cylinder head, the oil pump and pick-up tube, the oil pan, the water pump, the timing belt and timing cover, and the valve cover (see Chapter 2A or 2B). In order to save time and avoid problems, the external components must be installed in the following general order:

> *Thermostat and housing cover*
> *Water pump*
> *Intake and exhaust manifolds*
> *Carburetor or fuel injection components*
> *Emission control components*
> *Spark plug wires and spark plugs*
> *Ignition coils*
> *Oil filter*
> *Engine mounts and mount brackets*
> *Clutch and flywheel*
> * (manual transmission)*
> *Driveplate (automatic transmission)*

13 Initial start-up and break-in after overhaul

Warning: *Have a fire extinguisher handy when starting the engine for the first time.*

1 Once the engine has been installed in the vehicle, double-check the engine oil and coolant levels.

2 With the spark plugs out of the engine and the ignition system and fuel pump disabled, crank the engine until oil pressure registers on the gauge or the light goes out.

3 Install the spark plugs, hook up the plug wires and restore the ignition system and fuel pump functions.

4 Start the engine. It may take a few moments for the fuel system to build up pressure, but the engine should start without a great deal of effort.

5 After the engine starts, it should be allowed to warm up to normal operating temperature. While the engine is warming up, make a thorough check for fuel, oil and coolant leaks.

6 Shut the engine off and recheck the engine oil and coolant levels.

7 Drive the vehicle to an area with minimum traffic, accelerate from 30 to 50 mph, then allow the vehicle to slow to 30 mph with the throttle closed. Repeat the procedure 10 or 12 times. This will load the piston rings and cause them to seat properly against the cylinder walls. Check again for oil and coolant leaks.

8 Drive the vehicle gently for the first 500 miles (no sustained high speeds) and keep a constant check on the oil level. It is not unusual for an engine to use oil during the break-in period.

9 At approximately 500 to 600 miles, change the oil and filter.

10 For the next few hundred miles, drive the vehicle normally. Do not pamper it or abuse it.

11 After 2000 miles, change the oil and filter again and consider the engine broken in.

COMMON ENGINE OVERHAUL TERMS

B

Backlash - The amount of play between two parts. Usually refers to how much one gear can be moved back and forth without moving the gear with which it's meshed.

Bearing Caps - The caps held in place by nuts or bolts which, in turn, hold the bearing surface. This space is for lubricating oil to enter.

Bearing clearance - The amount of space left between shaft and bearing surface. This space is for lubricating oil to enter.

Bearing crush - The additional height which is purposely manufactured into each bearing half to ensure complete contact of the bearing back with the housing bore when the engine is assembled.

Bearing knock - The noise created by movement of a part in a loose or worn bearing.

Blueprinting - Dismantling an engine and reassembling it to EXACT specifications.

Bore - An engine cylinder, or any cylindrical hole; also used to describe the process of enlarging or accurately refinishing a hole with a cutting tool, as to bore an engine cylinder. The bore size is the diameter of the hole.

Boring - Renewing the cylinders by cutting them out to a specified size. A boring bar is used to make the cut.

Bottom end - A term which refers collectively to the engine block, crankshaft, main bearings and the big ends of the connecting rods.

Break-in - The period of operation between installation of new or rebuilt parts and time in which parts are worn to the correct fit. Driving at reduced and varying speed for a specified mileage to permit parts to wear to the correct fit.

Bushing - A one-piece sleeve placed in a bore to serve as a bearing surface for shaft, piston pin, etc. Usually replaceable.

C

Camshaft - The shaft in the engine, on which a series of lobes are located for operating the valve mechanisms. The camshaft is driven by gears or sprockets and a timing chain. Usually referred to simply as the cam.

Carbon - Hard, or soft, black deposits found in combustion chamber, on plugs, under rings, on and under valve heads.

Cast iron - An alloy of iron and more than two percent carbon, used for engine blocks and heads because it's relatively inexpensive and easy to mold into complex shapes.

Chamfer - To bevel across (or a bevel on) the sharp edge of an object.

Chase - To repair damaged threads with a tap or die.

Combustion chamber - The space between the piston and the cylinder head, with the piston at top dead center, in which air-fuel mixture is burned.

Compression ratio - The relationship between cylinder volume (clearance volume) when the piston is at top dead center and cylinder volume when the piston is at bottom dead center.

Connecting rod - The rod that connects the crank on the crankshaft with the piston. Sometimes called a con rod.

Connecting rod cap - The part of the connecting rod assembly that attaches the rod to the crankpin.

Core plug - Soft metal plug used to plug the casting holes for the coolant passages in the block.

Crankcase - The lower part of the engine in which the crankshaft rotates; includes the lower section of the cylinder block and the oil pan.

Crank kit - A reground or reconditioned crankshaft and new main and connecting rod bearings.

Crankpin - The part of a crankshaft to which a connecting rod is attached.

Crankshaft - The main rotating member, or shaft, running the length of the crankcase, with offset throws to which the connecting rods are attached; changes the reciprocating motion of the pistons into rotating motion.

Cylinder sleeve - A replaceable sleeve, or liner, pressed into the cylinder block to form the cylinder bore.

D

Deburring - Removing the burrs (rough edges or areas) from a bearing.

Deglazer - A tool, rotated by an electric motor, used to remove glaze from cylinder walls so a new set of rings will seat.

E

Endplay - The amount of lengthwise movement between two parts. As applied to a crankshaft, the distance that the crankshaft can move forward and back in the cylinder block.

F

Face - A machinist's term that refers to removing metal from the end of a shaft or the face of a larger part, such as a flywheel.

Fatigue - A breakdown of material through a large number of loading and unloading cycles. The first signs are cracks followed shortly by breaks.

Feeler gauge - A thin strip of hardened steel, ground to an exact thickness, used to check clearances between parts.

Free height - The unloaded length or height of a spring.

Freeplay - The looseness in a linkage, or an assembly of parts, between the initial application of force and actual movement. Usually perceived as slop or slight delay.

Freeze plug - See Core plug.

G

Gallery - A large passage in the block that forms a reservoir for engine oil pressure.

Glaze - The very smooth, glassy finish that develops on cylinder walls while an engine is in service.

H

Heli-Coil - A rethreading device used when threads are worn or damaged. The device is installed in a retapped hole to reduce the thread size to the original size.

I

Installed height - The spring's measured length or height, as installed on the cylinder head. Installed height is measured from the spring seat to the underside of the spring retainer.

J

Journal - The surface of a rotating shaft which turns in a bearing.

K

Keeper - The split lock that holds the valve spring retainer in position on the valve stem.

Key - A small piece of metal inserted into matching grooves machined into two parts fitted together - such as a gear pressed onto a shaft - which prevents slippage between the two parts.

Knock - The heavy metallic engine sound, produced in the combustion chamber as a result of abnormal combustion - usually detonation. Knock is usually caused by a loose or worn bearing. Also referred to as detonation, pinging and spark knock. Connecting rod or main bearing knocks are created by too much oil clearance or insufficient lubrication.

L

Lands - The portions of metal between the piston ring grooves.

Lapping the valves - Grinding a valve face and its seat together with lapping compound.

Lash - The amount of free motion in a gear train, between gears, or in a mechanical assembly, that occurs before movement can begin. Usually refers to the lash in a valve train.

Lifter - The part that rides against the cam to transfer motion to the rest of the valve train.

M

Machining - The process of using a machine to remove metal from a metal part.

Main bearings - The plain, or babbit, bearings that support the crankshaft.

Main bearing caps - The cast iron caps, bolted to the bottom of the block, that support the main bearings.

O

O.D. - Outside diameter.

Oil gallery - A pipe or drilled passageway in the engine used to carry engine oil from one area to another.

Oil ring - The lower ring, or rings, of a piston; designed to prevent excessive amounts of oil from working up the cylinder walls and into the combustion chamber. Also called an oil-control ring.

Oil seal - A seal which keeps oil from leaking out of a compartment. Usually refers to a dynamic seal around a rotating shaft or other moving part.

O-ring - A type of sealing ring made of a special rubberlike material; in use, the O-ring is compressed into a groove to provide the sealing action.

Overhaul - To completely disassemble a unit, clean and inspect all parts, reassemble it with the original or new parts and make all adjustments necessary for proper operation.

P

Pilot bearing - A small bearing installed in the center of the flywheel (or the rear end of the crankshaft) to support the front end of the input shaft of the transmission.

Pip mark - A little dot or indentation which indicates the top side of a compression ring.

Piston - The cylindrical part, attached to the connecting rod, that moves up and down in the cylinder as the crankshaft rotates. When the fuel charge is fired, the piston transfers the force of the explosion to the connecting rod, then to the crankshaft.

Piston pin (or wrist pin) - The cylindrical and usually hollow steel pin that passes through the piston. The piston pin fastens the piston to the upper end of the connecting rod.

Piston ring - The split ring fitted to the groove in a piston. The ring contacts the sides of the ring groove and also rubs against the cylinder wall, thus sealing space between piston and wall. There are two types of rings: Compression rings seal the compression pressure in the combustion chamber; oil rings scrape excessive oil off the cylinder wall.

Piston ring groove - The slots or grooves cut in piston heads to hold piston rings in position.

Piston skirt - The portion of the piston below the rings and the piston pin hole.

Plastigage - A thin strip of plastic thread, available in different sizes, used for measuring clearances. For example, a strip of plastigage is laid across a bearing journal and mashed as parts are assembled. Then parts are disassembled and the width of the strip is measured to determine clearance between journal and bearing. Commonly used to measure crankshaft main-bearing and connecting rod bearing clearances.

Press-fit - A tight fit between two parts that requires pressure to force the parts together. Also referred to as drive, or force, fit.

Prussian blue - A blue pigment; in solution, useful in determining the area of contact between two surfaces. Prussian blue is commonly used to determine the width and location of the contact area between the valve face and the valve seat.

R

Race (bearing) - The inner or outer ring that provides a contact surface for balls or rollers in bearing.

Ream - To size, enlarge or smooth a hole by using a round cutting tool with fluted edges.

Ring job - The process of reconditioning the cylinders and installing new rings.

Runout - Wobble. The amount a shaft rotates out-of-true.

S

Saddle - The upper main bearing seat.

Scored - Scratched or grooved, as a cylinder wall may be scored by abrasive particles moved up and down by the piston rings.

Scuffing - A type of wear in which there's a transfer of material between parts moving against each other; shows up as pits or grooves in the mating surfaces.

Seat - The surface upon which another part rests or seats. For example, the valve seat is the matched surface upon which the valve face rests. Also used to refer to wearing into a good fit; for example, piston rings seat after a few miles of driving.

Short block - An engine block complete with crankshaft and piston and, usually, camshaft assemblies.

Static balance - The balance of an object while it's stationary.

Step - The wear on the lower portion of a ring land caused by excessive side and back-clearance. The height of the step indicates the ring's extra side clearance and the length of the step projecting from the back wall of the groove represents the ring's back clearance.

Stroke - The distance the piston moves when traveling from top dead center to bottom dead center, or from bottom dead center to top dead center.

Stud - A metal rod with threads on both ends.

T

Tang - A lip on the end of a plain bearing used to align the bearing during assembly.

Tap - To cut threads in a hole. Also refers to the fluted tool used to cut threads.

Taper - A gradual reduction in the width of a shaft or hole; in an engine cylinder, taper usually takes the form of uneven wear, more pronounced at the top than at the bottom.

Throws - The offset portions of the crankshaft to which the connecting rods are affixed.

Thrust bearing - The main bearing that has thrust faces to prevent excessive endplay, or forward and backward movement of the crankshaft.

Thrust washer - A bronze or hardened steel washer placed between two moving parts. The washer prevents longitudinal movement and provides a bearing surface for thrust surfaces of parts.

Tolerance - The amount of variation permitted from an exact size of measurement. Actual amount from smallest acceptable dimension to largest acceptable dimension.

U

Umbrella - An oil deflector placed near the valve tip to throw oil from the valve stem area.

Undercut - A machined groove below the normal surface.

Undersize bearings - Smaller diameter bearings used with re-ground crankshaft journals.

V

Valve grinding - Refacing a valve in a valve-refacing machine.

Valve train - The valve-operating mechanism of an engine; includes all components from the camshaft to the valve.

Vibration damper - A cylindrical weight attached to the front of the crankshaft to minimize torsional vibration (the twist-untwist actions of the crankshaft caused by the cylinder firing impulses). Also called a harmonic balancer.

W

Water jacket - The spaces around the cylinders, between the inner and outer shells of the cylinder block or head, through which coolant circulates.

Web - A supporting structure across a cavity.

Woodruff key - A key with a radiused backside (viewed from the side).

Chapter 3
Cooling, heating and air conditioning systems

Contents

Specifications

Refrigerant type	R-134a

Torque specifications

Ft-lbs (unless otherwise indicated)

Note: *One foot-pound (ft-lb) of torque is equivalent to 12 inch-pounds (in-lbs) of torque. Torque values below approximately 15 foot-pounds are expressed in inch-pounds, because most foot-pound torque wrenches are not accurate at these smaller values.*

Air conditioning compressor bolts	18
Air conditioning condenser bolts	97 in-lbs
Cooling module	
Bolts	30
Nuts	27 in-lbs
Refrigerant line-to-compressor and condenser inlet and outlet nuts	133 in-lbs
Thermostatic expansion valve bolts	71 in-lbs
Engine oil cooler threaded tube (gasoline engine)	43
Engine oil cooler (diesel engine)	
Nut	18
Bolts	89 in-lbs
Fan clutch hub-to-water pump nut (gasonline engine)	98
Fan clutch hub-to-cooling fan drive nut (diesel engine)	98
Expansion tank bolts	168 in-lbs
Radiator	
Gasoline engine	26
Diesel engine	
Primary radiator bracket bolts	26
Secondary radiator bolts	97 in-lbs
Thermostat housing bolts	
Primary thermostat housing bolts	89 in-lbs
Secondary thermostat housing bolts (diesel engine)	53 in-lbs

Torque specifications (continued)   Ft-lbs (unless otherwise indicated)

Note: *One foot-pound (ft-lb) of torque is equivalent to 12 inch-pounds (in-lbs) of torque. Torque values below approximately 15 foot-pounds are expressed in inch-pounds, because most foot-pound torque wrenches are not accurate at these smaller values.*

Water pump bolts	
Gasoline engine	
Step 1	Hand tighten
Step 2	177 in-lbs
Step 3	Tighten an additional 45 degrees
Diesel engine	
Primary water pump bolts	
Small bolts	89 in-lbs
Large bolts	18
Secondary water pump bolts	18
Oil cooler	
Gasoline engine (threaded insert)	43
Diesel engine	
Mounting bolts	89 in-lbs
Mounting nut (if equipped)	18

1 General Information

Engine cooling system

1   The cooling system consists of a radiator, an expansion tank, a pressure cap (located on the expansion tank), a single thermostat on gasoline models and a dual thermostat on diesel models, a cooling fan, cooling fan stator (diesel models) and clutch, and a belt-driven water pump.

2   Diesel models are equipped with a secondary cooling system which serves the EGR system cooler, the Charge Air Cooler (intercooler), the transmission fluid cooler, and the fuel cooler. It uses a separate water pump, and a separate radiator (mounted in front of the primary radiator) and expansion tank. The secondary radiator is equipped with two thermostats (one mounted on each side).

3   The cooling fan and clutch are mounted on the front of the water pump on gasoline models and on the cooling fan drive diesel models. On all models except the 6.7L diesel, the fan incorporates a fluid-drive fan clutch, which saves horsepower and reduces noise. When the engine is cold, the fluid in the clutch offers little resistance and allows the fan to freewheel. As the engine heats up and reaches a predetermined temperature, the fluid in the clutch thickens and drives the fan. The system used on the 6.7L diesel engine is similar; however the fan is driven electrically.

4   The expansion tank (referred to by the manufacturer as a "degas bottle") functions somewhat differently than a conventional recovery tank. Designed to separate any trapped air in the coolant, it is pressurized by the radiator and has a pressure cap on top. The radiator on these models does not have a pressure cap. When the thermostat is closed, no coolant flows in the expansion tank, but when the engine is fully warmed up, coolant flows from the top of the radiator through a small hose that enters the top of the expansion tank, where the air separates and the coolant falls into a coolant reservoir in the bottom of the tank, which is fed to the cooling system through a larger hose connected to the lower radiator hose.

Warning: *Unlike a conventional coolant recovery tank, the pressure cap on the expansion tank should never be opened after the engine has warmed up, because of the danger of severe burns caused by steam or scalding coolant.*

5   Coolant in the right side of the radiator circulates through the lower radiator hose to the water pump, where it is forced through coolant passages in the cylinder block. The coolant then travels up into the cylinder head, circulates around the combustion chambers and valve seats, travels out of the cylinder head past the open thermostats into the upper radiator hose and back into the radiator.

6   When the engine is cold, the thermostats restrict the circulation of coolant to the engine. When the minimum operating temperature is reached, the thermostats begins to open, allowing coolant to return to the radiator.

Transmission cooling systems

7   Vehicles with an automatic transmission are equipped with a external transmission cooler. On gasoline models the cooler is located in the the radiator and on diesel models, it's located inside the engine compartment on the passenger's side of the frame. On gasoline models the transmission is connected to the cooler by two hoses: one delivers hot transmission fluid to the radiator and the other brings the cooled fluid back to the transmission. On diesel models, two lines connect the transmission to the cooler and the other two hoses connect the cooler to the secondary cooling system.

Engine oil cooling system

8   Besides the engine and transmission cooling systems described above, engine heat is also dissipated through an external oil cooler that's integrated into the lubrication system. The oil cooler helps keep engine oil temperatures within design limits under extreme load conditions.

9   The oil cooling system on gasoline engines consists of a housing mounted inline between the oil filter and the filter adapter, a heat exchanger inside the radiator and a pair of hoses that deliver hot oil from the housing to the radiator and bring the cooled oil back to the housing.

10   The oil cooling system on the diesel engine consists of a water-cooled heat exchanger housed on the side of the oil pan. The forward end of the exchanger is connected to the lower radiator hose connection; the rear fitting on the exchanger is connected to the cylinder block.

Heating system

11   The heating system consists of the heater controls, the heater core, the heater blower assembly (which houses the blower motor and the blower motor resistor), and the hoses connecting the heater core to the engine cooling system. Hot engine coolant is circulated through the heater core. When the heater mode is activated, a flap door opens to expose the heater box to the passenger compartment. A fan switch on the heater controls activates the blower motor, which forces air through the core, heating the air.

Air conditioning system

12   The air conditioning system consists of the condenser, which is mounted in front of the radiator, the evaporator case assembly under the dash, a compressor mounted on the engine, and the plumbing connecting all of the above components.

13   A blower fan forces the warmer air of the passenger compartment through the evaporator core (sort of a radiator-in-reverse), transferring the heat from the air to the refrigerant. The liquid refrigerant boils off into low pressure vapor, taking the heat with it when it leaves the evaporator.

2.2 The cooling system pressure tester is connected in place of the pressure cap on the coolant expansion tank, then pumped up to pressurize the system

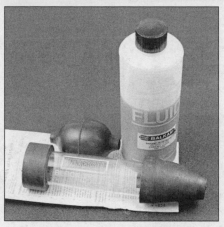

2.5a The combustion leak detector consists of a bulb, syringe and test fluid

2.5b Place the tester over the cooling system filler neck and use the bulb to draw a sample into the tester

2   Troubleshooting

Coolant leaks

1   A coolant leak can develop anywhere in the cooling system, but the most common causes are:

   a) A loose or weak hose clamp
   b) A defective hose
   c) A faulty pressure cap
   d) A damaged radiator
   e) A bad heater core
   f) A faulty water pump
   g) A leaking gasket at any joint that carries coolant

2   Coolant leaks aren't always easy to find. Sometimes they can only be detected when the cooling system is under pressure. Here's where a cooling system pressure tester comes in handy. After the engine has cooled completely, the tester is attached in place of the pressure cap on the expansion tank, then pumped up to the pressure value equal to that of the pressure cap rating (see illustration). Now, leaks that only exist when the engine is fully warmed up will become apparent. The tester can be left connected to locate a nagging slow leak.
Note: *Diesel engine models have a secondary expansion tank mounted to the fan shroud that is tested in the same way.*

Coolant level drops, but no external leaks

3   If you find it necessary to keep adding coolant, but there are no external leaks, the probable causes include:

   a) A blown head gasket
   b) A leaking intake manifold gasket (only on engines that have coolant passages in the manifold) cracked cylinder head or cylinder block

4   Any of the above problems will also usually result in contamination of the engine oil,

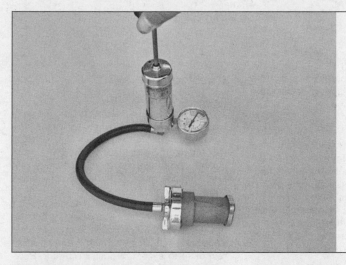

2.8 Checking the cooling system pressure cap with a cooling system pressure tester

which will cause it to take on a milkshake-like appearance. A bad head gasket or cracked head or block can also result in engine oil contaminating the cooling system.

5   Combustion leak detectors (also known as block testers) are available at most auto parts stores. These work by detecting exhaust gases in the cooling system, which indicate a compression leak from a cylinder into the coolant. The tester consists of a large bulb-type syringe and bottle of test fluid (see illustration). A measured amount of the fluid is added to the syringe. The syringe is placed over the cooling system filler neck and, with the engine running, the bulb is squeezed and a sample of the gases present in the cooling system are drawn up through the test fluid (see illustration). If any combustion gases are present in the sample taken, the test fluid will change color.

6   If the test indicates combustion gas is present in the cooling system, you can be sure that the engine has a blown head gasket or a crack in the cylinder head or block, and will require disassembly to repair.

Pressure cap

Warning: *Wait until the engine is completely cool before beginning this check.*

7   The cooling system is sealed by a spring-loaded cap, which raises the boiling point of the coolant. If the cap's seal or spring are worn out, the coolant can boil and escape past the cap. With the engine completely cool, remove the cap and check the seal; if it's cracked, hardened or deteriorated in any way, replace it with a new one.

8   Even if the seal is good, the spring might not be; this can be checked with a cooling system pressure tester (see illustration). If the cap can't hold a pressure within approximately 1-1/2 lbs of its rated pressure (which is marked on the cap), replace it with a new one.

9   The cap is also equipped with a vacuum relief spring. When the engine cools off, a vacuum is created in the cooling system. The vacuum relief spring allows air back into the system, which will equalize the pressure and prevent damage to the radiator (the radiator tanks could collapse if the vacuum is great enough). If, after turning the engine off and

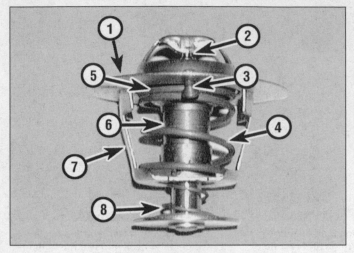

2.10 Typical thermostat:

1	Flange	5	Valve seat
2	Piston	6	Valve
3	Jiggle valve	7	Frame
4	Main coil spring	8	Secondary coil spring

2.28 The water pump weep hole is generally located on the underside of the pump

allowing it to cool down you notice any of the cooling system hoses collapsing, replace the pressure cap with a new one.

Thermostat

10   Before assuming the thermostat (see illustration) is responsible for a cooling system problem, check the coolant level (see Chapter 1), drivebelt tension (see Chapter 1) and temperature gauge (or light) operation.

11   If the engine takes a long time to warm up (as indicated by the temperature gauge or heater operation), the thermostat is probably stuck open. Replace the thermostat with a new one.

Note: *Diesel engine models are equipped with a dual thermostat and a pair of secondary thermostats for the charge air cooler (CAC) system.*

12   If the engine runs hot or overheats, a thorough test of the thermostat should be performed.

13   Definitive testing of the thermostat can only be made when it is removed from the vehicle. If the thermostat is stuck in the open position at room temperature, it is faulty and must be replaced.

Caution: *Do not drive the vehicle without a thermostat. The computer may stay in open loop and emissions and fuel economy will suffer.*

14   To test a thermostat, suspend the (closed) thermostat on a length of string or wire in a pot of cold water.

15   Heat the water on a stove while observing the thermostat. The thermostat should fully open before the water boils.

16   If the thermostat doesn't open and close as specified, or sticks in any position, replace it.

Cooling fan

Electric cooling fan

17   If the engine is overheating and the cooling fan is not coming on when the engine temperature rises to an excessive level, unplug the fan motor electrical connector(s) and connect the motor directly to the battery with fused jumper wires. If the fan motor doesn't come on, replace the motor.

18   If the radiator fan motor is okay, but it isn't coming on when the engine gets hot, the fan relay might be defective. A relay is used to control a circuit by turning it on and off in response to a control decision by the Powertrain Control Module (PCM). These control circuits are fairly complex, and checking them should be left to a qualified automotive technician. Sometimes, the control system can be fixed by simply identifying and replacing a bad relay.

19   Locate the fan relays in the engine compartment fuse/relay box.

20   Test the relay (see Chapter 12).

21   If the relay is okay, check all wiring and connections to the fan motor. Refer to the wiring diagrams at the end of Chapter 12. If no obvious problems are found, the problem could be the Engine Coolant Temperature (ECT) sensor or the Powertrain Control Module (PCM). Have the cooling fan system and circuit diagnosed by a dealer service department or repair shop with the proper diagnostic equipment.

Belt-driven cooling fan

22   Disconnect the cable from the negative terminal of the battery and rock the fan back and forth by hand to check for excessive bearing play.

23   With the engine cold (and not running),

turn the fan blades by hand. The fan should turn freely.

24   Visually inspect for substantial fluid leakage from the clutch assembly. If problems are noted, replace the clutch assembly.

25   With the engine completely warmed up, turn off the ignition switch and disconnect the negative battery cable from the battery. Turn the fan by hand. Some drag should be evident. If the fan turns easily, replace the fan clutch.

Water pump

26   A failure in the water pump can cause serious engine damage due to overheating.

Drivebelt-driven water pump

27   There are two ways to check the operation of the water pump while it's installed on the engine. If the pump is found to be defective, it should be replaced with a new or rebuilt unit.

28   Water pumps are equipped with weep (or vent) holes (see illustration). If a failure occurs in the pump seal, coolant will leak from the hole.

29   If the water pump shaft bearings fail, there may be a howling sound at the pump while it's running. Shaft wear can be felt with the drivebelt removed if the water pump pulley is rocked up and down (with the engine off).

Timing chain or timing belt-driven water pump

30   Water pumps driven by the timing chain or timing belt are located underneath the timing chain or timing belt cover.

31   Checking the water pump is limited because of where it is located. However, some basic checks can be made before deciding to remove the water pump. If the pump is found

to be defective, it should be replaced with a new or rebuilt unit.

32   One sign that the water pump may be failing is that the heater (climate control) may not work well. Warm the engine to normal operating temperature, confirm that the coolant level is correct, then run the heater and check for hot air coming from the ducts.

33   Check for noises coming from the water pump area. If the water pump impeller shaft or bearings are failing, there may be a howling sound at the pump while the engine is running.

Note: *Be careful not to mistake drivebelt noise (squealing) for water pump bearing or shaft failure.*

34   It you suspect water pump failure due to noise, wear can be confirmed by feeling for play at the pump shaft. This can be done by rocking the drive sprocket on the pump shaft up and down. To do this you will need to remove the tension on the timing chain or belt as well as access the water pump.

All water pumps

35   In rare cases or on high-mileage vehicles, another sign of water pump failure may be the presence of coolant in the engine oil. This condition will adversely affect the engine in varying degrees.

Note: *Finding coolant in the engine oil could indicate other serious issues besides a failed water pump, such as a blown head gasket or a cracked cylinder head or block.*

36   Even a pump that exhibits no outward signs of a problem, such as noise or leakage, can still be due for replacement. Removal for close examination is the only sure way to tell. Sometimes the fins on the back of the impeller can corrode to the point that cooling efficiency is diminished significantly.

Heater system

37   Little can go wrong with a heater. If the fan motor will run at all speeds, the electrical part of the system is okay. The three basic heater problems fall into the following general categories:

a) Not enough heat
b) Heat all the time
c) No heat

38   If there's not enough heat, the control valve or door is stuck in a partially open position, the coolant coming from the engine isn't hot enough, or the heater core is restricted. If the coolant isn't hot enough, the thermostat in the engine cooling system is stuck open, allowing coolant to pass through the engine so rapidly that it doesn't heat up quickly enough. If the vehicle is equipped with a temperature gauge instead of a warning light, watch to see if the engine temperature rises to the normal operating range after driving for a reasonable distance.

39   If there's heat all the time, the control valve or the door is stuck wide open.

40   If there's no heat, coolant is probably not reaching the heater core, or the heater core is plugged. The likely cause is a collapsed or plugged hose, core, or a frozen heater control valve. If the heater is the type that flows coolant all the time, the cause is a stuck door or a broken or kinked control cable.

Air conditioning system

41   If the cool air output is inadequate:

a) Inspect the condenser coils and fins to make sure they're clear.
b) Check the compressor clutch for slippage.
c) Check the blower motor for proper operation.
d) Inspect the blower discharge passage for obstructions.
e) Check the system air intake filter for clogging.

42   If the system provides intermittent cooling air:

a) Check the circuit breaker, blower switch and blower motor for a malfunction.
b) Make sure the compressor clutch isn't slipping.
c) Inspect the plenum door to make sure it's operating properly.
d) Inspect the evaporator to make sure it isn't clogged.
e) If the unit is icing up, it may be caused by excessive moisture in the system, incorrect super heat switch adjustment or low thermostat adjustment.

43   If the system provides no cooling air:

a) Inspect the compressor drivebelt. Make sure it's not loose or broken.
b) Make sure the compressor clutch engages. If it doesn't, check for a blown fuse.
c) Inspect the wire harness for broken or disconnected wires.
d) If the compressor clutch doesn't engage, bridge the terminals of the air conditioning pressure switch(es) with a jumper wire; if the clutch now engages, and the system is properly charged, the pressure switch is bad.
e) Make sure the blower motor is not disconnected or burned out.
f) Make sure the compressor isn't partially or completely seized.
g) Inspect the refrigerant lines for leaks.
h) Check the components for leaks.
i) Inspect the receiver-drier/accumulator or expansion valve/tube for clogged screens.

44   If the system is noisy:

a) Look for loose panels in the passenger compartment.
b) Inspect the compressor drivebelt. It may be loose or worn.
c) Check the compressor mounting bolts. They should be tight.
d) Listen carefully to the compressor. It may be worn out.
e) Listen to the idler pulley and bearing and the clutch. Either may be defective.
f) The winding in the compressor clutch coil or solenoid may be defective.
g) The compressor oil level may be low.

h) The blower motor fan bushing or the motor itself may be worn out.
i) If there is an excessive charge in the system, you'll hear a rumbling noise in the high pressure line, a thumping noise in the compressor, or see bubbles or cloudiness in the sight glass.
j) If there's a low charge in the system, you might hear hissing in the evaporator case at the expansion valve, or see bubbles or cloudiness in the sight glass.

---

3   Air conditioning and heating system - check and maintenance

---

Air conditioning system

Warning: *The air conditioning system is under high pressure. Do not loosen any hose fittings or remove any components until after the system has been discharged. Air conditioning refrigerant should be properly discharged into an EPA-approved recovery/recycling unit at a dealer service department or an automotive air conditioning repair facility. Always wear eye protection when disconnecting air conditioning system fittings.*

Caution: *All models covered by this manual use environmentally friendly R-134a. This refrigerant (and its appropriate refrigerant oils) are not compatible with R-12 refrigerant system components and must never be mixed or the components will be damaged.*

Caution: *When replacing entire components, additional refrigerant oil should be added equal to the amount that is removed with the component being replaced. Read the can before adding any oil to the system, to make sure it is compatible with the R-134a system.*

1   The following maintenance checks should be performed on a regular basis to ensure that the air conditioning continues to operate at peak efficiency.

a) Inspect the condition of the compressor drivebelt. If it is worn or deteriorated, replace it (see Chapter 1).
b) Check the drivebelt tension (see Chapter 1).
c) Inspect the system hoses. Look for cracks, bubbles, hardening and deterioration. Inspect the hoses and all fittings for oil bubbles or seepage. If there is any evidence of wear, damage or leakage, replace the hose(s).
d) Inspect the condenser fins for leaves, bugs and any other foreign material that may have embedded itself in the fins. Use a fin comb or compressed air to remove debris from the condenser.
e) Make sure the system has the correct refrigerant charge.
f) If you hear water sloshing around in the dash area or have water dripping on the carpet, check the evaporator housing drain tube (see illustration 3.25) and insert a piece of wire into the opening to check for blockage.

2   It's a good idea to operate the system

3.8 Insert a thermometer in the center vent, turn on the air conditioning system and wait for it to cool down; depending on the humidity, the output air should be 35 to 40 degrees cooler than the ambient air temperature

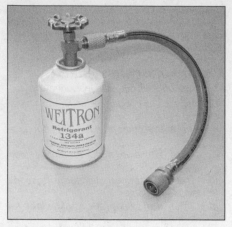

3.10 R-134a automotive air conditioning charging kit

3.12 Location of the low-side charging port

for about ten minutes at least once a month. This is particularly important during the winter months because long term non-use can cause hardening, and subsequent failure, of the seals. Note that using the Defrost function operates the compressor.

3   If the air conditioning system is not working properly, proceed to Step 6 and perform the general checks outlined below.

4   Because of the complexity of the air conditioning system and the special equipment necessary to service it, in-depth troubleshooting and repairs beyond checking the refrigerant charge and the compressor clutch operation are not included in this manual. However, simple checks and component replacement procedures are provided in this chapter. For more complete information on the air conditioning system, refer to the *Haynes Automotive Heating and Air Conditioning Manual*.

5   The most common cause of poor cooling is simply a low system refrigerant charge. If a noticeable drop in system cooling ability occurs, one of the following quick checks will help you determine if the refrigerant level is low.

Checking the refrigerant charge

6   Warm the engine up to normal operating temperature.

7   Place the air conditioning temperature selector at the coldest setting and put the blower at the highest setting.

8   Insert a thermometer in the center air distribution duct (see illustration) while operating the air conditioning system at its maximum setting - the temperature of the output air should be 35 to 40 degrees F below the ambient air temperature (down to approximately 40 degrees F). If the ambient (outside) air temperature is very high, say 110 degrees F, the duct air temperature may be as high as 60 degrees F, but generally the air conditioning is 35 to 40 degrees F cooler than the ambient air.

9   Further inspection or testing of the system requires special tools and techniques and is beyond the scope of the home mechanic.

Adding refrigerant

Caution: *Make sure any refrigerant, refrigerant oil or replacement component you purchase is designated as compatible with R-134a systems.*

10   Purchase an R-134a automotive charging kit at an auto parts store (see illustration). A charging kit includes a can of refrigerant, a tap valve and a short section of hose that can be attached between the tap valve and the system low side service valve.

Caution: *Never add more than one can of refrigerant to the system. If more refrigerant than that is required, the system should be evacuated and leak tested.*

11   Back off the valve handle on the charging kit and screw the kit onto the refrigerant can, making sure first that the O-ring or rubber seal inside the threaded portion of the kit is in place.

Warning: *Wear protective eyewear when dealing with pressurized refrigerant cans.*

12   Remove the dust cap from the low-side charging port and attach the hose's quick-connect fitting to the port (see illustration). The fittings on the charging kit are designed to fit only on the low side of the system.

Warning: *DO NOT hook the charging kit hose to the system high side!*

13   Warm up the engine and turn on the air conditioning. Keep the charging kit hose away from the fan and other moving parts.

Note: *The charging process requires the compressor to be running. If the clutch cycles off, you can put the air conditioning switch on High and leave the car doors open to keep the clutch on and compressor working. The compressor can be kept on during the charging by removing the connector from the pressure switch and bridging it with a paper clip or jumper wire during the procedure.*

14   Turn the valve handle on the kit until the stem pierces the can, then back the handle out to release the refrigerant. You should be able to hear the rush of gas. Keep the can upright at all times, but shake it occasionally. Allow stabilization time between each addition.

Note: *The charging process will go faster if you wrap the can with a hot-water-soaked rag to keep the can from freezing up.*

15   If you have an accurate thermometer, you can place it in the center air conditioning duct inside the vehicle and keep track of the output air temperature. A charged system that is working properly should cool down to approximately 40 degrees F. If the ambient (outside) air temperature is very high, say 110 degrees F, the duct air temperature may be as high as 60 degrees F, but generally the air conditioning is 35 to 40 degrees F cooler than the ambient air.

16   When the can is empty, turn the valve handle to the closed position and release the connection from the low-side port. Reinstall the dust cap.

17   Remove the charging kit from the can and store the kit for future use with the piercing valve in the UP position, to prevent inadvertently piercing the can on the next use.

Heating systems

18   If the carpet under the heater core is damp, or if antifreeze vapor or steam is coming through the vents, the heater core is leaking. Remove it (see Section 13) and install a new unit (most radiator shops will not repair a leaking heater core).

19   If the air coming out of the heater vents isn't hot, the problem could stem from any of the following causes:

a) *The thermostat is stuck open, preventing the engine coolant from warming up enough to carry heat to the heater core. Replace the thermostat (see Section 4).*

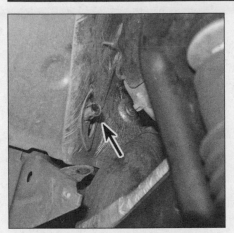

3.24 The evaporator drain hose is located on the passenger's side of the firewall (seen here from the right wheelwell with the inner fender splash shield removed)

4.4 Use a small hooked tool and pull the upper radiator hose clip outward to free the hose fitting

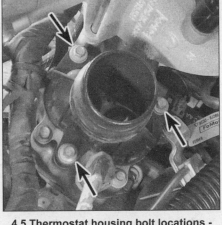

4.5 Thermostat housing bolt locations - diesel model shown

b) There is a blockage in the system, preventing the flow of coolant through the heater core. Feel both heater hoses at the firewall. They should be hot. If one of them is cold, there is an obstruction in one of the hoses or in the heater core, or the heater control valve is shut. Detach the hoses and back flush the heater core with a water hose. If the heater core is clear but circulation is impeded, remove the two hoses and flush them out with a water hose.

c) If flushing fails to remove the blockage from the heater core, the core must be replaced (see Section 13).

Eliminating air conditioning odors

20   Unpleasant odors that often develop in air conditioning systems are caused by the growth of a fungus, usually on the surface of the evaporator core. The warm, humid environment there is a perfect breeding ground for mildew to develop.

21   The evaporator core on most vehicles is difficult to access, and factory dealerships have a lengthy, expensive process for eliminating the fungus by opening up the evaporator case and using a powerful disinfectant and rinse on the core until the fungus is gone. You can service your own system at home, but it takes something much stronger than basic household germ-killers or deodorizers.

22   Aerosol disinfectants for automotive air conditioning systems are available in most auto parts stores, but remember when shopping for them that the most effective treatments are also the most expensive. The basic procedure for using these sprays is to start by running the system in the RECIRC mode for ten minutes with the blower on its highest speed. Use the highest heat mode to dry out the system and keep the compressor from engaging by disconnecting the wiring connector at the compressor.

23   The disinfectant can usually comes with

a long spray hose. Insert the nozzle into the vents closest to the evaporator core, and spray according to the manufacturer's recommendations. Try to cover the whole surface of the evaporator core, by aiming the spray up, down and sideways. Follow the manufacturer's recommendations for the length of spray and waiting time between applications.

24   Once the evaporator has been cleaned, the best way to prevent the mildew from coming back again is to make sure your evaporator housing drain tube is clear (see illustration).

Automatic heating and air conditioning systems

25   Some vehicles are equipped with an optional automatic climate control system. This system has its own computer that receives input from various sensors in the heating and air conditioning system. This computer, like the PCM, has self-diagnostic capabilities to help pinpoint problems or faults within the system. Vehicles equipped with automatic heating and air conditioning

4.6 On diesel models, note the position of the bleed hole; it should be facing toward the engine

systems are very complex and considered beyond the scope of the home mechanic. Vehicles equipped with automatic heating and air conditioning systems should be taken to a dealer service department or other qualified facility for repair.

4   Thermostat - replacement

Warning: *The engine must be completely cool when this procedure is performed.*
Note: *Don't drive the vehicle without a thermostat! The computer may stay in open loop and emissions and fuel economy will suffer.*
Note: *On diesel engine models there are four thermostats used. The primary thermostats are located at the end of the upper radiator hose. The two secondary thermostats are located in the side tanks of the secondary cooling system radiator (low temperature thermostat on the left side, high temperature thermostat on the right side).*

Primary thermostat

1   Drain the cooling system so that the coolant level is below the thermostat (see Chapter 1).

2   On gasoline models, remove the air filter housing outlet duct (see Chapter 4B).

3   On gasoline models, loosen the upper radiator hose clamp and pull off the hose from the thermostat housing.

4   On diesel models. pull the upper radiator hose clip outward until the end of the clip has dropped into the detent on the quick-connector (see illustration), then disconnect the upper radiator hose.

5   Remove the thermostat housing bolts (see illustration) and lift off the housing.

6   Note the orientation of the thermostat (see illustration), then remove it along with the O-ring. Discard the old O-ring. Make sure that nothing falls into the thermostat opening; it's a good idea to stuff a rag into the opening until you're ready to install the thermostat.

4.18a Remove the cover bolts . . .

4.18b . . . then carefully pry the cover out of the housing

4.19 Remove and discard the O-ring from the cover

4.20 Place the thermostat on the cover, then install the thermostat (with the spring toward the housing) and cover into the housing

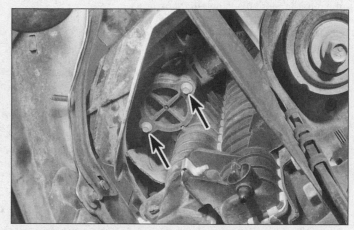

4.25 Remove the low temperature thermostat cover bolts

7   Be sure to use a replacement thermostat that's identical to the old thermostat. Compare the old and new thermostats carefully before installing the new unit.

Note: *On diesel engines, there are two thermostats used within one housing. The thermostats are combined into one unit and are not serviced separately.*

8   Install the thermostat. Make sure that the spring end faces down, toward the engine.

9   Install the new O-ring in the recess in the thermostat housing.

10   Carefully position the thermostat housing, install the bolts and tighten them to the torque listed in this Chapter's Specifications. Do NOT overtighten the thermostat housing bolts; doing so will crack or distort the housing.

11   On gasoline models, inspect the hose and the clamp (see Chapter 1). Replace the hose and/or clamp if damaged and reattach the radiator hose to the housing and tighten the clamp.

12   On diesel models, push the upper hose quick-connect fitting on to the thermostat housing connection, then slide the spring clip inward until it locks in place.

13   Fill the cooling system with the correct amount and type of coolant (see Chapter 1).

14   Start and run the engine until it reaches normal operating temperature, then check the coolant level and look for leaks.

Secondary thermostats (diesel models only)

15   Drain the secondary cooling system (see Chapter 1).

High-temperature thermostat

16   Remove the grille (see Chapter 11).

17   Remove the plastic push-pins holding the air deflector over the thermostat cover.

18   Place a drain pan underneath the thermostat. Remove the thermostat cover bolts, then remove the cover and thermostat (see illustrations).

Warning: *Be prepared for coolant spillage!*

19   Remove the thermostat O-ring from the cover (see illustration).

20   Install the thermostat (see illustration), making sure that the spring end faces up, toward the housing.

21   Install the bolts and tighten them to the torque listed in this Chapter's Specifications.

22   Remaining installation is the reverse of removal.

23   Fill the secondary cooling system (see Chapter 1).

24   Start and run the engine until it reaches normal operating temperature, then check the secondary coolant level and look for leaks.

Low-temperature thermostat

25   Place a drain pan underneath the thermostat. Working from under the vehicle, remove the thermostat cover bolts (see illustration), then remove the cover and thermostat.

Warning: *Be prepared for coolant spillage!*

26   Remove the thermostat O-ring (see illustration 4.19).

27   Install the thermostat. Making sure that the spring end faces up, toward the housing (see illustration 4.20).

28   Install the bolts and tighten them to the torque listed in this Chapter's Specifications.

29   Fill the secondary cooling system (see Chapter 1).

30   Start and run the engine until it reaches normal operating temperature, then check the secondary coolant level and look for leaks.

5.13 Disconnect the coolant lines (1) then remove the mounting bolts (2) and remove the tank

6.7a Remove the upper shroud fasteners (one of two) . . .

6.7b . . . and lift the upper shroud up and out

5   Expansion tank - removal and installation

Warning: *The engine must be completely cool before beginning this procedure.*
1   Drain the cooling system (see Chapter 1) below the level of the expansion tank.
Caution: *Once the turkey baster has been used for siphoning coolant do not use it with any type of food or food product.*

Gasoline models

2   Disconnect the (MAF) sensor electrical connector (see Chapter 6A) then remove the top half of the air filter housing (see Chapter 1).
3   Using hose clamp pliers, clamp the hose to the expansion tank.
4   Remove the hose retaining clip and disconnect the quick-connect hose from the expansion tank.

5   Remove the expansion tank mounting bolts then pull the expansion tank out of the grommets. Remove the tank from the vehicle.
6   Installation is the reverse of removal.
7   Refill the cooling system (see Chapter 1).

Diesel models

8   Drain the secondary cooling system (see Chapter 1).

Primary expansion tank

9   The primary expansion tank is integral with the secondary battery tray. See Chapter 5, Section 4 for the removal and installation procedure.

Secondary expansion tank

10   Remove the air filter housing lid (see Chapter 1) and the air intake duct.
11   Remove the hose retaining clip and dis-

connect the quick-connect hose from the bottom of the expansion tank.
12   Using pliers, squeeze the clamps on the hoses and slide them back, then disconnect the hoses from the expansion tank.
13   Remove the secondary expansion tank mounting bolts (see illustration) and separate the tank from the upper fan shroud. Remove the expansion tank.
14   Installation is the reverse of removal.
15   Refill the secondary cooling system (see Chapter 1).

6   Cooling fan and shroud - removal and installation

Warning: *Wait until the engine is completely cool before beginning this procedure.*

Removal

1   Disconnect the cable(s) from the negative battery terminals (see Chapter 5).
2   On diesel models, drain the primary cooling system (see Chapter 1). If you're going to remove the lower fan shroud, also drain the secondary cooling system.
3   On gasoline models, disconnect the expansion tank hose from the shroud then remove the vacuum tank reservoir (see Chapter 9) from the shroud, if equipped.
4   On diesel models, remove the secondary coolant expansion tank (see Section 5).
5   Remove the power steering fluid reservoir bolt, then pull the reservoir up from the shroud and secure it out of the way.
6   Detach the wiring harness from the upper shroud.
7   On diesel engine models, remove the upper shroud fasteners and remove the shroud (see illustrations).

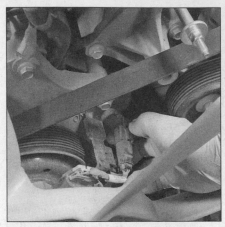

6.8a Disconnect the electrical connector from the harness . . .

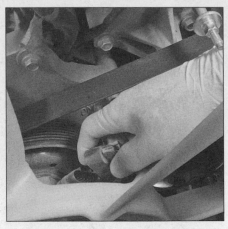

6.8b . . . and from the fan clutch

6.9 On models equipped with an electric fan clutch, remove the harness bracket bolts and bracket (the harness travels under the front of the engine; left-side bracket shown)

6.14a Special fan clutch tools like these are available at most auto parts stores

6.14b Using the special tools, hold the clutch drive and remove the fan clutch from the drive

6.15 Lift the fan and clutch up and out, making sure not to contact the radiator

8   Disconnect the fan clutch electrical harness (see illustrations), if equipped.

9   Remove the fan clutch electrical harness bracket bolts and bracket (see illustration), if equipped.

Gasoline models

10   Disconnect the impact sensor wiring harness from the bottom of the fan shroud. Press the five retaining tabs inward at the bottom of the shroud, then move the lower or inner section of the shroud upward and off of the main shroud.

11   Using a pair of special wrenches, unscrew the fan clutch from the water pump (see illustrations 6.14a and 6.14b). The fan clutch is usually very tight.

12   Position the fan and fan clutch assembly inside the fan shroud, then unbolt the fan shroud and lift the fan assembly and the shroud up and out of the engine compartment together.

13   Installation is the reverse of removal.

Diesel models

14   Using a pair of special wrenches, unscrew the fan clutch from the clutch drive (see illustrations). The fan clutch is usually very tight.

15   Remove the fan and clutch (see illustration).

16   To remove the lower shroud, remove the radiator (see Section 7).

17   Detach the wiring harnesses from the lower shroud.

18   Detach the shroud from the cooling fan stator.

19   Remove the lower fan shroud for the vehicle.

20   Installation is the reverse of removal. When installing the fshroud, make sure that the locator tabs on the bottom edge of the shroud fit into their respective slots.

7   Radiator - removal and installation

Warning: *The air conditioning system is under high pressure. DO NOT loosen any fittings or remove any components until after the system has been discharged. Air conditioning refrigerant should be properly discharged into an EPA-approved container at a dealer service department or an automotive air conditioning repair facility. Always wear eye protection when disconnecting air conditioning system fittings.*

Warning: *The engine must be completely cool when this procedure is performed.*

Note: *Diesel engine models use two radiators - primary and secondary - while gasoline models only use one.*

1   If you're working on a diesel model, have the air conditioning system refrigerant recovered by a licensed air conditioning technician.

7.3 Lower air deflector push-pin retainer locations

7.21 Disconnect the by-pass hose clamp (1) and the lower hose spring clip (2)

7.22 Radiator bracket bolt locations (right-side shown, left-side identical)

7.26a Disconnect the hoses and remove the mounting bolts from the right side . . .

1   Quick-connect spring clamp and connector
2   Hose clamps
3   Right-side mounting bolts

7.26b . . . then disconnect the hose and remove the mounting bolts from the left side

1   Hose clamp
2   Left-side mounting bolts

2   Drain the cooling system (see Chapter 1).

3   Remove the lower air deflector push-pin retainers (see illustration) and move the deflector out of the way.

4   Remove the grille (see Chapter 11).

5   Remove the cooling fan and shroud (see Section 6).

Gasoline models

6   Remove the air conditioning condenser mounting bolts and reposition the condenser (see Section 17).

7   Remove the transmission cooler mounting bolts and secure the cooler out of the way (see Chapter 7A).

8   Disconnect the upper and lower radiator hoses from the radiator.

9   Disconnect and plug the transmission cooler lines from the radiator (see Chapter 7A).

10   Disconnect and plug the power steering cooler hoses (see Chapter 10).

11   Remove the cooling fan and clutch assembly (see Section 6).

12   Remove the radiator mounting bolts and remove the radiator from the vehicle.

13   Installation is the reverse of removal. Tighten the mounting bolts to the torque listed in this Chapter's Specifications.

14   Fill the cooling system with the proper mixture of antifreeze, and also check the automatic transmission and power steering fluid levels.

15   Start the engine and allow it to reach normal operating temperature, then check for leaks.

Diesel models

Primary radiator

16   Remove the air filter housing and intake duct (see Chapter 4B).

17   Remove the secondary coolant expansion tank (see Section 5).

18   Remove the secondary battery (see Chapter 5).

19   Remove the charge air cooler (CAC)

upper mounting bolts (see Chapter 4B).

20   Remove the condenser (see Section 17).

21   Disconnect the lower hose spring clip and by-pass hose clamp (see illustration) and disconnect the hoses from the radiator.

22   Remove the insulators from the radiator support brackets (see illustration) and inspect them for cracks, tears and deterioration. If the upper insulators are damaged or worn, replace them.

23   With the help of an assistant, lift the radiator up and off of the brackets, then remove the radiator.

Secondary radiator

24   Remove the air filter housing (see Chapter 4B).

25   Remove the condenser (see Section 17).

26   Disconnect the hoses from the radiator and remove the mounting bolts (see illustrations).

27   Remove the secondary radiator from the vehicle.

8.15a To detach the left side of the cooling module, remove the two bolts (1) and lower nut at the bottom (B)

8.15b To detach the right side of the cooling module, remove the two bolts (1) and lower nut at the bottom

Primary or secondary radiator

28   Installation is the reverse of removal. Tighten the mounting bolts to the torque listed in this Chapter's Specifications.
29   Fill the system with the proper mixture of antifreeze (see Chapter 1).
30   Start the engine and allow it to reach normal operating temperature, then check for leaks.
31   Have the air conditioning system evacuated, leak checked and recharged by the shop that discharged it.

8   Cooling module - removal and installation

Warning: *The air conditioning system is under high pressure. DO NOT loosen any fittings or remove any components until after the system has been discharged. Air conditioning refrigerant should be properly discharged into an EPA-approved container at a dealer service department or an automotive air conditioning repair facility. Always wear eye protection when disconnecting air conditioning system fittings.*
Warning: *The engine must be completely cool when this procedure is performed.*
Note: *The cooling module consists of the radiator, air conditioning condenser, power steering fluid cooler, transmission fluid cooler, and secondary radiator (diesel models). These components can all be removed as a unit for certain procedures where improved access to the front of the engine is required.*
1   Have the air conditioning system discharged by a licensed air conditioning technician.
2   Drain the cooling system(s) (see Chapter 1).
3   Remove the air filter housing and outlet pipe.

4   Remove both headlight housings (see Chapter 12).
5   Remove the air deflector push-pins and remove the air deflectors.
6   Remove the grille and bracket (see Chapter 11).
7   Disconnect the lines to the condenser (see illustration 17.8a).
8   Disconnect the wiring harness retainers from the upper and lower shrouds.
9   On diesel models, disconnect the EGR cooler-to-secondary expansion tank hose from the tanks. Disengage the hose retainers and secure the hose out of the way.
10   On 4WD gasoline models equipped with shift-on-the-fly, disconnect the electrical connector and remove the connector bracket bolt from the shroud. Then disconnect the vacuum reservoir electrical connector, the reservoir hoses and the reservoir bolts, then remove the reservoir from the shroud.
11   On models without shift-on-the-fly, disconnect the dummy connector and remove the connector bracket bolt from the shroud.
12   Remove the bolts and secure the power steering reservoir out of the way.
13   On diesel models, remove the secondary coolant expansion tank (see Section 5) and disconnect the hoses to the secondary radiator (see Section 7).
14   Disconnect the electrical connectors at the shroud, to the charge air cooler (CAC) (see Chapter 4B) and impact sensors (see Chapter 12).
15   Remove the cooling module nuts and bolts (see illustrations).
16   With the help of an assistant, remove the cooling module.
17   Installation is the reverse of removal. When installing the module, tighten the fasteners to the torque listed in this Chapter's Specifications.
18   Fill the system with the proper mixture of antifreeze, and also check the automatic

transmission and power steering fluid levels.
19   Start the engine and allow it to reach normal operating temperature, then check for leaks.

9   Water pump - removal and installation

Removal

Warning: *Wait until the engine is completely cool before beginning this procedure.*
1   Disconnect the cable(s) from the negative battery terminal(s), (see Chapter 5).
2   Drain the cooling system (see Chapter 1).
Note: *If you're working on a diesel model, it is only necessary to drain the cooling system that the particular water pump being replaced is associated with (primary or secondary).*
3   Remove the cooling fan, the fan shroud (see Section 6) and fan stator (refer to Section 11) on diesel engines.

Gasoline engine

4   If the water pump pulley bolts aren't too tight, you might be able to loosen them now, using the drivebelt to hold the pulley. If that doesn't work, you'll have to hold the pulley with a strap wrench after you remove the drivebelt.
5   Remove the drivebelt (see Chapter 1).
6   Remove the pulley bolts and remove the pulley. If you were unable to loosen the water pump pulley bolts with the drivebelt installed, use a strap wrench to hold the pulley as you loosen the bolts.
7   Remove the water pump retaining bolts and remove the water pump. Take note of the installed positions of the various bolts.
Note: *If the water pump sticks, dislodge it with a soft-face hammer or a hammer and a block of wood.*

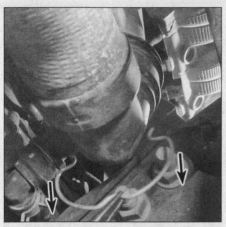

9.13 Remove the lower radiator hose clip

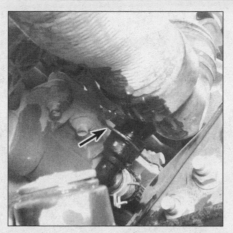

9.14 Pull the clip outward, then pull the hose fitting off

9.17 Carefully clean the gasket mating surface

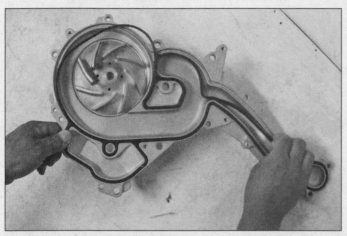

9.18 Make sure the gasket is completely installed into the grooves on the pump

9.19 Place a screwdriver between the bolt heads to hold the pulley and loosen the opposite bolts

8   Before installation, remove and clean all gasket or sealant material from the water pump and cylinder block.

9   Inspect the O-ring and sealing surface of water pump housing in the block for dirt and/or debris. Clean them thoroughly before reassembly.

10   Lubricate a new O-ring seal with clean antifreeze and install it to the water pump.

Diesel engine

Primary pump

11   Remove the four cooling fan drive drive bolts and remove the assembly.

12   Disconnect the camshaft position (CMP) sensor electrical connector (see Chapter 6B) and the harness retainers, then secure the harness out of the way.

13   Pull the lower radiator hose clip downwards until the clip for quick-connector (see illustration) can be removed.

14   Pull the heater hose clips outward until the end of the clip drops into the detent on the quick-connector (see illustration). Disconnect

the hose connection from the bottom of the lower radiator hose connector, then disconnect the radiator hose from the pump.

15   Remove the battery cable bracket bolts and nuts, then move the cable along with the bracket out of the way.

16   Remove the six large bolts and 12 small bolts around the perimeter of the pump, noting their positions (see illustration 9.22) and remove the pump from the front cover.

17   Before installation, clean off the gasket mating surfaces (see illustration).

18   Install a new gasket into the groove along the pump sealing surface (see illustration).

Secondary pump

Note: *The manufacturer states that the secondary water pump, the pulley and housing must be replaced as an assembly. Some aftermarket manufacturers make just the pump body and impeller; check on the availability of parts before beginning this procedure.*

19   Remove the water pump pulley mounting bolts (see illustration).

20   Pull the secondary radiator hose clips outward until the end of the clip has dropped into the detent on the quick-connector (see illustration), then disconnect the radiator hoses.

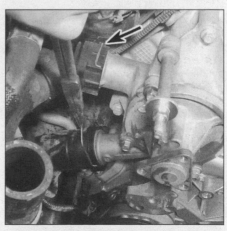

9.20 Use a small hooked tool or pliers and pull the hose clip outward

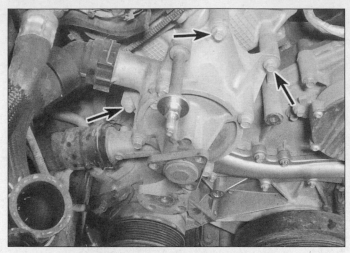

9.21 Remove the housing bolts and remove the pump assembly - 3 of 4 shown

9.22 On diesel models, hand tighten all bolts then torque the smaller bolts first, followed by the larger bolts

11.3 Remove the stator mounting nuts

11.4 Move the stator around the fan and off of the lower shroud

21   Remove the housing mounting bolts and remove the assembly (see illustration).

All engines

22   Install the water pump (see illustration) or water pump housing and tighten the bolts by hand evenly, then tighten the bolts to the torque listed in this Chapter's Specifications.

23   The remainder of installation is the reverse of removal.

24   Refill the cooling system (see Chapter 1).

25   Start the engine and make sure there are no leaks. Check the level frequently during the first few weeks of operation to ensure there are no leaks and that the coolant level is stable.

10   Coolant temperature sending unit - general information

1   The coolant temperature indicator system consists of a temperature gauge mounted in the dash and a Cylinder Head Temperature (CHT) sensor (gasoline models) or an Engine Coolant Temperature (ECT) sensor (diesel models) (see Chapter 6B).

2   These sensors also serve as the sensor(s) for the Powertrain Control Module (PCM).

3   If an overheating indication occurs, check the coolant level in the system. Make sure that the wiring between the gauge and the sensor(s) is secure and all fuses are intact.

11   Cooling fan stator (diesel engine models) - removal and installation

Warning: *Wait until the engine is completely cool before beginning this procedure.*

1   Remove the upper fan shroud (see Section 6).

2   Disengage the automatic transmission cooler-to-expansion tank hose retainer from the stator.

3   Remove the stator-to-cooling fan shroud nuts (see illustration) and pull the stator off of the studs.

4   Rotate the stator while twisting it to free it from the lower section of the fan shroud (see illustration).

5   Installation is the reverse of removal.

12.4 Depress the connector lock and disconnect the connector from the module

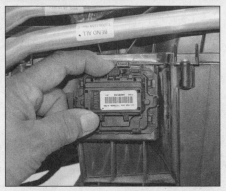

12.5a To replace the blower motor speed control module, press the retaining tab downward . . .

12.5b . . . then pull the module from the case

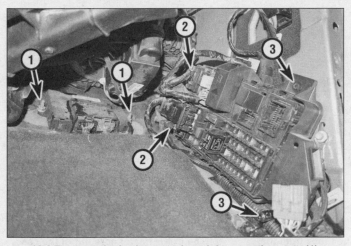

12.9 Remove the body control module mounting nuts (1), disconnect the electrical connectors from the interior fuse panel (2) and remove the fuse panel (3) fasteners

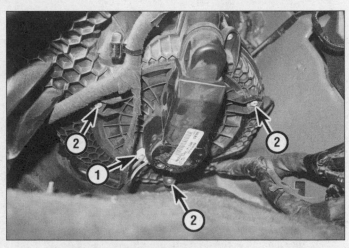

12.11 Disconnect the electrical connector from the blower motor (1) then remove the mounting screws (2)

12 Air conditioning and heater blower motor and blower motor speed control module - replacement

Blower motor speed control module

Warning: *The air conditioning system is under high pressure. DO NOT loosen any fittings or remove any components until after the system has been discharged. Air conditioning refrigerant should be properly discharged into an EPA-approved container at a dealer service department or an automotive air conditioning repair facility. Always wear eye protection when disconnecting air conditioning system fittings.*

1   Have the air conditioning system discharged by a licensed air conditioning technician.
2   Disconnect the cable(s) from the negative battery terminal(s) (see Chapter 5).
3   Disconnect the cable(s) from the nega-

tive battery terminal(s) (see Chapter 5). Remove the heater/evaporator case (see Section 13).
4   Disconnect the blower motor speed control module electrical connector (see illustration).
5   Press the module retaining tab down (see illustration) and remove the resistor from the case (see illustration).
6   Installation is the reverse of removal.

Blower motor

7   Disconnect the cable(s) from the negative battery terminal(s) (see Chapter 5).
8   Remove the passenger's side kick-panel (see Chapter 11).
9   Remove the body control module and the interior fuse panel (see illustration).
10   On 4WD models, remove the transfer case control module (TCCM), (see Chapter 7B).
11   Disconnect the electrical connector from the blower motor (see illustration).
12   Remove the blower motor (see illustration).

Note: *On some models it may be necessary to remove the four screws for the housing extension and the blower motor screws, then remove the extension and blower motor together to clear the floorboard.*
13   Installation is the reverse of removal.

12.12 Lower the motor, then rotate the motor assembly out of the housing

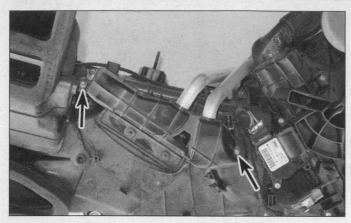

13.3a Remove the heater core cover screws . . .

13.3b . . . then separate the cover from the case - heater/evaporator case removed for clarity

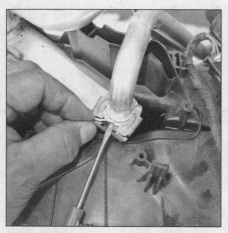

13.4a Use a small screwdriver to pry up the clip lock, then remove the clip and . . .

13.4b . . . separate the tubes from the heater core

13.5 Lift the heater core out of the case and replace the O-rings on the heater core tubes

13 Heater core - removal and installation

Warning: *The vehicles covered by this manual are equipped with a Supplemental Restraint System (SRS), more commonly known as airbags. Always disable the airbag system before working in the vicinity of any airbag system components to avoid the possibility of accidental deployment of the airbag, which could cause personal injury (see Chapter 12).*

Note: *The heater and evaporator cores are mounted in a large housing under the instrument panel. The instrument panel must be removed for access to the housing, and this is a difficult procedure for the home mechanic, involving many hard-to-find/reach fasteners and electrical connectors.*

Removal

1   Drain the cooling system (see Chapter 1). On diesel models, only drain the primary system.
2   Remove the instrument panel (see Chapter 11).
3   Remove the heater core cover screws and cover (see illustrations).
4   Disconnect the heater core tube clips at the heater core and disconnect the tubes from the heater core (see illustrations). Always replace the O-ring and clips with new ones.
5   Remove the heater core from the case (see illustration).

Installation

6   Installation is the reverse of removal, noting for the following:
a) *Replace the heater core tube O-rings and the tube clips with new ones.*
b) *Fill the cooling system with the correct type and amount of coolant (see Chapter 1).*
c) *Start the engine and check for leaks.*

14 Air conditioning and heater control assembly - removal and installation

Warning: *The vehicles covered by this manual are equipped with a Supplemental Restraint System (SRS), more commonly known as airbags. Always disable the airbag system before working in the vicinity of any airbag system components to avoid the possibility of accidental deployment of the airbag, which could cause personal injury (see Chapter 12).*

1   Disconnect the cable(s) from the negative battery terminal(s) (see Chapter 5).
2   Using a trim tool, pry out the power outlet panel and USB panel from the bottom corners of the center trim panel (see illustrations), then disconnect the electrical connectors.
3   Remove the instrument center trim panel fasteners (see illustration). Starting at the bottom, pry the panel outward (see illustration) and pull the panel out far enough to disconnect the electrical connectors to the electrical components (see Chapter 12), if necessary.
4   Disengage the lever-type electrical connector from the heating and air conditioning control module (see illustration).
5   Remove the heater and air conditioning control assembly retaining screws (see illustration).
6   Remove the control module from the panel.
7   Installation is the reverse of removal.
8   Verify that all controls operate correctly.

14.2a Pry out the power outlet panel . . .

14.2b . . . then the USB panel and disconnect the electrical connectors

14.3a Center trim panel mounting screw locations

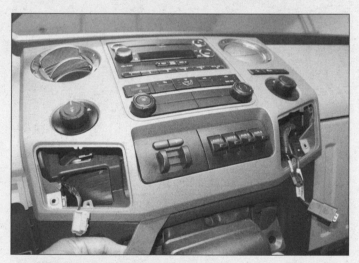

14.3b Start prying from the bottom of the panel to remove the panel

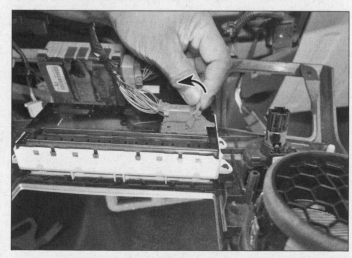

14.4 Depress the clip and rotate the lever back over the connector to release it from the module

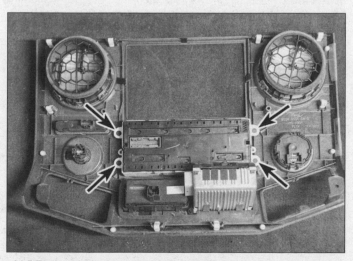

14.5 To detach the heater and air conditioning control assembly, remove these four screws

15.3 Unscrew the drier cap to access the snap-ring and inner receiver-drier plug

16.6 Unplug the electrical connector from the compressor clutch field coil (diesel model shown)

16.7 To disconnect the inlet and outlet lines from the compressor, remove the manifold nuts; be sure to replace the manifold O-rings before reattaching it (diesel model shown)

16.8 To remove the compressor, remove these bolts (diesel model shown)

15 Air conditioning receiver-drier - removal and installation

Warning: *The air conditioning system is under high pressure. DO NOT loosen any fittings or remove any components until after the system has been discharged. Air conditioning refrigerant should be properly discharged into an EPA-approved container at a dealer service department or an automotive air conditioning repair facility. Always wear eye protection when disconnecting air conditioning system fittings.*

Note: *These vehicles are not equipped with a traditional receiver-drier. Instead, a desiccant cartridge is built into the left side of the condenser. It should be replaced whenever the compressor is replaced or whenever a component has been replaced because of a leak in the system.*

1   Have the air conditioning system discharged by a licensed air conditioning technician.

2   Remove the condenser (see Section 17).

3   Remove the plastic receiver-drier cap (see illustration). Once the cap is removed, use a small screwdriver to remove the snap-ring, then unscrew the drier plug and pull the receiver/drier cartridge. Cap the opening.

4   Pull the desiccant cartridge from the tube.

5   Installation is otherwise the reverse of removal.

6   Take the vehicle to the shop that discharged it and have the system evacuated and recharged.

16 Air conditioning compressor - removal and installation

Removal

Warning: *The air conditioning system is under high pressure. DO NOT loosen any fittings or remove any components until after the system has been discharged. Air conditioning refrigerant should be properly discharged into an EPA-approved container at a dealer service department or an automotive air conditioning repair facility. Always wear eye protection when disconnecting air conditioning system fittings.*

1   Have the air conditioning system discharged by a licensed air conditioning technician.

2   Remove the drivebelt (see Chapter 1).

3   Raise the vehicle and support it securely on jackstands.

4   Remove the right-side inner fender splash shield (Chapter 11, Section 15).

5   Detach the steering damper from the steering linkage (see Chapter 10).

6   Disconnect the electrical connector from the compressor clutch field coil (see illustration).

7   Disconnect the compressor inlet and outlet lines from the compressor (see illustration). Remove and discard the old O-rings.

8   Remove the compressor mounting bolts (see illustration) and remove the compressor.

Installation

9   If a new compressor is being installed, follow the directions with the compressor regarding the draining of excess oil prior to installation.

10   The clutch may have to be transferred from the original to the new compressor.

11   Before reconnecting the inlet and outlet lines to the compressor, replace all manifold O-rings and lubricate them with refrigerant oil.

12   Have the system evacuated, recharged and leak tested by the shop that discharged it.

13   Installation is otherwise the reverse of removal.

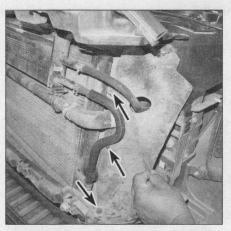

17.4 Remove the air deflector push pins - left side shown, right side similar

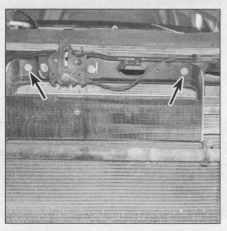

17.5 Power steering cooler mounting bolts

17.7a Disconnect the secondary radiator hose-to-condenser retainer . . .

17.7b . . . then loosen the secondary radiator hose clamp (1) and disengage the radiator-to-condenser retainer (2) and separate the hose

17.8a To disconnect the refrigerant inlet and outlet lines from the condenser, remove these nuts

17   Air conditioning condenser - removal and installation

Warning: *The air conditioning system is under high pressure. DO NOT loosen any fittings or remove any components until after the system has been discharged. Air conditioning refrigerant should be properly discharged into an EPA-approved container at a dealer service department or an automotive air conditioning repair facility. Always wear eye protection when disconnecting air conditioning system fittings.*

Warning: *Wait until the engine is completely cool before beginning this procedure.*

1   Have the air conditioning system discharged by a licensed air conditioning technician.

2   On diesel engine models, drain the secondary cooling system (see Chapter 1).

3   Remove the grille (see Chapter 11).

4   Remove the air deflector push-pin retainers (see illustration) and move the deflectors out of the way.

5   Remove the power steering cooler mounting bolts (see illustration) and secure the cooler out of the way. It is not necessary to disconnect the lines to the cooler.

6   On diesel engine models, remove the horns (see Chapter 12).

7   Loosen the secondary radiator hose clamp, then disconnect the hose retainers in front of the condenser and separate the hose (see illustrations).

8   Disconnect the refrigerant inlet and outlet line nuts, then detach the lines from the condenser (see illustrations). Remove and discard the old O-rings.

17.8b Pull the fittings off and replace the O-rings

17.9 Condenser mounting bolt locations

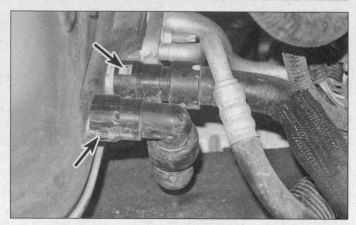

18.4 Working through the fenderwell, depress the retainers and detach the heater hoses from the heater core pipes

18.7a Remove the heater core/evaporator core case fasteners from the firewall from the right-side . . .

18.7b . . . lower section . . .

difficult procedure for the home mechanic, involving many hard-to-find/reach fasteners and electrical connectors. It is recommended that evaporator core replacement for these models be performed at a dealership.

Warning: *The air conditioning system is under high pressure. DO NOT loosen any fittings or remove any components until after the system has been discharged. Air conditioning refrigerant should be properly discharged into an EPA-approved container at a dealer service department or an automotive air conditioning repair facility. Always wear eye protection when disconnecting air conditioning system fittings.*

Warning: *Wait until the engine is completely cool before beginning this procedure.*

Removal

1   Have the air conditioning system discharged by a licensed air conditioning technician.

2   Remove the instrument panel (see Chapter 11).

3   Remove the right-side inner fender splash shield (see Chapter 11, Section 17).

4   Disconnect the heater hoses from the heater core (see illustration).

5   Remove the TXV refrigerant line manifold mounting nut (see Section 20) and move the lines out of the way.

6   On diesel models, disconnect the electric heater cables from the battery and battery tray.

7   Remove the heater core/evaporator core case nuts and bolts from the engine compartment side of the firewall (see illustrations).

8   Remove the heater core/evaporator core case fasteners from the interior side of the firewall (see illustrations) and remove the case.

9   Remove the right-hand temperature blend door motor (see illustrations).

10   Remove the instrument panel seal (see illustration), if not already removed, remove the TXV valve (see Section 20).

11   Remove the heater core tube and evaporator tube support (see illustrations).

18.7c . . . and center section

9   Remove the condenser mounting bolts (see illustration) and remove the condenser.

10   Remove the condenser. If you're going to reinstall the same condenser, store it with the line fittings facing up to prevent oil from draining out.

11   If you're going to install a new condenser, pour one ounce of refrigerant oil of the correct type into it prior to installation.

12   Before reconnecting the refrigerant lines to the condenser, be sure to coat a pair of new O-rings with refrigerant oil, install them in the refrigerant line fittings and then tighten the condenser inlet and outlet nuts to the torque listed in this Chapter's Specifications.

13   Installation is otherwise the reverse of removal.

14   On diesel models, refill the secondary cooling system (see Chapter 1).

15   Have the system evacuated, recharged and leak tested by the shop that discharged it.

---

18   Air conditioning evaporator core - removal and installation

---

Note: *The heater and evaporator cores are mounted in a large housing under the instrument panel. The instrument panel must be removed for access to the housing, and this is a*

18.8a Remove the case fasteners from inside the vehicle . . .

18.8b . . . then remove the heater core/evaporator core housing

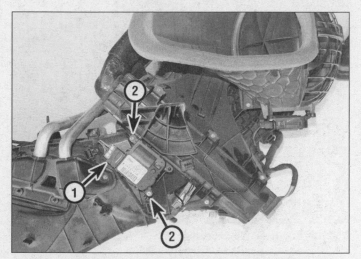

18.9a Disconnect the electrical connector (1) to the blend door motor and remove the motor fasteners (2) . . .

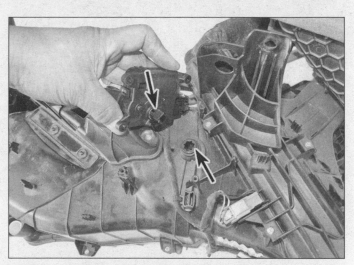

18.9b . . . and lift the blend door motor off, noting how the blend door motor shaft engages with the blend door

18.10 Lift the instrument panel seal off of the evaporator and heater core lines

18.11a Remove the support mounting screws . . .

18.11b . . . then rotate the support around and off of the tubes

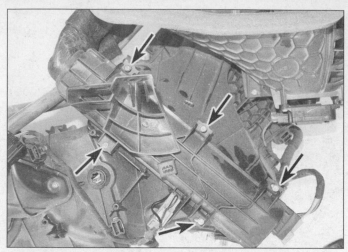

18.12 Remove the evaporator core cover fasteners and cover

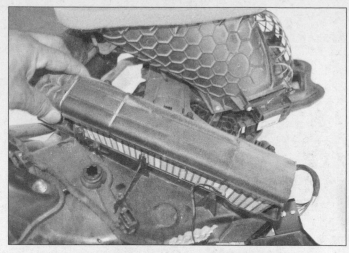

18.13 Lift the evaporator core from the case

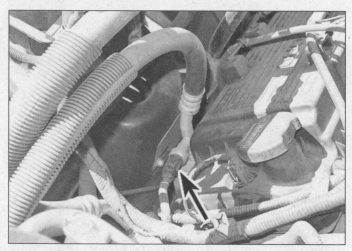

19.1 To remove the air conditioning pressure cycling switch, unplug the electrical connector and unscrew the switch from the refrigerant line (gasoline engine model shown, diesel similar)

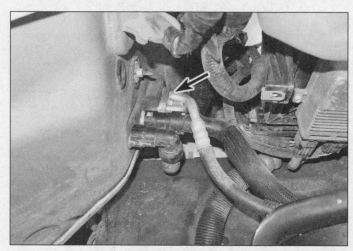

20.2 The Thermostatic Expansion Valve is located on the firewall - as viewed from below

12   Remove the evaporator core cover (see illustration).

13   Remove the evaporator core from the case (see illustration).

14   Remove the evaporator temperature sensor from the evaporator core.

15   Installation is the reverse of removal.

16   Refill the cooling system (see Chapter 1).

17   Have the system evacuated, recharged and leak tested by the shop that discharged it.

19   Air conditioning pressure cycling switch - replacement

Note: *A Schrader valve prevents refrigerant loss during pressure cycling switch replacement.*

Note: *The pressure cycling switch threaded into a fitting on the compressor-to-condenser discharge line at the right side of the engine compartment, near the right-front corner of the engine.*

1   Unplug the electrical connector from the pressure cycling switch (see illustration).

2   Unscrew the pressure cycling switch from the refrigerant line.

3   Lubricate the switch O-ring with clean refrigerant oil of the correct type.

4   Screw the new switch onto the refrigerant line fitting until hand tight, then tighten it securely.

5   Reconnect the electrical connector.

20   Air conditioning thermostatic expansion valve (TXV) - general information

Warning: *The air conditioning system is under high pressure. DO NOT loosen any hose fittings or remove any components until the system has been discharged. Air conditioning refrigerant must be properly discharged into an EPA-approved recovery/recycling unit by a dealer service department or an automotive air conditioning repair facility. Always wear eye protection when disconnecting air conditioning system fittings.*

1   There are several ways that air conditioning systems convert the high-pressure liquid refrigerant from the compressor to lower-pressure vapor. The conversion takes place at the air conditioning evaporator; the evaporator is chilled as the refrigerant passes through, cooling the airflow through the evaporator for delivery to the vents. The conversion is usually accomplished by a sudden change in the tubing size. Many vehicles have a removable controlled orifice in one of the refrigerant lines at the firewall.

2   The models covered by this manual use a Thermostatic Expansion Valve (TXV) that accomplishes the same thing as a controlled orifice (see illustration).

3   Have the air conditioning system discharged by a licensed air conditioning technician.

20.6 Remove the air conditioning line mounting nut

20.7 Valve mounting bolt locations - evaporator case removed for clarity

4 Loosen the right-front wheel lug nuts, then raise the vehicle and support it securely on jackstands.

5 Remove the right side inner fender splash shield (see Chapter 11, Section 15).

6 Disconnect the refrigerant lines from the TXV at the firewall (see illustration).

7 Remove the refrigerant line mounting nut, separate the lines and remove the O-rings (see illustration).

8 Installation is the reverse of removal.

9 Have the system evacuated, recharged and leak tested by the shop that discharged it.

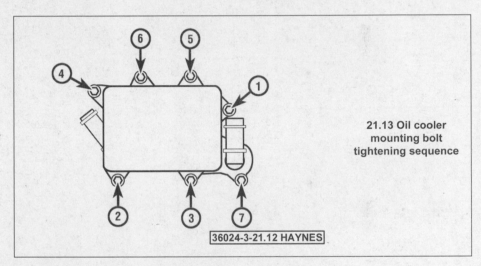

21.13 Oil cooler mounting bolt tightening sequence

36024-3-21.12 HAYNES

21 Oil cooler - removal and installation

Warning: *Wait until the engine is completely cool before beginning this procedure.*

1 Raise the vehicle and support it securely on jackstands.

2 Drain the engine coolant (see Chapter 1).

Gasoline engines

3 Remove the oil filter (see Chapter 1).

4 Disconnect the two coolant hoses from the oil cooler.

5 The oil cooler is attached to the engine by the same large threaded tube that the oil filter is screwed onto. Remove the oil cooler threaded tube and remove the oil cooler assembly.

6 Installation is the reverse of removal. Be sure to use a new O-ring gasket between the oil cooler and the engine.

7 Refill the cooling system and check the engine oil level (see Chapter 1).

Diesel engine

8 Disconnect the two coolant hoses from the oil cooler.

9 On 2011 models, remove the lower oil pan (see Chapter 2B, Section 14) and remove the oil cooler mounting nut that is accessible once the oil pan has been removed.

Note: *On 2012 late-build and later models, the internal nut is no longer used on the oil cooler and the lower oil pan does not have to be removed.*

10 Remove the oil cooler mounting bolts and remove the oil cooler.

11 Remove and discard the old gaskets.

12 On 2011 and 2012 early-build models, install the nut and tighten it to the torque listed in this Chapter's Specifications.

13 Be sure to use new gaskets and tighten the oil cooler nut, then the bolts (in sequence) (see illustration) to the torque listed in this Chapter's Specifications.

14 Installation is otherwise the reverse of removal.

15 Refill the cooling system and engine oil (if the pan was removed), or check the engine oil level (see Chapter 1).

Notes

Chapter 4 Part A
Fuel and exhaust systems - gasoline engines

Contents

Specifications

Fuel system

Fuel system pressure	
Key on, engine off (KOEO)	51 to 62 psi
Engine running	51 to 62 psi
Fuel system hold pressure (after 5 minutes)	Less than 5 psi loss
Injector resistance (approximate)	Unspecified*

Although unspecified, testing can be done for comparison purposes and to check for an open or shorted coil.

Torque specifications   Ft-lbs (unless otherwise indicated)

Note: *One foot-pound (ft-lb) of torque is equivalent to 12 inch-pounds (in-lbs) of torque. Torque values below approximately 15 foot-pounds are expressed in inch-pounds, because most foot-pound torque wrenches are not accurate at these smaller values.*

Throttle body bolts	
Step 1	106 in-lbs
Step 2	Tighten an additional 60 degrees
Fuel rail mounting bolts	
Step 1	89 in-lbs
Step 2	Tighten an additional 90 degrees
Fuel tank strap bolts	
Side tank (midship)	30
Rear tank	53
Fuel tank skid plate bolts (rear tank)	66
Fuel tank shield bolts (midship tank)	16

1 General information and precautions

Fuel system warnings

Warning: *Gasoline is extremely flammable and repairing fuel system components can be dangerous. Consider your automotive repair knowledge and experience before attempting repairs, which may be better suited for a professional mechanic.*

a) *Don't smoke or allow open flames or bare light bulbs near the work area*

b) *Don't work in a garage with a gas-type appliance (water heater, clothes dryer)*

c) *Use fuel-resistant gloves. If any fuel spills on your skin, wash it off immediately with soap and water*

d) *Clean up spills immediately*

e) *Do not store fuel-soaked rags where they could ignite*

f) *Prior to disconnecting any fuel line, you must relieve the fuel pressure (see Section 3)*

g) *Wear safety glasses*

h) *Have a proper fire extinguisher on hand*

Fuel system

1 This Chapter covers the removal and installation procedures for the important parts of the air intake, fuel and exhaust systems. Because emission control systems are integral parts of the engine management system, there are many cross-references to Chapter 6A. Information on the engine management system, information sensors and output actuators is also in Chapter 6A.

2 The air intake system consists of the air filter housing, the air intake duct, the throttle body, and the intake manifold. Incoming air passes through the air filter element, the Mass Air Flow (MAF) sensor, the air intake duct, the throttle body, the intake manifold plenum and the intake manifold runners before being mixed with fuel sprayed into the intake ports by the fuel injectors.

3 The Sequential Fuel Injection (SFI) system consists of the fuel tank, a 2-speed electric fuel pump/fuel level sending unit module mounted inside the tank, the fuel pressure regulator (integral with the fuel pump module), a fuel pump flow control module, the fuel rail, the fuel injectors, and the metal and flexible fuel lines that connect the various components of the SFI system.

4 The lifetime fuel filter is part of the fuel pump module and is not serviceable.

5 Fuel is circulated from the fuel pump to the fuel rail through fuel lines running along the underside of the vehicle. Various sections of the fuel line are either rigid metal or nylon, or flexible fuel hose. The various sections of the fuel hose are connected either by quick-connect fittings or threaded metal fittings.

Exhaust system

6 The exhaust system consists of the exhaust manifold, catalytic converter, muffler, tailpipe and all connecting pipes, flanges and clamps. The catalytic converter is an emission control device added to the exhaust system to reduce pollutants.

2 Troubleshooting

Fuel pump

1 The fuel pump is located inside the fuel tank. Sit inside the vehicle with the windows closed, turn the ignition key to On (not Start) and listen for the sound of the fuel pump as it's briefly activated. You will only hear the sound for a second or two, but that sound tells you that the pump is working. Alternatively, have an assistant listen at the fuel filler cap.

2 If the pump does not come on, check the fuel pump fuses and fuel pump relay (see illus- tration). If the fuses and relay are okay, check the wiring back to the fuel pump. If the fuses, relay and wiring are okay, the fuel pump is probably defective. If the pump runs continu- ously with the ignition key in the On position, the Powertrain Control Module (PCM) is prob- ably defective. Have the PCM checked by a professional mechanic.

Fuel pump inertia switch

3 If the fuel pump does not come on, also verify that the fuel pump inertia switch has not been tripped. The fuel pump inertia switch turns off the fuel pump in the event of a collision or similar incident. The switch is located behind the passenger's side end of the dash (see illustra- tion). When the switch is triggered, the plunger at the top is in the up position. To reset the switch, press the plunger down until it clicks.

Fuel injection system

Note: *The following procedure is based on the assumption that the fuel pump is working and the fuel pressure is adequate (see Section 4).*

4 Check all electrical connectors that are related to the system. Check the ground wire connections for tightness.

5 Verify that the battery is fully charged (see Chapter 5).

6 Inspect the air filter element (see Chap- ter 1).

7 Check all fuses related to the fuel sys- tem (see Chapter 12).

8 Check the air induction system between the throttle body and the intake manifold for air leaks. Also inspect the condition of all vac- uum hoses connected to the intake manifold and to the throttle body.

9 Remove the air intake duct from the throttle body and look for dirt, carbon, varnish, or other residue in the throttle body, particu- larly around the throttle plate. If it's dirty, clean it with carb cleaner, a toothbrush and a clean shop towel.

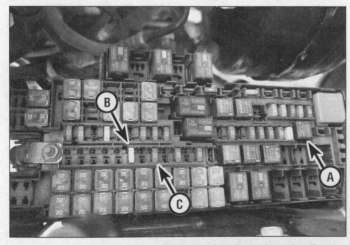

2.2 Fuel system related underhood fuse/relay box details

A   *Fuel pump relay (supplies power to the fuel pump control module)*
B   *Fuel pump relay fuse F66 (20 amp; always hot)*
C   *Fuel pump relay fuse F68 (10 amp; always hot)*

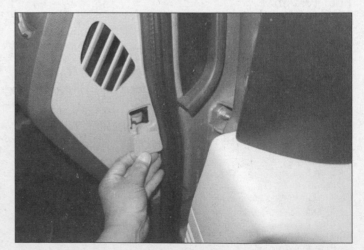

2.3 Pry off the small trim cover and depress the button on the top of the inertia switch

2.10 An automotive stethoscope is used to listen to the fuel injectors in operation

3.1 The FPCM is located on the left frame rail, under the cab

10   With the engine running, place an automotive stethoscope against each injector, one at a time, and listen for a clicking sound that indicates operation (see illustration).
Warning: *Stay clear of the drivebelt and any rotating or hot components.*
11   If you can hear the injectors operating, but the engine is misfiring, the electrical circuits are functioning correctly, but the injectors might be dirty or clogged. Try a commercial injector cleaning product (available at auto parts stores). If cleaning the injectors doesn't help, replace the injector(s).
12   If an injector is not operating (it makes no sound), disconnect the injector electrical connector and measure the resistance across the injector terminals with an ohmmeter. Compare this measurement to the other injectors. If the resistance of the non-operational injector is quite different from the other injectors, replace it.
13   If the injector is not operating, but the resistance reading is within the range of resistance of the other injectors, the PCM or the circuit between the PCM and the injector might be faulty.

---

3   Fuel pressure relief procedure

Warning: *Gasoline is extremely flammable, so take extra precautions when you work on any part of the fuel system. Don't smoke or allow open flames or bare light bulbs near the work area, and don't work in a garage where a gas-type appliance (such as a water heater or a clothes dryer) is present. Since gasoline is carcinogenic, wear fuel-resistant gloves when there's a possibility of being exposed to fuel, and, if you spill any fuel on your skin, rinse it off immediately with soap and water. Mop up any spills immediately and do not store fuel-soaked rags where they could ignite. The fuel system is under constant pressure, so, if any fuel lines are to be disconnected, the fuel pressure in the system must be relieved first. When you perform any kind of work on the fuel*

system, wear safety glasses and have a Class B type fire extinguisher on hand.
Note: *After the fuel pressure has been relieved, it's a good idea to lay a shop towel over any fuel connection to be disassembled, to absorb the residual fuel that may leak out when servicing the fuel system.*
1   Unplug the Fuel Pump Control Module (FPCM) electrical connector (see illustration).
2   Start the engine and allow it to run until it stops. This should take only a few seconds.
3   The fuel system pressure is now relieved. Disconnect the cable from the negative terminal of the battery before performing any work on the fuel system.
4   When you're finished working on the fuel system, simply plug the electrical connector back into the FPCM.

---

4   Fuel pump/fuel pressure - check

Warning: *Gasoline is extremely flammable, so take extra precautions when you work on any part of the fuel system. See the Warning in Section 3.*
Note: *After the fuel pressure has been relieved, it's a good idea to lay a shop towel over any fuel connection to be disassembled, to absorb the residual fuel that may leak out when servicing the fuel system.*

General checks

Note: *The inertia switch is an electrical device wired into the fuel pump circuit that will shut down power to the fuel pump in an accident (see Section 3). Be sure to check that the inertia switch is activated and in working order if the fuel pump is not receiving the proper voltage.*
Note: *The fuel pump relay is equipped with a primary and secondary voltage circuit. The primary circuit is controlled by the PCM and the secondary circuit is linked directly to battery voltage from the ignition switch. With the ignition switch On (engine not running), the PCM will ground the relay for one second. During cranking, the PCM grounds the fuel pump re-*

lay as long as the reference signal from the ignition system is received. If there are no reference pulses, the fuel pump will shut off after two or three seconds.
1   If you suspect insufficient fuel delivery, check the following items first:
a) *Check the battery and make sure it's fully charged (see Chapter 5).*
b) *Check the fuel pump fuse.*
c) *Check the fuel filter for restriction.*
d) *Inspect all fuel lines to ensure that the problem is not simply a leak in a line.*
2   Verify the fuel pump actually runs. Place the transmission in Park (automatic) or Neutral (manual) and apply the parking brake. Have an assistant turn the ignition switch to On - you should hear a brief whirring noise as the pump comes on and pressurizes the system. If there is no response from the fuel pump (makes no sound), check the fuel pump electrical circuit. If the fuel pump runs, but a fuel system problem is suspected, continue with the fuel pump pressure check.
Note: *The fuel pump is easily heard through the gas tank filler neck.*
3   If the pump does not turn on (makes no sound) with the ignition switch in the On position, check the ignition fuse located in the engine compartment fuse center. Also, check the fuel pump relay. Refer to Chapter 12 for testing relays.
Note: *The fuel pump relay is located in the fuse/relay center in the engine compartment (see Section 2).*
4   If the relays are good and the fuel pump does not operate, check the fuel pump circuit. If the wiring and the connectors are good, either the FPCM (see Section 9) or the fuel pump (see Section 8) is faulty.
Note: *The PCM commands the fuel pump control module (FPCM), which in turn operates the fuel pump at either 8 or 12 volts. If you suspect that the fuel pump control module is faulty, have the system diagnosed at a dealer or a qualified repair shop. Refer to Chapter 6A for information on replacement of this component.*

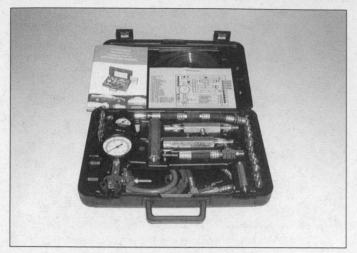

4.6 A typical fuel pressure gauge, with hoses and fittings suitable for tee-ing into the fuel system between the fuel delivery line and the fuel rail

4.7 Location of the fuel line quick-connect fitting on the left-side fuel rail

Fuel pump pressure check

Note: *Before proceeding, obtain a fuel pressure gauge capable of measuring fuel pressure well above the specified operating range of the fuel system you're going to test, and you'll also need fittings suitable for tee-ing the gauge into the fuel system between the fuel delivery line and the fuel rail.*

5   Relieve the fuel system pressure (see Section 3).

6   In addition to a fuel pressure gauge capable of reading fuel pressure up to 70 psi, you'll need a hose and an adapter suitable for tee-ing into the fuel system at the quick-connect fitting between the fuel delivery hose and the fuel rail (see illustration).

7   Disconnect the quick-connect fitting at the connection between the fuel delivery hose and the fuel rail (if you're unfamiliar with quick-connect fittings, refer to Section 5). Tee in the fuel pressure gauge between the fuel delivery hose and the fuel rail (see illustration).

8   Turn off all the accessories, then start the engine and let it idle. The fuel pressure should be within the operating range listed in this Chapter's Specifications. If the pressure reading is within the specified range, the system is operating correctly.

9   If the fuel pressure is higher than specified, then the pump, the Fuel Pump Driver Module (FPDM), the Powertrain Control Module (PCM) or the circuit connecting these components is probably defective. But checking this circuit is beyond the scope of the home mechanic, so have the circuit checked by a professional.

10   If the fuel pressure is lower than specified, inspect the fuel delivery lines and hoses for an obstruction or a kink. Also inspect all fuel delivery line and hose quick-connect fittings for leaks. If the lines, hoses, and connec-

tions are all in good shape, remove the fuel pump/fuel level sensor assembly (see Section 8) and inspect the fuel pump inlet strainer for restrictions. If everything else is okay, replace the fuel pump.

11   Turn the ignition switch to Off, wait five minutes and recheck the pressure on the gauge. Compare the reading with the hold pressure listed in this Chapter's Specifications. If the hold pressure is less than specified:

a) *The fuel delivery line or a quick-connect fitting might be leaking.*

b) *A fuel injector (or injectors) may be leaking.*

c) *The fuel pump might be defective.*

12   After the testing is complete, relieve the fuel pressure (see Section 3), remove the fuel pressure gauge and reconnect the fuel delivery line to the fuel rail.

---

5   Fuel lines and fittings - general information and disconnection

---

Warning: *Gasoline is extremely flammable. See* Fuel system warnings *in Section 1.*

1   Relieve the fuel pressure before servicing fuel lines or fittings (see Section 3), then disconnect the cable from the negative battery terminal (see Chapter 5) before proceeding.

2   The fuel supply line connects the fuel pump in the fuel tank to the fuel rail on the engine. The Evaporative Emission (EVAP) system lines connect the fuel tank to the EVAP canister and connect the canister to the intake manifold.

3   Whenever you're working under the vehicle, be sure to inspect all fuel and evaporative emission lines for leaks, kinks, dents and other damage. Always replace a dam-

aged fuel or EVAP line immediately.

4   If you find signs of dirt in the lines during disassembly, disconnect all lines and blow them out with compressed air. Inspect the fuel strainer on the fuel pump pick-up unit for damage and deterioration.

Steel tubing

5   It is critical that the fuel lines be replaced with lines of equivalent type and specification.

6   Some steel fuel lines have threaded fittings. When loosening these fittings, hold the stationary fitting with a wrench while turning the tube nut.

Plastic tubing

7   When replacing fuel system plastic tubing, use only original equipment replacement plastic tubing.

Caution: *When removing or installing plastic fuel line tubing, be careful not to bend or twist it too much, which can damage it. Also, plastic fuel tubing is NOT heat resistant, so keep it away from excessive heat.*

Flexible hoses

8   When replacing fuel system flexible hoses, use only original equipment replacements.

9   Don't route fuel hoses (or metal lines) within four inches of the exhaust system or within ten inches of the catalytic converter. Make sure that no rubber hoses are installed directly against the vehicle, particularly in places where there is any vibration. If allowed to touch some vibrating part of the vehicle, a hose can easily become chafed and it might start leaking. A good rule of thumb is to maintain a minimum of 1/4-inch clearance around a hose (or metal line) to prevent contact with the vehicle underbody.

Disconnecting Fuel Line Fittings

Two-tab type fitting; depress both tabs with your fingers, then pull the fuel line and the fitting apart

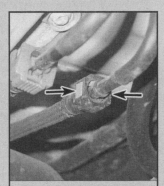

On this type of fitting, depress the two buttons on opposite sides of the fitting, then pull it off the fuel line

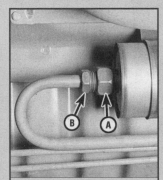

Threaded fuel line fitting; hold the stationary portion of the line or component (A) while loosening the tube nut (B) with a flare-nut wrench

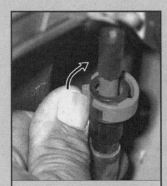

Plastic collar-type fitting; rotate the outer part of the fitting

Metal collar quick-connect fitting; pull the end of the retainer off the fuel line and disengage the other end from the female side of the fitting . . .

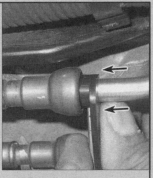

. . . insert a fuel line separator tool into the female side of the fitting, push it into the fitting and pull the fuel line off the pipe

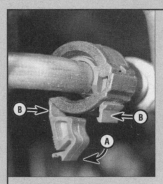

Some fittings are secured by lock tabs. Release the lock tab (A) and rotate it to the fully-opened position, squeeze the two smaller lock tabs (B) . . .

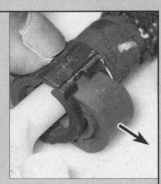

. . . then push the retainer out and pull the fuel line off the pipe

Spring-lock coupling; remove the safety cover, install a coupling release tool and close the tool around the coupling . . .

. . . push the tool into the fitting, then pull the two lines apart

Hairpin clip type fitting: push the legs of the retainer clip together, then push the clip down all the way until it stops and pull the fuel line off the pipe

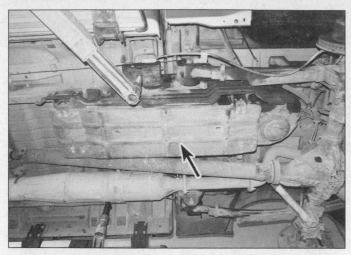

6.7 Fuel tank shield (midship-mounted tank)

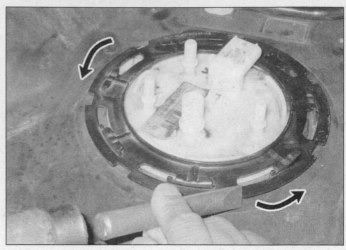

8.2 Loosen the fuel pump module lock ring by tapping it counterclockwise with a hammer and brass punch

6   Fuel tank - removal and installation

Warning: *Gasoline is extremely flammable, so take extra precautions when you work on any part of the fuel system. See the Warning in Section 3.*
Note: *Don't begin this procedure until the gauge indicates that the tank is empty or nearly empty. If the tank must be removed when it's full (for example, if the fuel pump malfunctions), siphon any remaining fuel from the tank prior to removal.*

Removal

1   Relieve the fuel system pressure:

a) On gasoline models, see Section 3.
b) On diesel models, see Chapter 4B, Section 3.

2   Remove the fuel filler cap to relieve fuel tank pressure.
3   Disconnect the cable(s) from the negative terminal(s) of the battery(ies) (see Chapter 5, Section 3).
4   If the tank is full or nearly full, siphon the fuel into an approved container using a siphoning kit available at most auto parts stores.
Warning: *DO NOT start the siphoning action by mouth! Use a siphoning kit (available at most auto parts stores).*
5   Raise the vehicle and support it securely on jackstands.
6   Depending on the model and optional equipment, the fuel tank may be located either forward of the rear axle (midship) or behind the rear axle (aft).
7   On side mount (midship) tanks, remove the fuel tank shield (see illustration). On rear mount tanks, the skid plate is removed with the tank.
8   Remove the fuel tank filler hose and vapor hose from the fuel filler neck and the fuel tank.

9   Place a floor jack under the tank. Raise the jack until it's supporting the tank.
10   Disconnect the fuel lines and EVAP vapor lines (see Section 5).
Note: *Disconnect the fuel tank pressure sensor, if equipped.*
11   Disconnect the electric fuel pump and sending unit electrical connector with a screwdriver (see Section 8). It may be necessary to lower the fuel tank slightly to access the components located directly on top of the fuel tank.
12   Remove the bolts that retain the fuel tank straps. The tank may be equipped with either 4 or 6 tank straps.
13   Lower the tank far enough to unplug any vapor lines or wire harness brackets that may be difficult to reach when the fuel tank is in the vehicle.
14   Slowly lower the jack while steadying the tank. Remove the tank from the vehicle.
15   If you're replacing the tank, or having it cleaned or repaired, refer to Section 7.
16   Refer to Section 8 to remove and install the fuel pump (gasoline) or sending unit (diesel).

Installation

17   Installation is the reverse of removal. Clean engine oil can be used as an assembly aid when pushing the fuel filler neck back into the tank.
18   Carefully angle the fuel tank filler neck into the filler pipe assembly and lift the tank into place. Torque the tank straps to the specifications in this Chapter's Specifications.
19   Install the shield to the tank.
20   Turn the ignition on and check for leaks.

7   Fuel tank - cleaning and repair

1   There are two different types of fuel tanks on these models: steel and plastic.
2   Fuel tanks may be steam-cleaned to

remove sediment or rust in the bottom of the tank. Remove the fuel tank sending unit/fuel pump and vapor valve prior to cleaning. Allow plenty of time for the tank to air dry before returning it to service.
3   Repairs to the steel fuel tank or filler pipe should be performed by a professional with the proper training to carry out this critical and potentially dangerous job. Even after cleaning and flushing, explosive fumes can remain and could explode during repair of the tank.
4   The plastic (polyethylene) fuel tank cannot be repaired. No reliable repair procedures are available to correct leaks or damage. Fuel tank replacement is the only approved service.
5   If the fuel tank is removed from the vehicle, it should not be placed in an area where sparks or open flames could ignite the fumes coming out of the tank. Be especially careful inside garages where a gas-type appliance is located, because the appliance could cause an explosion.
6   Whenever the fuel tank is steam-cleaned or otherwise serviced, the vapor valve assembly should be replaced. All grommets and seals must be replaced to prevent possible leakage.

8   Fuel pump/fuel pump module - removal and installation

Warning: *Gasoline is extremely flammable, so take extra precautions when you work on any part of the fuel system. See the Warning in Section 3.*
1   Remove the fuel tank (refer to Section 6).
2   Remove the fuel pump assembly from the fuel tank by turning the retaining ring counterclockwise, until it's loose (see illustration). A special tool designed for this purpose is available, but a hammer and brass punch

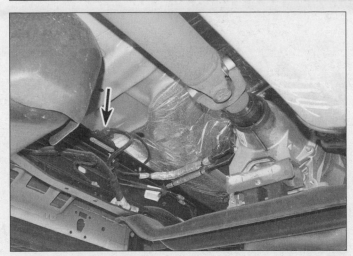

9.1 The FPCM is located on the left-side frame rail, under the cab

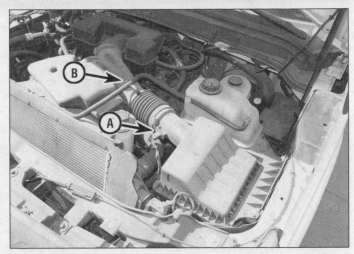

12.2 Location of MAF (A) and coolant reservoir hose (B)

will work. Remove the retaining ring.

Warning: *Use only a brass punch to avoid creating sparks.*

Note: *A large pair of channel lock pliers will also sometimes work.*

3   Remove the old seal ring, if equipped and discard it.

4   Separate the fuel pump from the assembly.

5   Clean the fuel pump mounting flange, the tank mounting surface and seal ring groove.

6   Installation is the reverse of removal. Apply a thin coat of heavy grease to the new seal ring to hold it in place during assembly.

7   Use care when inserting the fuel pump module into the tank. After installation, turn the ignition on and check for leaks.

9   Fuel Pump Control Module (FPCM) - replacement

Note: *On models with a side mounted fuel tank (pickups), the FPCM is located on top of the frame rail near the middle of the vehicle. On models with a rear mounted tank (cab and chassis), the FPCM is located near the charcoal canister.*

1   Locate the FPCM and disconnect the electrical connector (see illustration).

2   Remove the fasteners attaching the FPCM to the vehicle and remove the FPCM.

3   Installation is reverse of removal. No programming is necessary.

10   Fuel pressure regulator - removal and installation

Note: *All models are equipped with a returnless fuel system. The fuel pressure regulator is part of the fuel pump module. If the fuel pressure regulator is diagnosed as faulty, the fuel pump module will need to be replaced. Refer to Section 8 for more information.*

11   Fuel level sending unit - replacement

Warning: *Gasoline is extremely flammable, so take extra precautions when you work on any part of the fuel system. See the Warning in Section 3.*

Note: *Diesel models are equipped with a separate fuel level sensor mounted in the fuel tank and an externally mounted fuel pump. The fuel tank must be removed to access the fuel level sending unit assembly.*

Note: *Later models have a fuel level sender that is part of the fuel pump module. The sender is not replaceable by itself.*

1   Remove the fuel tank (see Section 6).

2   Remove the fuel pump/fuel level sending unit assembly (see Section 8). Carefully angle the assembly out of the opening without damaging the fuel level float located at the bottom of the assembly.

3   Remove the electrical connectors for the fuel level sending unit.

4   Disassemble the fuel level sending unit from the fuel pump.

5   Installation is the reverse of removal. Be sure to install a new rubber gasket.

6   Turn the ignition on and check for leaks.

12   Air filter housing and expansion resonator - removal and installation

Warning: *Wait until the engine is completely cool before beginning this procedure.*

Note: *The air filter housing base and coolant reservoir are one unit.*

1   Refer to Chapter 1 and remove the air filter.

2   Disconnect the MAF sensor and detach the coolant reservoir hose from the bracket on the air intake (see illustration).

3   Loosen the air intake tube clamp at the

expansion resonator before the throttle body and disconnect the air intake tube.

4   Disconnect the brake booster vacuum line, crankcase vent tube and detach the vacuum hose from the rear of the expansion resonator.

5   To remove the expansion resonator, remove the two bolts attaching the resonator to the intake manifold, then loosen the air intake tube clamp at the throttle body and disconnect the resonator.

6   To remove the air filter housing base/coolant expansion tank, drain the engine coolant to a point below the tank (see Chapter 1), then detach the coolant hoses at the tank. Remove the fasteners and remove the air filter housing base and coolant expansion tank.

7   Installation is the reverse of removal. Top off the coolant as necessary.

13   Throttle body - removal and installation

Removal

1   Remove the air intake expansion resonator (see Section 12).

2   Detach the Electronic Throttle Control (ETC) electrical connector from the throttle body.

3   On all models, remove the four throttle body mounting bolts and remove the throttle body.

4   Discard the O-ring gasket.

5   Clean the sealing surfaces. If scraping is necessary, be careful not to damage the sealing surfaces or allow material to drop into the intake manifold. Install a new O-ring gasket.

Installation

6   Installation is the reverse of removal. Be sure to tighten the throttle body mounting nuts to the torque listed in this Chapter's Specifications.

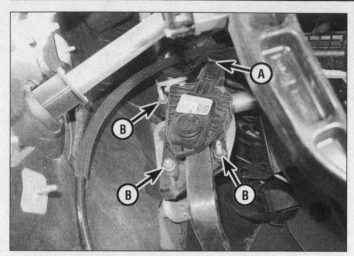

14.3 Slide out the locking tab (A) and disconnect the electrical connector, then remove the APP sensor mounting fasteners (B, adjustable pedal assembly shown)

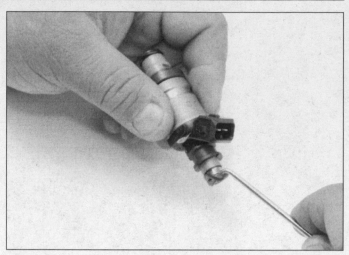

15.10a Remove the O-ring from the top of the fuel injector . . .

15.10b . . . then remove the lower O-ring from the injector

14   Accelerator Pedal Position (APP) sensor - replacement

Note: *All models are "drive-by-wire" and do not have an accelerator cable. The APP sensor communicates the desired pedal position to the PCM, the PCM then actuates the throttle body to the desired throttle angle.*

1   The APP sensor is used to monitor the driver's demand for power. The PCM adjusts the amount of fuel injected into the cylinders according to the position of the pedal. A problem in the APP sensor circuit will set a diagnostic trouble code (see Chapter 6A, Section 2 or Chapter 6B, Section 2).
2   Disconnect the APP sensor electrical connector.
3   Remove the fasteners attaching the sensor to its bracket and remove the APP sensor and accelerator pedal (see illustration).
4   Installation is reverse of removal.

15   Fuel rails and injectors - removal and installation

Warning: *Gasoline is extremely flammable, so take extra precautions when you work on any part of the fuel system. See the Warning in Section 3.*

Removal

1   Relieve the fuel pressure (see Section 3).
2   Disconnect the cable from the negative terminal of the battery (see Chapter 5, Section 3).
3   Remove the air filter housing (see Section 12).
4   Disconnect the electrical connectors and spark plug wires from the ignition coils.
5   Disconnect the fuel injector electrical connectors.
6   Disconnect the fuel line from the fuel rail (see Section 4, illustration 4.7, and Section 5).
7   Remove the fuel rail mounting bolts.
8   Carefully remove the fuel rail with the fuel injectors attached as an assembly. Use a rocking, side-to-side motion while lifting to remove the injectors from the fuel rail.
9   Remove the injector clips and pull the injectors from the fuel rail.

Installation

Caution: *Even if you only removed the fuel rail assembly to replace a single injector or a leaking O-ring, it's a good idea to remove all of the injectors from the fuel rail and replace all of the O-rings at the same time.*
10   Remove the O-rings from the injecors (see illustrations).
11   Inspect the injector plastic "hat" (covering the injector pintle) and washer for signs of deterioration. Replace as required. If the hat

is missing, look for it in the intake manifold.
12   Ensure that the injector caps are clean and free of contamination.
13   Install the injectors to the fuel rail. Ensure the injector clips retain the injectors securely.
14   Secure the fuel rail assembly with the retaining bolts and tighten them to the torque listed in this Chapter's Specifications.
15   The remainder of installation is the reverse of removal.
16   Turn the ignition on and check for leaks.

16   Exhaust system servicing - general information

Warning: *Inspection and repair of exhaust system components should be done only after enough time has elapsed after driving the vehicle to allow the system components to cool completely. Also, when working under the vehicle, make sure it is securely supported on jackstands.*

1   The exhaust system consists of the exhaust manifold, the catalytic converter, the resonator, exhaust pipe, muffler and all brackets, hangers and clamps. The exhaust system is attached to the body with mounting brackets and rubber hangers (see illustrations). If any of the parts are damaged or deteriorated, excessive noise and vibration will be transmitted to the body.
2   Conducting regular inspections of the exhaust system will keep it safe and quiet. Look for any damaged or bent parts, open seams, holes, loose connections, excessive corrosion or other defects which could allow exhaust fumes to enter the vehicle. Also check the catalytic converter when you inspect the exhaust system (see Chapter 6A). Deteriorated exhaust system components should not be repaired; they should be replaced with new parts.

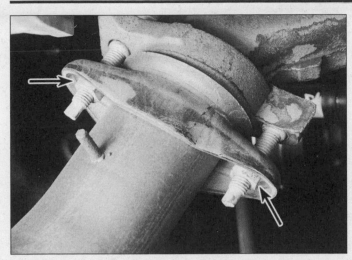

16.1a Be sure to apply penetrating fluid to the exhaust flange nuts before attempting to remove them

16.1b Check the condition of the rubber hangers that support the exhaust system

3   If the exhaust system components are extremely corroded or rusted together, they will probably have to be cut from the exhaust system. The convenient way to accomplish this is to have a muffler repair shop remove the corroded sections with a cutting torch. If, however, you want to save money by doing it yourself (and you don't have a welding outfit with a cutting torch), simply cut off the old components with a hacksaw. If you have compressed air, special pneumatic cutting chisels can also be used. If you decide to tackle the job at home, be sure to wear safety goggles to protect your eyes from metal chips and work gloves to protect your hands.

4   Here are some simple guidelines to follow when repairing the exhaust system:

a) *Work from the back to the front when removing exhaust system components.*

b) *Apply penetrating oil to the exhaust system component fasteners to make them easier to remove.*

c) *Use new gaskets, hangers and clamps when installing exhaust systems components.*

d) *Apply anti-seize compound to the threads of all exhaust system fasteners at reassembly.*

e) *Be sure to allow sufficient clearance between newly installed parts and all points on the underbody to avoid overheating the floor pan and possibly damaging the interior carpet and insulation. Pay particularly close attention to the catalytic converter and heat shield.*

Notes

Chapter 4 Part B
Fuel and exhaust systems - diesel engine

Contents

Specifications

Fuel system

Fuel system (low side) pressure	
Key on, engine off (KOEO)	57 to 73 psi *
Engine running	57 to 73 psi
High-pressure fuel pump gear backlash	0.0007-0.0074 in (0.020-0.190 mm)

* 30 seconds after turning the engine off, the fuel pressure will begin to bleed down to 0 psi. This is normal.

Torque specifications   Ft-lbs (unless otherwise indicated)

Note: *One foot-pound (ft-lb) of torque is equivalent to 12 inch-pounds (in-lbs) of torque. Torque values below approximately 15 foot-pounds are expressed in inch-pounds, because most foot-pound torque wrenches are not accurate at these smaller values.*

Fuel conditioning module bolts	177 in-lbs
High-pressure fuel line fittings	
Step 1	89 in-lbs
Step 2	Tighten an additional 65 degrees
High-pressure fuel pump	
Drive gear nut (right hand thread)	59
Pump spacers	18
Fuel injector hold-down clamp bolt	
Step 1	22
Step 2	Tighten an additional 90 degrees
Fuel rail mounting bolts	18
Fuel rail pressure sensor	52
Fuel pressure control valve	
Step 1	44
Step 2	Loosen 90 degrees
Step 3	63
Glow plugs	
Step 1	35
Step 2	Tighten an additional 45 to 55 degrees
Intercooler fasteners	106 in-lbs
Turbocharger-to-pedestal bolts	41
Throttle body bolts	89 in-lbs

1   General Information and precautions

Fuel system warnings

Warning: *Diesel fuel is flammable and repairing fuel system components can be dangerous. Consider your automotive repair knowledge and experience before attempting repairs, which may be better suited for a professional mechanic.*

a) *Don't smoke or allow open flames or bare light bulbs near the work area.*

b) *Don't work in a garage with a gas-type appliance (water heater, clothes dryer).*

c) *Use fuel-resistant gloves. If any fuel spills on your skin, wash it off immediately with soap and water.*

d) *Clean up spills immediately.*

e) *Do not store fuel-soaked rags where they could ignite.*

f) *Prior to disconnecting any fuel line, you must relieve the fuel pressure (see Section 3).*

g) *Wear safety glasses.*

h) *Have a proper fire extinguisher on hand.*

General information

1   The fuel system is what most sets apart diesel engines from their gasoline powered cousins. Simply stated, fuel is injected directly into the combustion chambers; the more fuel injected, the more power the engine produces. Unlike gasoline engines, diesels have no throttle plate to limit the entry of air into the intake manifold. The only control is the amount of fuel injected; an unrestricted supply of air is always available through the intake.

2   There are two major sub-systems in the fuel injection system; the low-pressure (also known as the supply) portion and the high-pressure injection (delivery) portion.

3   Fuel from the tank passes through the chassis-mounted diesel fuel conditioning module into the engine-mounted secondary fuel filter module. The low-pressure fuel then enters the high-pressure fuel pump. The fuel temperature sensor provides input to the PCM to control fuel system operating parameters. The fuel pressure switch ensures the minimum fuel supply pressure required by the high-pressure pump is provided.

4   Fuel to the high-pressure fuel injection pump flows through the cam chamber where it cools and lubricates the pump. A mechanical pressure regulator in the pump maintains the steady internal pressure. The fuel volume control valve mounted on top of the high pressure pump allows a PCM-controlled portion of the low-pressure fuel to flow to the the pump, where it is compressed and sent to the fuel rails as high pressure. The high-pressure pump generates pressure up to 29,000 psi (2,000 bar).

5   High-pressure fuel is supplied to the left fuel rail by two fuel rail supply tubes, and to the right fuel rail through a single fuel rail supply tube. The Fuel Rail Pressure (FRP) sensor measures the fuel pressure in left fuel rail while the fuel pressure control valve, mounted on the rear of the left fuel rail, controls the fuel pressure.

6   The PCM uses various inputs to determine high-pressure fuel system operating mode. Under certain conditions (including startup) the system operates with the fuel volume control valve fully open, and fuel pressure is controlled by the fuel pressure control valve. In other operating conditions, the fuel volume control valve meters fuel so that only the amount of fuel required for a given rail pressure is sent to the rails. In this mode the fuel pressure control valve performs the fuel trim function, bleeding off only small amounts of fuel.

7   Fuel injectors are controlled by the PCM using an internal injector driver. There can be up to 5 injection events per combustion cycle of the engine. The injector nozzle contains 8 fuel delivery orifices. The injector injects fuel directly into the combustion chamber at pressures up to 29,000 psi (2,000 bar). Each injector is assigned an Injector Quantity Adjustment (IQA) code when manufactured. The PCM is programmed with the IQA code of the fuel injector for each cylinder. If a fuel injector is replaced, the PCM will require programming with the new IQA code for that replacement injector.

8   Return fuel from the high-pressure pump combines with the return fuel from the fuel rail and flows to the fuel cooler, mounted on the left frame rail, then to the diesel fuel conditioning module. The return fuel from the injectors flows to a tee containing an orifice that controls the backpressure on the injectors. During engine cranking, fuel pressure from the diesel fuel conditioning module is applied to the injector return connectors through this return hose to create the backpressure necessary for the injectors to function. During normal operation, injector return fuel flowing though the orifice creates the required backpressure.

Diesel fuel contamination

9   Before you replace an injector or some other expensive component, find out what caused the failure. If water contamination is present, buying a new or rebuilt injector or other component won't do much good. The following procedure will help you determine if water contamination is present:

a) *Remove the engine fuel filter and inspect the contents for the presence of water or gasoline (see Chapter 1).*

b) *If the vehicle has been stalling, performance has been poor or the engine has been knocking loudly, suspect fuel contamination. Gasoline or water must be removed by flushing (see below).*

c) *If you find a lot of water in the fuel filter, remove the fuel return line and check for water there. If the pump has water in it, flush the system.*

d) *Small quantities of surface rust won't create a problem. If contamination is excessive, the vehicle will probably stall.*

f) *Sometimes contamination in the system becomes severe enough to cause damage to the system. If the damage reaches this stage, have the damaged parts replaced by an authorized diesel truck repair facility.*

Storage

10   Good quality diesel fuel contains inhibitors to stop the formation of rust in the fuel lines and the injectors, so as long as there are no leaks in the fuel system, it's generally safe from water contamination. Diesel fuel is usually contaminated by water as a result of careless storage. There's not much you can do about the storage practices of service stations where you buy diesel fuel, but if you keep a small supply of diesel fuel on hand at home, as many diesel owners do, follow these simple rules:

a) *Diesel fuel "ages" and goes stale. Don't store containers of diesel fuel for long periods of time. Use it up regularly and replace it with fresh fuel.*

b) *Keep fuel storage containers out of direct sunlight. Variations in heat and humidity promote condensation inside fuel containers.*

c) *Don't store diesel fuel in galvanized containers. It may cause the galvanizing to flake off, contaminating the fuel and clogging filters when the fuel is used.*

d) *Label containers properly, as containing diesel fuel.*

Fighting fungi and bacteria with biocides

11   If there's water in the fuel, fungi and/or bacteria can form in warm or humid weather. Fungi and bacteria plug fuel lines, fuel filters and injection nozzles; they can also cause corrosion in the fuel system.

12   If you've had problems with water in the fuel system and you live in a warm or humid climate, have your dealer correct the problem. Then, use a diesel fuel biocide to sterilize the fuel system in accordance with the manufacturer's instructions. Biocides are available from your dealer, service stations and auto parts stores. Consult your dealer for advice on using biocides in your area and for recommendations on which ones to use.

Cleaning the fuel system

Warning: *Diesel fuel is flammable, so take extra precautions when you work on any part of the fuel system. Don't smoke or allow open flames or bare light bulbs near the work area, and don't work in a garage where a gas-type appliance (such as a water heater or a clothes dryer) is present. Since diesel fuel is carcinogenic, wear latex gloves when there's a possibility of being exposed to fuel, and, if you spill any fuel on your skin, rinse it off immediately with soap and water. Mop up any spills immediately and do not store diesel fuel-soaked rags where they could ignite. When you perform any kind of work on the fuel system, wear safety glasses and have a Class B type fire*

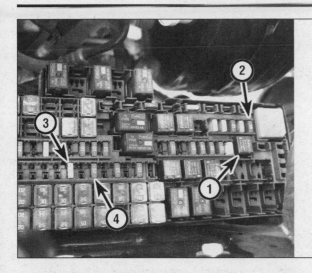

2.2 Fuel system-related underhood fuse/relay box details (diesel)

1   *Fuel pump (low pressure) relay*
2   *Fuel pressure control valve/ fuel volume control valve fuse F37 (10 amp; always hot)*
3   *Fuel pump relay fuse F66 (20 amp; always hot)*
4   *Fuel pump motor diode fuse F68 (10 amp; always hot)*

2.3 Pry off the small trim cover and depress the button on the top of the inertia switch

extinguisher on hand.
Note: *If gasoline has been accidentally pumped into the fuel tank, it should be drained immediately. Gasoline in the fuel in small amounts - up to 30 percent - isn't usually noticeable. At higher ratios, the engine may make a knocking noise, which will get louder as the ratio of gasoline increases.*

Water in the fuel system

13   Disconnect the cables from the negative battery terminals (see Chapter 5).
14   Drain the diesel fuel from the fuel tank into an approved container and dispose of it properly.
15   Remove the fuel tank gauge sending unit (see Chapter 4A).
16   Thoroughly clean the fuel tank. If it's rusted inside, send it to a repair shop or replace it.
17   Reinstall the fuel tank but don't connect the fuel lines to the fuel tank.
18   Disconnect the main fuel line from the fuel pump. Disconnect the supply line from the fuel filter/water separator (see Section 9) Using low air pressure, blow out the line toward the rear of the vehicle.
Warning: *Wear eye protection when using compressed air.*
19   Temporarily disconnect the fuel return line at the fuel filter/water separator and at the fuel tank, using low air pressure, blow out the line toward the rear of the vehicle.
20   Reconnect the main fuel and return lines at the tank. Fill the tank to a fourth of its capacity with clean diesel fuel. Install the cap on the fuel filler neck.
21   Discard the fuel filter (see Chapter 1).
22   Connect the fuel line to the fuel pump.
23   Reconnect the battery cables.
24   Purge the fuel pump by cranking the engine until clean fuel is pumped out. Catch the fuel in a closed metal container.
25   Install a new fuel filter (see Chapter 1).
26   Install a hose from the fuel return line (at the fuel filter/water separator) to a metal container with a capacity of at least two gallons.
27   Crank the engine until clean fuel appears

at the return line. Don't crank the engine for more than 30 seconds at a time. If it's necessary to crank it again, allow a three-minute interval before resuming.
28   Drain the fuel tank into an approved container and fill the tank with clean, fresh diesel fuel (see Chapter 4A).
29   Remove the fuel lines from the fuel filter/ water separator at the cylinder heads.
30   Connect a short pipe and hose from the ends of the fuel lines to a metal container.
31   Crank the engine to purge the fuel pump and fuel filter. Don't crank the engine more than 30 seconds. Allow two or three minutes between cranking intervals for the starter to cool.
32   Remove the short pipe and hose and connect the fuel lines to the cylinder heads.
33   Remove the fuel drain plugs from the rear of the cylinder heads (refer to Section 16) and allow the diesel fuel and air to drain.
34   Install the fuel drain plugs on the back of the cylinder heads (see Section 16).
35   Try to start the engine. If it doesn't start, purge the air from the system again by removing the drain plugs at the back of the cylinder heads (see Section 16).
Warning: *Avoid sources of ignition and have a fire extinguisher handy. Clean all the diesel fuel that may have spilled during the bleeding procedure.*
36   Start the engine and run it at idle for 15 minutes.

"Water-In-Fuel" (WIF) warning system

37   The system detects the presence of water in the fuel filter/water separator when it reaches excessive amounts. Water is detected by a probe located in the fuel filter that completes a circuit through a wire to a light in the instrument cluster that reads "Water In Fuel."
38   This system includes a bulb-check feature - when the ignition is turned on, the bulb glows momentarily, then fades away.
39   If the light comes on immediately after you've filled the tank or let the vehicle sit for an extended period of time, drain the water

from the system immediately. Do not start the engine. There might be enough water in the system to shut the engine down before you've driven even a short distance. If, however, the light comes on during a cornering or braking maneuver, there's less water in the system; the engine probably won't shut down immediately, but you still should drain the water soon.
40   Water is heavier than diesel fuel, so it sinks to the bottom of the fuel tank. A bleed-off pipe on the fuel filter/water separator, which deposits the excess water onto the street enables you to siphon most of the water from the fuel filter/water separator without having to remove the tank. However, if the fuel tank is full of excessive amounts of water, siphoning won't remove all of the water; you'll still need to remove the tank and thoroughly clean it.
Warning: *Do not start a siphon by mouth - use a siphoning kit (available at most auto parts stores).*

2   Troubleshooting

Fuel pump

1   The low pressure fuel pump is part of the fuel conditioning module. Sit inside the vehicle with the windows closed, turn the ignition key to On (not Start) and listen for the sound of the fuel pump as it's briefly activated. You will only hear the sound for a second or two, but that sound tells you that the pump is working. Alternatively, have an assistant listen at the center of the vehicle.
2   If the pump does not come on, check the fuel pump fuses and fuel pump (FP) relay (see illustration). If the fuse and relay are okay, check the wiring back to the fuel pump. If the fuse, relay and wiring are okay, the fuel pump is probably defective. If the pump runs continuously with the ignition key in the On position, the Powertrain Control Module (PCM) is probably defective. Have the PCM checked by a professional mechanic.

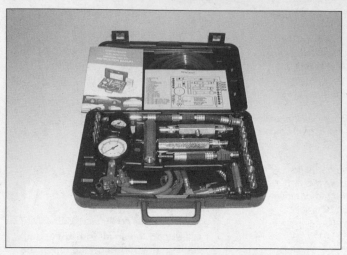

4.6a This fuel pressure testing kit contains all the necessary fittings and adapters, along with the fuel pressure gauge, to test most automotive systems

4.6b Location of the fuel inlet line quick-connect fitting at the secondary fuel filter

Fuel pump inertia switch

3   If the fuel pump does not come on, also verify that the fuel pump inertia switch has not been tripped. The fuel pump inertia switch turns off the fuel pump in the event of a collision or similar incident. The switch is located behind the passenger side end of the dash (see illustration). When the switch is triggered, the plunger at the top is in the up position. To reset the switch, press the plunger down until it clicks.

3   Fuel pressure relief procedure

Warning: *Diesel fuel is flammable, so take extra precautions when you work on any part of the fuel system. See the Warning in Section 1.*
Note: *After the fuel pressure has been relieved, it's a good idea to lay a shop towel over any fuel connection to be disassembled, to absorb the residual fuel that may leak out when servicing the fuel system.*
1   Fuel system pressure cannot be relieved manually.
2   When servicing the low pressure fuel system, turn the ignition off and wait for the fuel conditioning module pump to stop running. After the pump stops running, wait two minutes for the fuel pressure to bleed off before servicing the low pressure fuel system.
3   The high-pressure fuel system (the part of the system from the output side of the camshaft-driven high-pressure fuel pump to the fuel rails) loses pressure as the engine cools down. Wait until the engine is cool (or five minutes after shutting off the engine if the engine is already cool) before servicing any high-pressure fuel system component.

4   Fuel pressure check

Warning: *Diesel fuel is flammable, so take extra precautions when you work on any part of the fuel system. See the Warning in Section 1.*
Warning: *DO NOT attempt to check the pressure on the high side of the system (from the camshaft-driven high-pressure fuel pump to the fuel rails). The pressure on the high-pressure side of the system can be as high as 29,000 psi.*
Note: *This procedure checks the delivery pressure from the fuel conditioning module pump to the high-pressure fuel pump.*
Note: *After the fuel pressure has been relieved, it's a good idea to lay a shop towel over any fuel connection to be disassembled, to absorb the residual fuel that may leak out when servicing the fuel system.*

General checks

Note: *The inertia switch is an electrical device wired into the fuel pump circuit that will shut down power to the fuel pump in an accident (see Section 2). Be sure to check that the inertia switch is activated and in working order if the fuel pump is not receiving the proper voltage.*
Note: *The fuel pump relay is equipped with a primary and secondary voltage circuit. The primary circuit is controlled by the PCM and the secondary circuit is linked directly to battery voltage from the ignition switch. With the ignition switch On (engine not running), the PCM will ground the relay for one second. During cranking, the PCM grounds the fuel pump relay as long as the reference signal from the camshaft position sensor is received (see Chapter 6A). If there are no reference pulses, the fuel pump will shut off after two or three seconds.*

1   If you suspect insufficient fuel delivery, check the following items first:
   a) *Check the battery and make sure it's fully charged (see Chapter 5).*
   b) *Check the fuel pump fuses (see Chapter 4A, Section 2).*
   c) *Check the fuel filter for restriction.*
   d) *Inspect all fuel lines to ensure that the problem is not simply a leak in a line.*
2   Verify the fuel pump actually runs. Place the transmission in Park (automatic) or Neutral (manual) and apply the parking brake. Have an assistant turn the ignition switch to On - you should hear a brief whirring noise (for approximately two seconds) as the pump comes on and pressurizes the system. If there is no response from the fuel pump (makes no sound), check the fuel pump electrical circuit. If the fuel pump runs, but a fuel system problem is suspected, continue with the fuel pump pressure check.
3   If the pump does not turn on (makes no sound) with the ignition switch in the ON position, check the fuel pump related fuses located in the engine compartment fuse center (see Section 2). Also, check the fuel pump relay.
Note: *The fuel pump relay is located in the fuse/relay center in the engine compartment (see illustration 2.2).*
4   If the relay is good and the fuel pump does not operate, check the fuel pump circuit. If the wiring and the connectors are good, replace the fuel pump.

Pressure check

Note: *In order to perform the fuel pressure test, you will need a fuel pressure gauge capable of measuring high fuel pressure. The fuel gauge must be equipped with the proper fitting required to attach it to the fuel pump and the inlet line.*

5.3 Outlet line at the secondary fuel filter

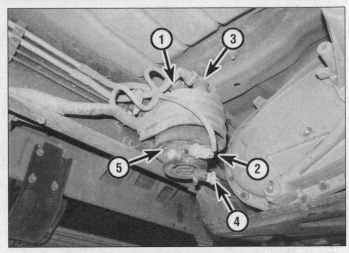

8.4 Fuel conditioning module components

1	Electrical connector	4	Water drain
2	Water-in-fuel sensor	5	Primary fuel filter housing
3	Fuel lines		

5   Relieve the fuel system pressure (see Section 3).

6   Remove the fuel inlet line from the secondary fuel filter and attach a fuel pressure gauge (see illustrations).

7   Turn all the accessories Off and switch the ignition key On (engine not running). The fuel pump should run for approximately two seconds. Note the reading on the gauge and compare to this Chapter's Specifications. If the fuel pressure is lower than specified, check the fuel line from the tank for restrictions. If no restriction is found, check the fuel pump wiring for high resistance. If no problems are found, replace the fuel pump.

8   Approximately 30 seconds after turning the engine off, the fuel pressure will begin to bleed down to 0 psi - this is normal.

9   If the fuel pressure recorded is lower than specified, check the water separator/ Diesel Fuel Conditioner Module (DFCM) filter for clogging (see Chapter 1, Section 19). If no restrictions are found, have the fuel pump and lines diagnosed by a dealer service department or other qualified truck repair facility.

Note: *High fuel pump pressure is more difficult to diagnose. The fuel return lines, the pressure regulator, the fuel pump or the fuel filter/ water separator assembly may be damaged, clogged or the fuel lines may be bent. Have the system checked by a qualified technician.*

5   Fuel system bleeding procedure

Warning: *Diesel fuel is flammable, so take extra precautions when you work on any part of the fuel system. See the Warning in Section 1.*

1   To bleed the system, perform the following procedure three times: Turn the ignition to On and wait 30 seconds for the fuel conditioning module to operate, then turn the ignition Off.

2   Verify that the vehicle operates normally. If it is determined that air is still in the system,

perform the next steps.

3   Disconnect the outlet fuel line from the secondary fuel filter (see illustration).

4   Connect a hose to the secondary fuel filter outlet fitting and insert the other end in a container to catch fuel.

5   Perform the following three times: Turn the ignition to On and wait 30 seconds for the fuel conditioning module to operate, then turn the ignition Off.

6   Reconnect the fuel supply line to the secondary fuel filter. Turn the ignition on and check for leaks.

6   Fuel tank - removal and installation

1   The procedure for diesel models is very similar to the procedure for gasoline models. See Chapter 4A, Section 6.

7   Fuel pickup assembly/fuel level sending unit - replacement

1   See Chapter 4A, Section 8 for this procedure.

8   Fuel conditioning module/fuel filter replacement

Warning: *Diesel fuel is flammable, so take extra precautions when you work on any part of the fuel system. See the Warning in Section 1.*

Note: *After the fuel pressure has been relieved, it's a good idea to lay a shop towel over any fuel connection to be disassembled, to absorb the residual fuel that may leak out when servicing the fuel system.*

Note: *The fuel pump is part of the fuel condi-*

tioning module. This module also contains the primary fuel filter, the water separator, a drain valve and a pressure regulator.

1   Relieve the fuel system pressure (see Section 3).

2   Disconnect the cable(s) from the negative battery terminal(s) (see Chapter 5, Section 3).

Fuel conditioning module/ primary fuel filter

3   Raise the vehicle and support it on jackstands.

4   Disconnect the water in fuel sensor and fuel conditioning module electrical connectors (see illustration).

5   Place a drain pan under the fuel conditioning module and unscrew the bottom of the module three or four turns.

Note: *There is a check valve in the fuel conditioning module that stops the fuel from being siphoned out of the tank. It works when the filter cover is loosened or removed.*

6   Loosen the water drain and drain the fuel into the container.

7   To replace the primary fuel filter, unscrew the cover and remove the filter.

8   To remove the fuel conditioning module, disconnect the fuel lines quick-disconnect fittings from the module.

9   Remove the fuel conditioning module nuts and remove the fuel conditioning module and bracket from the frame.

10   Installation is the reverse of removal. Apply clean engine oil to the O-rings prior to installation.

11   Tighten the filter cover until it contacts the mechanical stops.

12   Turn the ignition on and check for any leaks.

Secondary fuel filter

13   See Chapter 1, Section 19 for the secondary fuel filter replacement procedure.

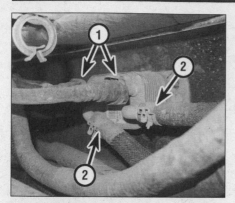

10.1 The fuel cooler is located on the inside of the left frame rail, above the mount for the left radius arm

1   Fuel lines
2   Coolant hoses

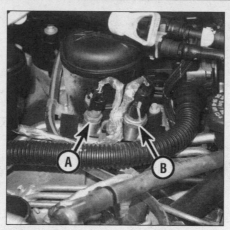

11.2 Locations of the fuel temperature sensor (A) and fuel pressure switch (B)

12.3 The Fuel Rail Pressure sensor is located near the rear of the alternator

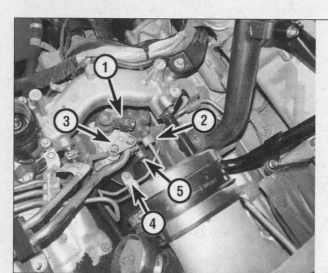

14.2 High-pressure fuel injection pump components

1   Fuel volume control valve
2   High-pressure fuel rail supply lines
3   Fuel pump spacers
4   Fuel supply lines
5   Fuel rail supply line bracket

1   Relieve the fuel pressure (see Section 3).
2   Remove the throttle body air inlet tube between the intercooler and the air filter housing (see Section 22).
3   Locate the FRP sensor and disconnect the electrical connector (see illustration).
4   Unscrew the FRP sensor from the fuel rail.
5   Installation is reverse of removal. Be sure to lubricate the O-ring with clean engine oil, then tighten the FRP sensor to the torque listed in this Chapter's Specifications.

13   Fuel pressure control valve - replacement

Note: *The fuel pressure control valve is located on the end of the driver's side fuel rail, toward the rear of the engine.*
1   Relieve fuel pressure (see Section 3).
2   Loosen the left-front wheel lug nuts, then raise and support the front of the vehicle on jackstands.
3   Remove the left front wheel and inner wheelhouse liner.
4   Locate and disconnect the fuel pressure control valve connector.
5   Loosen the fuel pressure control valve from the fuel rail and unscrew to remove.
6   Installation is reverse of removal. Installation is reverse of removal. Be sure to lubricate the O-ring with clean engine oil, then tighten the fuel pressure control valve to the torque listed in this Chapter's Specifications.

14   High-pressure fuel pump - removal and installation

Removal

1   Relieve the fuel system pressure (see Section 3).
2   Remove the fuel rail supply lines between the high-pressure fuel pump and the left and right bank fuel rails (see illustration).

9   Water separator - replacement

1   The water separator is part of the fuel conditioning module. See Section 8.

10   Fuel cooler - replacement

Warning: *Wait until the engine is completely cool before beginning this procedure.*
Note: *The fuel cooler is bolted inside the frame rails toward the rear of the vehicle and has two coolant hoses and two quick-disconnect fuel lines attached to it.*
1   Pinch or clamp off the coolant hoses to prevent excessive coolant loss (see illustration). Remove the clamps and disconnect the coolant hoses.
2   Relieve the fuel system pressure (refer to Section 3).
3   Disconnect the fuel line quick-connect fittings.
4   Remove the fasteners attaching the cooler to the frame rail and remove the cooler.
5   Installation is reverse of removal. Top off

the coolant as necessary. Turn the ignition on to operate the fuel pump and check for leaks.

11   Fuel temperature sensor/ pressure switch - replacement

Note: *The fuel temperature sensor and fuel pressure switch are located on the fuel supply line, in front of the secondary fuel filter.*
1   Relieve the fuel pressure (see Section 3).
2   Disconnect the electrical connector for the sensor/switch (see illustration).
3   Unscrew the sensor/switch and remove from the vehicle.
4   Installation is reverse of removal.

12   Fuel Rail Pressure (FRP) sensor - replacement

Note: *The FRP sensor is located on the end of the driver side fuel rail, towards the front of the engine.*

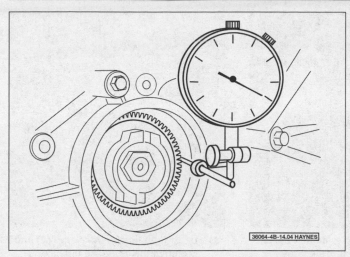

14.4 Using a dial indicator, check the high-pressure fuel pump drive gear backlash

14.7 Remove the fuel supply lines (A) and disconnect the volume control valve connector (B)

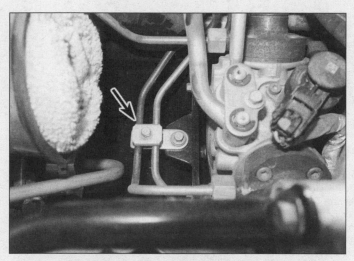

14.8 Remove the fuel rail supply line bracket

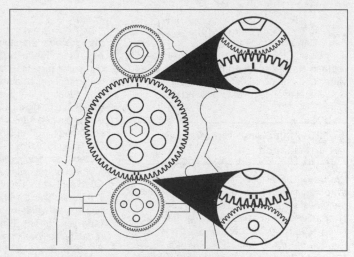

14.14 Install the drive gear and line up as shown

3   Remove the engine cooling fan (see Chapter 3) and the brake vacuum pump from the engine front cover to allow access to the high-pressure fuel pump drive gear.

4   Using a Dial Indicator with Holding Fixture (see illustration), check the high-pressure fuel pump drive gear backlash (see this Chapter's Specifications).

5   Rotate the crankshaft clockwise (looking at the front of the engine), until the single mark on the camshaft drive gear and the double mark on the high-pressure fuel pump drive gear are aligned as shown.

Note: *The keyway on the high-pressure fuel pump drive gear must be in the 12 o'clock position.*

6   Remove the high-pressure fuel pump drive gear nut.

7   Remove the fuel supply lines from the high-pressure fuel pump. Disconnect the volume control valve connector from the top of the pump (see illustration).

Note: *Slide the red connector lock out, then*

depress the tab to release the connector

8   Remove the fuel rail supply line bracket from the rear of the pump (see illustration).

9   Remove the three high-pressure pump spacers (see illustration 14.2) and slide the pump toward the rear of the engine until the gear comes in contact with the engine block.

10   Tap the high-pressure fuel pump shaft using a soft mallet to disengage the pump shaft from the gear.

Caution: *Do not pry on the pump to free the gear as this can damage the pump.*

11   Slide the high-pressure fuel pump rearward off of the studs until it can be removed through the opening in the top of the engine. Discard the O-ring on the front of the pump.

Installation

12   Replace the O-ring and lubricate with clean engine oil.

13   Install the high-pressure fuel pump and the spacers. Tighten the spacers to the torque listed in this Chapter's Specifications.

14   Install the drive gear and line up as shown (see illustration). Install the drive gear nut and tighten to the torque listed in this Chapter's Specifications.

15   The remainder of installation is the reverse of removal. Bleed the fuel system (see Section 5).

16   Turn the ignition on and check for leaks before starting the engine.

15   Fuel volume control valve - replacement

Note: *The fuel volume control valve is located on the top of the high pressure fuel pump, at the front of the engine.*

1   Relieve fuel pressure (see Section 3).

2   Locate the fuel volume control valve and disconnect the electrical connector (see illustration 14.2).

3   Remove the two bolts and remove the

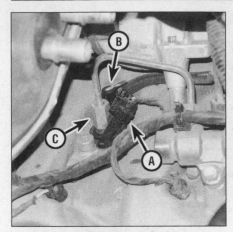

16.7 Disconnect the injector electrical connector (A), the fuel return line (B) and the high-pressure fuel line (C)

17.1a Remove the clip to detach the air intake pipe from the throttle body . . .

17.1b . . . then loosen the clamp (A) and remove the air intake pipe (B)

fuel volume control valve.

4   Check the condition of the O-ring and replace as necessary. Lubricate the O-ring with clean engine oil before installation.

5   Installation is reverse of removal.

16   Fuel injectors - removal and installation

Warning: *Diesel fuel is flammable, so take extra precautions when you work on any part of the fuel system. See the Warning in Section 1.*
Note: *Several special tools are needed to remove and install the fuel injectors. Read through the procedure and obtain the special tools (or their equivalents) before beginning.*
Caution: *Each fuel injector is assigned a fuel flow trim code during the injector manufacturing process and is assigned a specific Injector Quantity Adjustment (IQA) code. The PCM is programmed with the IQA code of the fuel injector for each cylinder. If an injector is replaced, retain the label that comes with the new injector and have the PCM programmed with the new IQA code for that replacement injector.*

Removal

1   Relieve the fuel system pressure (see Section 3).

2   Disconnect the cable from the negative battery terminal(s) (see Chapter 5, Section 3).

3   Depending on the injector(s) requiring replacement, remove the driver's or passenger's front inner wheelwell.
Note: *If the passenger's side injectors are to be replaced, remove the glow plug control module (see Section 18).*

4   Remove the engine sound shield bolts and sound shield for the side of the engine

the injector(s) are being replaced on.

5   Disconnect the fuel injector electrical connector (see illustration 16.7). Remove the wiring harness retainers from the fuel rails.

6   Firmly pull upward on the fuel injector return line fitting lock. It may be necessary to use pliers or a similar tool to unlock the return line fitting. Pull upward again on the return line fitting to disconnect it from the injector.
Note: *If servicing cylinder 8, disconnect the fuel pressure control valve electrical connector to allow removal of the injector.*

7   Unbolt and remove the high pressure fuel line between the injector and the fuel rail (see illustration). It is recommended to not reuse these lines once removed.

8   Remove and discard the fuel injector hold down clamp bolt.

9   Mark the cylinder that the injector is removed from. It is important to install the injector into the same cylinder as it was removed.
Caution: *Each fuel injector is assigned a fuel flow trim code during the injector manufacturing process and is assigned a specific Injector Quantity Adjustment (IQA) code. The PCM is programmed with the IQA code of the fuel injector for each cylinder. If an injector is replaced, retain the label that comes with the new injector and have the PCM programmed with the new IQA code for that replacement injector.*

10   Remove the injector from the valve cover. A special tool may be necessary to remove the injector as they are installed with close tolerance and may be difficult to remove.

11   With the injector removed, discard the O-ring and copper seal from the end of the injector and discard the O-ring for the return line.

12   Remove and discard the injector oil seal from the valve cover.

Installation

13   Install new injector oil seals into the valve cover using a 22mm socket.

14   Install new O-rings and a new copper washer on each injector.

15   Install the injectors along with the clamps and bolts. Lubricate the injector hold down clamp and bolt with clean engine oil. Torque the injector hold down clamp bolt to this Chapter's Specifications.

16   Install new high pressure fuel lines between the injector and throttle body. If more than one high pressure fuel line is removed, the fuel rail mounting bolts must be loosened before the new lines are installed. Torque the lines and fuel rail bolts to this Chapter's Specifications.

17   After installing the return lines with new O-rings, press down firmly to slide the lock down to secure the return lines to the fuel injector.

18   The remainder of installation is the reverse of removal. Top off the engine coolant as necessary.

19   If any fuel injectors have been replaced, the IQA code for the new injector will need to be programmed into the PCM.

20   Bleed the fuel system (see Section 5).

21   Run the engine and check for leaks.

17   Throttle body - removal and installation

Removal

1   Remove the air intake pipe connection by removing the clip, then loosen the clamp and remove the intake pipe from the throttle body (see illustrations).

2   Detach the Electronic Throttle Control (ETC) electrical connector.

17.3 Disconnect the electrical connector (A) and remove the throttle body bolts (B)

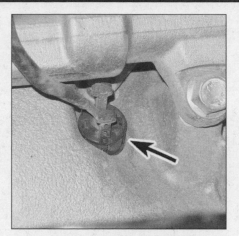

18.8 Squeeze the glow plug connector tabs to remove the connector

18.15 Disconnect the electrical connectors and remove the glow plug control module

3   Remove the four throttle body mounting bolts (see illustration) and remove the throttle body.

4   Discard the gasket.

Installation

5   Installation is the reverse of removal. Clean the sealing surfaces. If scraping is necessary, be careful not to damage the sealing surfaces or allow material to drop into the intake manifold. Install a new gasket. Be sure to tighten the throttle body mounting bolts to the torque listed in this Chapter's Specifications.

---

18   Glow plug system - general information and component replacement

Warning: *Diesel fuel is flammable, so take extra precautions when you work on any part of the fuel system. See the Warning in Section 1.*
Warning: *The glow plugs operate on 115-volts DC. Do not pierce the wiring insulation or use a conventional test light on the glow plug wiring, as you could receive a severe shock.*

General information

1   The glow plugs are designed to heat up extremely fast. When voltage is applied to the glow plug, rapid heating of the glow plug tip occurs, which in turn heats up the air inside the pre-chamber. As the fuel is injected into the pre-chamber, it is ignited. The rapid burning of the fuel forces it through the small opening in the pre-chamber into the cylinder where it mixes with more hot air and a second combustion takes place. To prevent overheating of the glow plug, a cycling device is used in the circuit.

Note: *If the tip is missing from a glow plug or is damaged, the fuel quality is probably incorrect. If the tip is distorted, the glow plugs are probably staying on too long. Glow plugs are*

not interchangeable between the early and later model systems.

2   The glow plug system is electronically controlled by the Glow Plug Control Module (GPCM). If coolant temperature is below 60°C (140°F), the GPCM energizes the glow plugs immediately after the key is turned to On (and turns on the WAIT-TO-START lamp). Depending on the Engine Coolant Temperature (ECT) sensor and battery voltage, the GPCM determines how long the glow plugs are to be energized.

3   The WAIT-TO-START lamp is located in the instrument cluster. This lamp informs the driver when the glow plugs are hot enough for the engine to be started. For a bulb check, the bulb will come on when the ignition switch is in the Start position. The on time of the glow plugs is independent of the WAIT-TO-START lamp on time.

4   The GPCM monitors and detects individual glow plug functions. This information is relayed to the PCM where it is accessible by scan tool. Glow plug system diagnostics are part of the On Board II (OBD-II) diagnostic system (see Chapter 6A).

5   The wiring harness to the glow plug control module incorporates a 125A mega fuse.

Component replacement
Glow plugs

6   Depending on the glow plug(s) requiring replacement, remove the driver's or passenger's front inner wheelwell.

7   Remove the engine sound shield bolts and sound shield for the side of the engine the glow plugs are being replaced on.

8   Pull the glow plug connector off of the glow plug using the tabs at the top of the connector sleeve (see illustration).

9   Unscrew the glow plugs.

10   Installation is the reverse of removal noting the following items:

11   Tighten the glow plugs to the torque listed in this Chapter's Specifications.

12   Pull the glow plug connector out of the sleeve. Connect the connector to the glow plug, then press the sleeve down to lock the connector onto the glow plug.

Glow plug control module

Note: *The glow plug control module is attached to the underside of the primary battery tray.*

13   Disconnect the cable from the negative battery terminal(s) (see Chapter 5, Section 3).

14   Remove the passenger's front inner wheelwell.

15   Disconnect the electrical connectors from the glow plug control module (see illustration).

16   Remove the control module mounting bolts and remove the control module.

17   Installation is the reverse of removal.

---

19   Turbocharger - general information and inspection

General information

1   The turbocharger pressurizes the air entering the combustion chamber by using an exhaust gas-driven turbine. Compressing more air into the combustion chamber increases power output, improves fuel efficiency and performance at high altitudes. The turbocharger identification number is located on an ID plate riveted to the housing. This number should be referred to if replacement is required.

Note: *The turbocharger is normally covered by the Federally mandated emissions warranty. Check with a dealer service department concerning coverage.*

2   A turbocharger improves engine performance, lowers the density of exhaust smoke, improves fuel economy, reduces engine noise and mitigates the effects of lower density air at higher altitude.

3   The single stage variable geometry tur-

20.4 Remove the bolts and position the transmission fluid dipstick tube aside

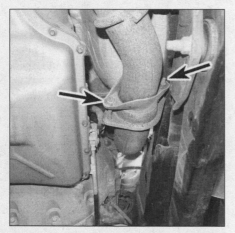

20.11 Remove the down pipe fasteners at the particulate filter

bocharger is electronically controlled by the turbocharger actuator through the PCM. The turbocharger actuator controls intake manifold pressure (similar to the function of a wastegate). The turbocharger uses moveable vanes in the turbine housing to modify the flow of the exhaust gases through the turbocharger. These vanes can be positioned to change the angle or direction and velocity of flow to the turbine wheel in relation to engine operating conditions. As power demand increases and decreases, exhaust gas velocity also increases and decreases as well as intake manifold boost pressure.

4   The turbo vanes are connected to the unison ring. When the unison ring moves, the vanes move. The unison ring is operated by the turbocharger actuator. Turbocharger control uses the Exhaust Pressure (EP) sensor to provide feedback to the PCM. In response to engine speed, engine load, manifold pressure and barometric pressure, the PCM controls the turbocharger actuator position to match manifold boost to engine demands.

5   Turbocharged models are equipped with an "intercooler," a heat exchanger through which the compressed air intake charge is routed to lower the temperature of the intake

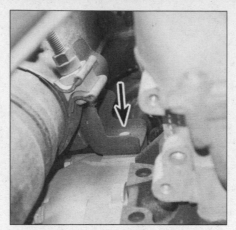

20.10 Remove the upper down pipe bracket bolt

charge. Cooler air is denser, which promotes combustion efficiency, increasing power and reducing emissions.

Inspection

6   Though it's a relatively simple device, the turbocharger is a precision component. Special tools are needed to disassemble and overhaul a turbocharger, so servicing should be left to a specialist. However, you can inspect some things yourself, such as a cracked turbo mounting flange, a worn out or overheated turbine/compressor shaft bearing or a defective wastegate actuator.

7   A turbocharger has its own distinctive sound, so a change in the quality or the quantity of noise can be a sign of potential problems. Before assuming that a funny sound is caused by a defective turbocharger, inspect the exhaust manifold for cracks and loose connections. For example, a high-pitched or whistling sound might indicate an intake air or exhaust gas leak. Inspect the turbocharger mounting flange at the exhaust manifold and make sure that the hose clamp that attaches the air intake duct to the turbocharger is tight.

8   If an unusual sound is coming from the turbocharger, remove the intake duct between the air cleaner housing and the turbocharger. Reach inside the housing and turn the compressor wheel to make sure it spins freely. If it doesn't, it's possible the turbo lubricating oil has sludged or coked up from overheating. Push in on the turbine wheel and check for binding. The turbine should rotate freely with no binding or rubbing on the housing. If it does, the turbine or compressor shaft bearing is worn out.

Warning: *Inspect the turbocharger with the engine off and cool to the touch. Touching or reaching inside a hot and/or operating turbocharger can cause serious injury.*

9   The turbocharger is lubricated by engine oil that has been pressurized, cooled and filtered. Oil is delivered to the turbocharger entering through a channel directly under the turbocharger from the engine block. Oil travels to the turbocharger's bearing housing, where it lubricates the shaft and bearings. A return

channel at the bottom of the turbocharger routes the engine oil back to the crankcase. Because the turbine and compressor wheels spin at speeds over 100,000 rpm, severe damage can result from the interruption or contamination of the oil supply to the turbocharger bearings. Burned oil on the turbine housing is a sign of a blocked return line. **Caution:** *Whenever a major engine bearing such as a main, connecting rod or camshaft bearing is replaced, flush the turbocharger oil passages with clean oil.*

20   Turbocharger - removal and installation

Warning: *Wait until the engine is completely cool before beginning this procedure.*
Caution: *The turbocharger is a precision component which has been assembled and balanced to very fine tolerances. Do not disassemble it or try to repair it. Turbochargers should only be overhauled or repaired by authorized turbocharger repair facilities. An incorrectly assembled turbocharger could result in damage to the turbocharger and/or the engine.*

Removal

1   Disconnect the cables from the negative battery terminals (see Chapter 5, Section 3).
2   Drain the engine coolant (see Chapter 1).
3   Remove the lower intake manifold from the engine (see Chapter 2B, Section 5).
4   Remove the two bolts and move the transmission dipstick tube out of the way (see illustration).
5   Disconnect the heater hose from the heater core inlet tube. Remove the bolt and remove the heater core inlet tube (inspect and replace the O-ring if needed).
6   Disconnect the electrical wiring from the turbocharger.
7   Remove the bolts and the turbocharger heat shield.
8   Mark the position of the clamp on the pipe for installation and loosen the upper down pipe clamp. It is recommended to not reuse the clamp.
9   Remove the passenger side inner wheel well.
10   Remove the the upper down pipe bracket bolt (see illustration).
11   Remove the two down pipe fasteners at the particulate filter (see illustration) and remove the down pipe from the vehicle.
12   Mark the position of the clamps on the pipes for installation and loosen the left and right turbocharger inlet pipe clamps and slide the clamps away from the turbocharger. It is recommended to not reuse the clamps.
13   Disconnect the turbocharger coolant outlet tube. Remove the heat shield if necessary.
14   Disconnect the oil supply line at the turbocharger.
15   Remove the four turbocharger mounting bolts.

21.2 Location of the wastegate control valve

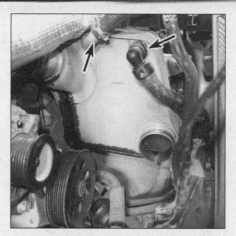

23.4 Disconnect the coolant hoses for the intercooler

23.6 Location of the Charge Air Cooler Temperature (CACT) sensor

16   Lift the turbocharger up and disconnect the oil drain tube. Remove the turbocharger.

Installation

17   Use a die to clean the studs or tap to clean bolt holes on the turbocharger and coat them with anti-seize compound.
18   Inspect and replace any gaskets or O-rings.
19   Install the turbocharger onto the engine.
20   The remainder of installation is the reverse of removal.
21   Tighten the bolts to the torque listed in this Chapter's Specifications.
22   Refill the cooling system (see Chapter 1).

21   Wastegate control valve - replacement

Note: *The wastegate control valve is located directly under the throttle body.*
1   Remove the throttle body (see Section 17).
2   Locate the wastegate control valve and disconnect the electrical connector (see illustration).
3   Disconnect the vacuum line connector at the valve.
4   Unscrew the two bolts and remove the wastegate control valve.
5   Installation is reverse of removal.

22   Air filter housing - replacement

1   Disconnect the MAF sensor electrical connector (see Chapter 6B).

2   Loosen the air intake tube clamp at the intake pipe and disconnect the air intake.
3   Disengage the air filter lid clips and remove the air filter lid and air intake duct and air filter element (see Chapter 1).
4   Remove the mounting bolt from the air filter housing base and lift the air filter housing base from the engine compartment.
5   Installation is the reverse of removal.

23   Charge Air Cooler (intercooler) - removal and installation

Warning: *Wait until the engine is completely cool before beginning this procedure.*
1   Disconnect the cables from the negative terminals of the batteries (see Chapter 5). Drain the primary and secondary cooling systems (see Chapter 3).
2   Remove the driver's side (secondary) battery (see Chapter 5).
3   Disconnect the upper radiator hose from the radiator and disconnect the vacuum tube quick-disconnect fitting.
4   Disconnect the coolant hoses from the intercooler (see illustration).
5   Loosen the clamp and disconnect the intercooler upper (inlet) tube from the intercooler.
6   Disconnect the Charge Air Cooler Temperature (CACT) sensor connector from the lower (outlet) tube (see illustration).
7   Loosen the clamps and disconnect the intercooler lower (outlet) tube from the intercooler and the upper outlet tube.
8   Remove the two bolts and remove the intercooler from the engine compartment (see illustration).

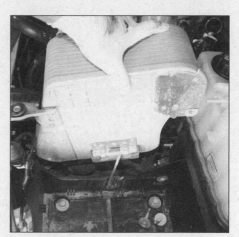

23.8 Remove the bolts and lift the intercooler out of the engine compartment

9   Inspect the intercooler for cracks and damage. Replace it or have it repaired if necessary.
10   Installation is the reverse of removal.
11   Tighten the intercooler mounting bolts to the torque listed in this Chapter's Specifications.
12   Refill the cooling systems (see Chapter 1).

24   Accelerator Pedal Position (APP) sensor - replacement

1   Refer to Chapter 4A for this procedure.

Notes

Chapter 5
Engine electrical systems

Contents

Specifications

Charging system

Charging voltage .... 13.5 to 15.0 volts

Torque specifications   Ft-lbs (unless otherwise indicated)

Note: *One foot-pound (ft-lb) of torque is equivalent to 12 inch-pounds (in-lbs) of torque. Torque values below approximately 15 foot-pounds are expressed in inch-pounds, because most foot-pound torque wrenches are not accurate at these smaller values.*

Starter motor mounting bolts 18
Alternator mounting bolts 35

1   General information and precautions

General information

Ignition system

1   The ignition system on gasoline engines consists of the ignition coils, the spark plugs, the spark plug wires (to the lower spark plugs), the Camshaft Position (CMP) sensor, the Crankshaft Position (CKP) sensor, the knock sensor, and the Powertrain Control Module (PCM).

2   The CKP, CMP and knock sensors are information sensors used by the PCM to control ignition timing and other engine operating parameters. The PCM also uses a number of other information sensors to make decisions regarding the correct ignition timing. These other sensors include the Throttle Position (TP) sensor, the Engine Coolant Temperature (ECT) sensor, the Mass Air Flow (MAF) sensor, the Intake Air Temperature (IAT) sensor, the Vehicle Speed Sensor (VSS) and the transmission gear position sensor or Transmission Range (TR) switch. For more information on these and other sensors, refer to Chapter 6A or 6B.

Charging system

3   The charging system on gasoline and diesel engines includes the alternator (with an integral voltage regulator) or alternators (diesel models with dual alternators), the Powertrain Control Module (PCM), the Body Control Module (BCM), a charge indicator light on the dash, the battery, a fuse or fusible link and the wiring connecting all of these components. The charging system supplies electrical power for the ignition system, the lights, the radio, etc. The alternator is driven by a drivebelt.

Starting system

4   The starting system on gasoline and diesel engines consists of the battery (or batteries), the ignition switch, the starter relay, the Powertrain Control Module (PCM), the Body Control Module (BCM), the Transmission Range (TR) switch, the starter motor and solenoid assembly, and the wiring connecting all of these components.

Precautions

5   Always observe the following precautions when working on the electrical system:

  a) *Be extremely careful when servicing engine electrical components. They are easily damaged if checked, connected or handled improperly.*

  b) *Never leave the ignition switched on for long periods of time when the engine is not running.*

  c) *Never disconnect the battery cables while the engine is running.*

  d) *Maintain correct polarity when connecting battery cables from another vehicle during jump starting - see the* Booster battery (jump) starting *Section at the front of this manual.*

  e) *Always disconnect the cable from the negative battery terminal before working on the electrical system, but read the battery disconnection procedure first (see Section 3).*

6   It's also a good idea to review the safety-related information regarding the engine electrical systems located in the *Safety first!* Section at the front of this manual before beginning any operation included in this chapter.

2   Troubleshooting

Ignition system

1   If a malfunction occurs in the ignition system, do not immediately assume that any particular part is causing the problem. First, check the following items:

  a) *Make sure that the cable clamps at the battery terminals are clean and tight.*

  b) *Test the condition of the battery (see Steps 19 through 22). If it doesn't pass all the tests, replace it.*

  c) *Check the ignition coil or coil pack connections.*

  d) *Check any relevant fuses in the engine compartment fuse and relay box (see Chapter 12). If they're burned, determine the cause and repair the circuit.*

Check

Warning: *Because of the high voltage generated by the ignition system, use extreme care when performing a procedure involving ignition components.*

Note: *The ignition system components on these vehicles are difficult to diagnose. In the event of ignition system failure that you can't diagnose, have the vehicle tested at a dealer service department or other qualified auto repair facility.*

Note: *For the following test, you'll need a spark tester (available at auto parts stores).*

2   If the engine turns over but won't start, verify that there is sufficient secondary ignition voltage to fire the spark plug as follows:

3   Disconnect a spark plug wire from a spark plug and install the tester (see illustration) between the spark plug wire boot and the spark plug.

All models

4   Crank the engine while watching the tester. If the tester flashes, sufficient voltage is reaching the spark plug to fire it.

Caution: *Do NOT crank the engine or allow it to run for more than five seconds; running the engine for more than five seconds may set a Diagnostic Trouble Code (DTC) for a cylinder misfire.*

5   Repeat this test on the remaining cylinders.

6   Proceed on this basis until you have verified that there's a good spark from each coil. If there is, then you have verified that the coils are functioning correctly.

7   If there is no spark, then either the coil is bad or the spark plug wire is bad.

8   Also inspect the coil electrical connector. Make sure that it's clean, tight and in good condition.

9   If all the coils are firing correctly, but the engine misfires when the spark plug wires are connected to the spark plugs, then one or more of the plugs might be fouled. Remove and check the spark plugs or install new ones (see Chapter 1).

10   Also inspect the boots carefully for corrosion (high resistance) or deterioration of the insulation (low resistance). If any of the boots look damaged or deteriorated, replace them as a set.

11   No further testing of the ignition system is possible without special tools. If the problem persists, have the ignition system tested by a dealer service department or other qualified repair shop.

Charging system

12   If a malfunction occurs in the charging system, do not automatically assume the alternator is causing the problem. First check the following items:

  a) *Check the drivebelt tension and condition, as described in Chapter 1. Replace it if it's worn or deteriorated.*

  b) *Make sure the alternator mounting bolts are tight.*

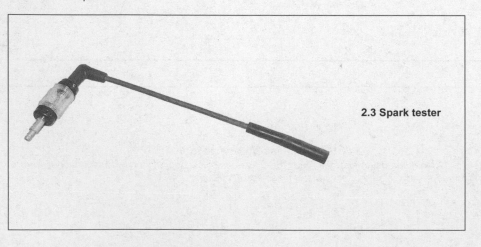

2.3 Spark tester

2.19 To test the open circuit voltage of the battery, touch the black probe of the voltmeter to the negative terminal and the red probe to the positive terminal of the battery; a fully charged battery should be at least 12.6 volts

2.21 Connect a battery load tester to the battery and check the battery condition under load following the tool manufacturer's instructions

c) *Inspect the alternator wiring harness and the connectors at the alternator and voltage regulator. They must be in good condition, tight and have no corrosion.*

d) *Check the fuses in the underhood fuse/ relay box. If any are burned, determine the cause, repair the circuit and replace the fuse (the vehicle will not start and/or the accessories will not work if the main fuse is blown).*

e) *Start the engine and check the alternator for abnormal noises (a shrieking or squealing sound indicates a bad bearing).*

f) *Check the battery. Make sure it's fully charged and in good condition (one bad cell in a battery can cause overcharging by the alternator).*

g) *Disconnect the battery cables (negative first, then positive). Inspect the battery posts and the cable clamps for corrosion. Clean them thoroughly if necessary (see Chapter 1). Reconnect the cables (positive first, negative last).*

Alternator - check

13 Use a voltmeter to check the battery voltage with the engine off. It should be at least 12.6 volts (see illustration 2.19).

14 Start the engine and check the battery voltage again. It should now be approximately 13.5 to 15 volts.

15 If the voltage reading is more or less than the specified charging voltage, the voltage regulator is probably defective, which will require replacement of the alternator (the voltage regulator is not replaceable separately). Remove the alternator and have it bench tested (most auto parts stores will do this for you).

16 The charging system (battery) light on the instrument cluster lights up when the ignition key is turned to On, but it should go out when the engine starts.

17 If the charging system light stays on after the engine has been started, there is a problem with the charging system. Before replacing the alternator, check the battery condition, alternator belt tension and electrical cable connections.

18 If replacing the alternator doesn't restore voltage to the specified range, have the charging system tested by a dealer service department or other qualified repair shop.

Battery - check

19 Check the battery state of charge. Visually inspect the indicator eye on the top of the battery (if equipped with one); if the indicator eye is black in color, charge the battery as described in Chapter 1. Next perform an open circuit voltage test using a digital voltmeter. With the engine and all accessories Off, touch the negative probe of the voltmeter to the negative terminal of the battery and the positive probe to the positive terminal of the battery (see illustration). The battery voltage should be 12.6 volts or slightly above. If the battery is less than the specified voltage, charge the battery before proceeding to the next test. Do not proceed with the battery load test unless the battery charge is correct.

Note: *The battery's surface charge must be removed before accurate voltage measurements can be made. Turn on the high beams for ten seconds, then turn them off and let the vehicle stand for two minutes.*

20 Disconnect the negative battery cable, then the positive cable from the battery.

21 Perform a battery load test. An accurate check of the battery condition can only be performed with a load tester (see illustration). This test evaluates the ability of the battery to operate the starter and other accessories during periods of high current draw. Connect the load tester to the battery terminals. Load test the battery according to the tool manufacturer's instructions. This tool increases the load demand (current draw) on the battery.

22 Maintain the load on the battery for 15 seconds and observe that the battery voltage does not drop below 9.6 volts. If the battery condition is weak or defective, the tool will indicate this condition immediately.

Note: *Cold temperatures will cause the minimum voltage reading to drop slightly. Follow the chart given in the manufacturer's instructions to compensate for cold climates. Minimum load voltage for freezing temperatures (32 degrees F) should be approximately 9.1 volts.*

Starting system

The starter rotates, but the engine doesn't

23 Remove the starter (see Section 8). Check the overrunning clutch and bench test the starter to make sure the drive mechanism extends fully for proper engagement with the flywheel ring gear. If it doesn't, replace the starter.

24 Check the flywheel ring gear for missing teeth and other damage. With the ignition turned off, rotate the flywheel so you can check the entire ring gear.

The starter is noisy

25 If the solenoid is making a chattering noise, first check the battery (see Steps 19 through 22). If the battery is okay, check the cables and connections.

26 If you hear a grinding, crashing metallic sound when you turn the key to Start, check for loose starter mounting bolts. If they're tight, remove the starter and inspect the teeth on the starter pinion gear and flywheel ring gear. Look for missing or damaged teeth.

27 If the starter sounds fine when you first turn the key to Start, but then stops rotating the engine and emits a zinging sound, the problem is probably a defective starter drive that's not staying engaged with the ring gear. Replace the starter.

The starter rotates slowly

28 Check the battery (see Steps 19 through 22).

29 If the battery is okay, verify all connections (at the battery, the starter solenoid and motor) are clean, corrosion-free and tight. Make sure the cables aren't frayed or damaged.

30 Check that the starter mounting bolts are tight so it grounds properly. Also check the pinion gear and flywheel ring gear for evidence of a mechanical bind (galling, deformed gear teeth or other damage).

The starter does not rotate at all

31 Check the battery (see Steps 19 through 22).

32 If the battery is okay, verify all connections (at the battery, the starter solenoid and motor) are clean, corrosion-free and tight. Make sure the cables aren't frayed or damaged.

33 Check all of the fuses in the underhood fuse/relay box.

34 Check that the starter mounting bolts are tight so it is grounded properly.

35 Check for voltage at the starter solenoid "S" terminal when the ignition key is turned to the start position. If voltage is present, replace the starter/solenoid assembly. If no voltage is present, the problem could be the starter relay, the Transmission Range (TR) sensor (see Chapter 7A), or with an electrical connector somewhere in the circuit (see the wiring diagrams). Also, on many modern vehicles, the Powertrain Control Module (PCM) and the Body Control Module (BCM) control the voltage signal to the starter solenoid; on such vehicles a special scan tool is required for diagnosis.

3 Battery - disconnection and reconnection

Note: *To disconnect the battery for service procedures requiring power to be cut from the vehicle, first open the driver's door to disable Retained Accessory Power (RAP), then loosen the cable end nut(s) and disconnect the cable(s) from the negative battery terminal(s). Isolate the cable end(s) to prevent it from coming into accidental contact with the battery terminal(s).*

1 The battery is located in the right side of the engine compartment on all vehicles covered by this manual; diesel models are equipped with a second battery located in the left side of the engine compartment. To disconnect the battery(ies) for service procedures that require battery disconnection, simply disconnect the cable(s) from the negative battery terminal(s) (see Section 4). Make sure that you isolate the cable(s) to prevent it from coming into contact with the battery negative terminal(s).

Warning: *When performing a procedure on a dual-battery equipped model that requires battery disconnection, disconnect the negative cable from BOTH batteries.*

Warning: *Always disconnect the negative battery cable first, then the positive battery cable second. When reconnecting the battery cables, always connect the positive battery cable first, then the negative battery cable second.*

2 Some vehicle systems (radio, alarm system, power door locks, etc.) require battery power all the time, either to enable their operation or to maintain control unit memory (Powertrain Control Module, automatic transaxle control module, etc.), which would be lost if the battery were to be disconnected. So before you disconnect the battery, note the following points:

a) *Before connecting or disconnecting the cable from the negative battery terminal, make sure that you turn the ignition key and the lighting switch to their Off positions. Failure to do so could damage semiconductor components.*

b) *On a vehicle with power door locks, it is a wise precaution to remove the key from the ignition and to keep it with you, so that it does not get locked inside if the power door locks should engage accidentally when the battery is reconnected!*

c) *After the battery has been disconnected, then reconnected (or a new battery has been installed) on vehicles with an automatic transmission, the Transmission Control Module (TCM) will need some time to relearn its adaptive strategy. As a result, shifting might feel firmer than usual. This is a normal condition and will not adversely affect the operation or service life of the transaxle. Eventually, the TCM will complete its adaptive learning process and the shift feel of the transaxle will return to normal.*

d) *The engine management system's PCM has some learning capabilities that allow it to adapt or make corrections in response to minor variations in the fuel system in order to optimize drivability and idle characteristics. However, the PCM might lose some or all of this information when the battery is disconnected. The PCM must go through a relearning process before it can regain its former drivability and performance characteristics. Until it relearns this lost data, you might notice a difference in drivability, idle and/ or (if you have an automatic) shift "feel." To facilitate this relearning process, refer to* Enabling the PCM to relearn *below.*

Memory savers

3 Devices known as memory savers (typically, small 9-volt batteries) can be used to avoid some of the above problems. A memory saver is usually plugged into the cigarette lighter, and then you can disconnect the vehicle battery from the electrical system. The memory saver will deliver sufficient current to maintain security alarm codes and - maybe, but don't count on it! - PCM memory. It will also run unswitched (always on) circuits such as the clock and radio memory, while isolating the car battery in the event that a short circuit occurs while the vehicle is being serviced.

Warning: *If you're going to work around any airbag system components, disconnect the battery and do not use a memory saver. If you do, the airbag could accidentally deploy and cause personal injury.*

Caution: *Because memory savers deliver current to operate unswitched circuits when the battery is disconnected, make sure that the circuit that you're going to service is actually open before working on it!*

Enabling the PCM to relearn

4 After the battery has been reconnected, perform the following procedure in order to facilitate PCM relearning:

5 Start the engine and allow it to warm up to its normal operating temperature.

6 Drive the vehicle at part-throttle, under moderate acceleration and idle conditions, until normal performance returns.

7 Park the vehicle and apply the parking brake with the engine running.

8 Depress the brake pedal and put the shift lever in Drive.

9 Allow the engine to idle for about two minutes, or until the idle stabilizes. Make sure that the engine is at its normal operating temperature.

4 Battery and battery tray - removal and installation

Note: *Battery straps and handlers are available at most auto parts stores for reasonable prices. They make it easier to remove and carry the battery.*

Note: *This procedure applies to primary or secondary batteries and trays.*

Removal

Battery

1 Disconnect the cable from the negative terminal of the battery (see illustrations). If a secondary battery is installed, disconnect the negative cable from the secondary battery as well.

2 Disconnect the cable from the positive terminal of the battery for the battery being removed.

3 Remove the battery hold-down bracket.

4 Lift out the battery. Be careful - it's heavy. If equipped, remove the battery cover from the battery.

Battery tray

Warning: *If you're removing the secondary (left-side) battery tray, wait until the engine is completely cool.*

5 To remove the battery tray for the primary (right-sde) battery, remove the air filter housing (see Chapter 4B, diesel models only), then disconnect the windshield washer fluid reservoir connector and pump hose. Remove the four bolts, then remove the battery tray

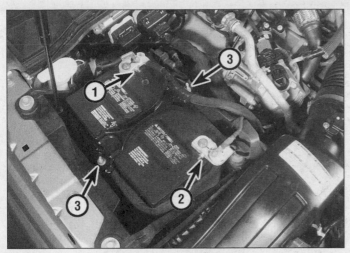

4.1a Battery details (right side)

1 *Negative cable clamp* 3 *Battery hold-down nuts*
2 *Positive cable clamp*

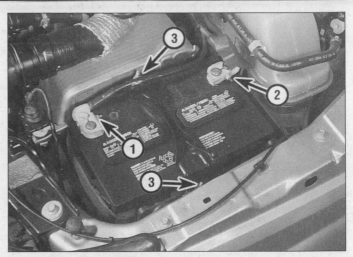

4.1b Battery details (left side)

1 *Negative cable clamp* 3 *Battery hold-down nuts*
2 *Positive cable clamp*

4.6a Disconnect the expansion tank hoses (lower hose shown) . . .

4.6b . . . remove the intercooler bolts (A, diesel models only), then remove the bolts (B) to remove the battery tray and coolant reservoir as one unit

and windshield washer fluid reservoir as one assembly.

6 To remove the battery tray for the secondary (left-side) battery (if equipped), drain the cooling system until the coolant level is below the coolant expansion tank (see Chapter 1) and disconnect the hoses from the expansion tank (see illustrations). On diesel models, unbolt the intercooler from the battery tray (see Chapter 4B, Section 23). Remove the four bolts and remove the battery tray and coolant expansion tank assembly.

Installation

7 If you are replacing the battery, make sure you get one that's identical, with the same dimensions, amperage rating, cold cranking rating, etc.

8 Installation is the reverse of removal. Connect the cable to the positive battery terminal(s) first, then connect the ground cable(s) to the negative battery terminal(s).

9 If the left-side battery tray was removed, refill the cooling system (see Chapter 1).

5 Battery cables - replacement

1 When removing the cables, always disconnect the cable(s) from the negative battery terminal first and hook it up last, or you might accidentally short out the battery with the tool you're using to loosen the cable clamps. Even if you're only replacing a cable for the positive terminal, be sure to disconnect the negative cable(s) from the battery first.

2 Disconnect the old cables from the battery or the mega-fuse terminal, then trace each of them to their opposite ends and disconnect them. Be sure to note the routing of each cable before disconnecting it to ensure correct installation.

3 If you are replacing any of the old cables, take them with you when buying new cables. It is vitally important that you replace the cables with identical parts.

4 Clean the threads of the solenoid or ground connection with a wire brush to remove rust and corrosion. Apply a light coat of battery terminal corrosion inhibitor or petroleum jelly to the threads to prevent future corrosion.

5 Attach the cable to the solenoid or ground connection and tighten the mounting nut/bolt securely.

6 Before connecting a new cable to the battery, make sure that it reaches the battery post without having to be stretched.

7 Connect the cable to the positive battery terminal first, then connect the ground cable to the negative battery terminal.

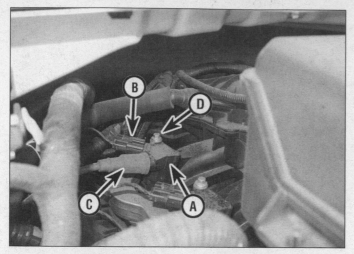

6.3 Ignition coil details

A   Ignition coil   C   Spark plug wire
B   Coil electrical connector   D   Coil bolt

7.6 Remove the four alternator mounting bolts (two bolts shown - gasoline models)

7.14 Disconnect the B+ terminal and unplug the connector on the rear of the alternator

7.15 Location of the alternator mounting bolts (diesel models)

6   Ignition coils - replacement

Note: *The 6.2L engine is equipped with 16 spark plugs. The upper spark plugs are coil-on-plug while the lower spark plugs have a spark plug wire that comes from the ignition coil.*

1   Disconnect the cable(s) from the negative battery terminal(s) (see Section 3).
2   Remove the air intake duct and expansion resonator (see Chapter 4A, Section 12).
3   Remove the spark plug wire from the coil being removed (see illustration).
4   Disconnect the ignition coil electrical connector (see illustration 6.3).
5   Remove the bolt securing the ignition coil (see illustration 6.3), then pull the coil from the upper spark plug using a twisting motion to remove it from the cylinder head.
6   Installation is the reverse of the removal procedure with the following additions:

a)   Prior to installing the coil, coat the interior of the rubber boot with silicone dielectric compound.
b)   Connect each coil electrical connector to its correct coil and make sure they are tight and secure.

7   Alternator - removal and installation

Gasoline engine

1   Disconnect the cable(s) from the negative battery terminal(s) (see Section 3).
2   Remove the expansion resonator from the throttle body and air filter housing (see Chapter 4A, Section 12).
3   Remove the drivebelt (see Chapter 1).
4   Remove the alternator harness clamp from the alternator stud.
5   Disconnect the electrical connectors from the alternator.
6   Remove the four bolts from the alterna-

tor bracket and remove the bracket (see illustration).
7   Remove the two lower mounting bolts and separate the alternator from the engine.
8   Installation is the reverse of removal.
9   Install the drivebelt and reconnect the cable to the negative terminal of the battery.

Diesel engine

10   Disconnect the cables from the negative battery terminals (see Section 3).
11   Remove the intercooler inlet and outlet tubes (see Chapter 4B, Section 23).
12   For a secondary (right-side) alternator, remove the air intake tube (see Chapter 4B, Section 22).
13   Remove the drivebelt (see Chapter 1).
14   Disconnect the wiring from the alternator (see illustration).
15   Remove the two bolts and separate the alternator from the engine (see illustration).
16   Installation is the reverse of removal.
17   Install the drivebelt and reconnect the cables to the negative battery terminals.

8.7 Locating the stud and ground strap (gasoline models)

8.8a Location of the short starter motor mounting bolts (diesel models)

8.8b Location of the long starter motor mounting bolt (diesel models) looking from under vehicle

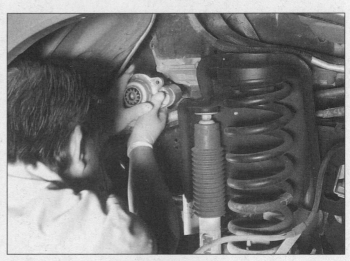

8.8c On diesel models, guide the starter up and over the frame rail

8   Starter motor - removal and installation

1   Disconnect the cable(s) from the negative battery terminal(s) (see Section 4). If you're working on a diesel model, loosen the right front wheel lug nuts.
2   Raise the vehicle and support it securely on jackstands. If you're working on a diesel model, remove the right-front wheel.

3   On diesel models, remove the inner fender splash shield (see Chapter 11).
4   If the starter has a heat shield, remove it. Remove the protective plastic cap and disconnect the wiring from the terminals on the starter solenoid.
5   Remove any harness clamps or straps and position the harness aside.
6   On diesel models, disconnect the oil pressure switch and oil temperature sensor connectors and if equipped, disconnect the

block heater from the engine block.
7   On gasoline models, remove the nut and ground strap from the starter bolt (see illustration).
8   On all models, remove the starter motor mounting bolts (see illustrations) and detach the starter from the engine.
9   Installation is the reverse of removal. Tighten the starter mounting bolts to the torque listed in this Chapter's Specifications.

Notes

Chapter 6 Part A
Emissions and engine control systems - gasoline engine

Contents

Specifications

Torque specifications

Ft-lbs (unless otherwise indicated)

Note: *One foot-pound (ft-lb) of torque is equivalent to 12 inch-pounds (in-lbs) of torque. Torque values below approximately 15 foot-pounds are expressed in inch-pounds, because most foot-pound torque wrenches are not accurate at these smaller values.*

Cylinder head temperature sensor	97 in-lbs
Knock sensor bolts	177 in-lbs
VCT variable force solenoid bolts	
Step 1	89 in-lbs
Step 2	Tighten additional 45 degrees
CMP sensor mounting bolt	89 in-lbs
CKP sensor mounting bolt	89 in-lbs

1   General Information

1   To prevent pollution of the atmosphere from incompletely burned and evaporating gases, and to maintain good driveability and fuel economy, a number of emission control systems are incorporated (see illustrations). They include the:

a) *On-Board Diagnostic (OBD) II system*
b) *Electronic Fuel Injection (EFI) system*
c) *Evaporative Emissions Control (EVAP) system*
d) *Positive Crankcase Ventilation (PCV) system*
e) *Catalytic converter(s)*

2   Before assuming that an emissions control system is malfunctioning, check the fuel and ignition systems carefully. The diagnosis of some emission control devices requires specialized tools, equipment and training. If checking and servicing become too difficult or if a procedure is beyond your ability, consult a dealer service department or other repair shop. Remember, the most frequent cause of emissions problems is simply a loose or broken wire or vacuum hose, so always check the hose and wiring connections first.

3   This doesn't mean, however, that emissions control systems are particularly difficult to maintain and repair. You can quickly and easily perform many checks and do most of the regular maintenance at home with common tune-up and hand tools.

Note: *Because of a Federally mandated warranty which covers the emissions control system components, check with your dealer about warranty coverage before working on any emissions-related systems. Once the warranty has expired, you may wish to perform some of the component checks and/or replacement procedures in this Chapter to save money.*

4   Pay close attention to any special precautions outlined in this Chapter. It should be noted that the illustrations of the various systems may not exactly match the system

installed on your vehicle because of changes made by the manufacturer during production or from year-to-year.

5   A Vehicle Emissions Control Information (VECI) label is attached to the underside of the hood (see illustration). This label contains information regarding the types of emissions control systems installed on the vehicle. When servicing the engine or emissions systems, the VECI label on your particular vehicle should always be checked for up-to-date information.

2   On Board Diagnosis (OBD) system and diagnostic trouble codes

General description

1   All models are equipped with the second generation OBD-II system. This system consists of an on-board computer known as the Powertrain Control Module (PCM), and information sensors, which monitor various functions of the engine and send data to the PCM. This system incorporates a series of diagnostic monitors that detect and identify fuel injection and emissions control system faults and store the information in the computer memory. This system also tests sensors and output actuators, diagnoses drive cycles, freezes data and clears codes.

2   The PCM is the brain of the electronically controlled fuel and emissions system. It receives data from a number of sensors and other electronic components (switches, relays, etc.). Based on the information it receives, the PCM generates output signals to control various relays, solenoids (fuel injectors) and other actuators. The PCM is specifically calibrated to optimize the emissions, fuel economy and driveability of the vehicle.

3   It isn't a good idea to attempt diagnosis or replacement of the PCM or emission control components at home while the vehicle

is under warranty. Because of a federally-mandated warranty which covers the emissions system components and because any owner-induced damage to the PCM, the sensors and/or the control devices may void this warranty, take the vehicle to a dealer service department if the PCM or a system component malfunctions.

Scan tool information

4   Because extracting the Diagnostic Trouble Codes (DTCs) from an engine management system is now the first step in troubleshooting many computer-controlled systems and components, a code reader, at the very least, will be required (see illustration). More powerful scan tools can also perform many of the diagnostics once associated with expensive factory scan tools (see illustration). If you're planning to obtain a generic scan tool for your vehicle, make sure that it's compatible with OBD-II systems. If you don't plan to purchase a code reader or scan tool and don't have access to one, you can have the codes extracted by a dealer service department or an independent repair shop.

Note: *Some auto parts stores even provide this service.*

Obtaining and clearing Diagnostic Trouble Codes (DTCs)

5   Before outputting any DTCs stored in the PCM, thoroughly inspect ALL electrical connectors and hoses. Make sure that all electrical connections are tight, clean and free of corrosion. Make sure that all hoses are correctly connected, fit tightly and are in good condition (no cracks or tears). Also, make sure that the engine is tuned up. A poorly running engine is probably one of the biggest causes of emission-related malfunctions. Often, simply giving the engine a good tune-up will correct the problem.

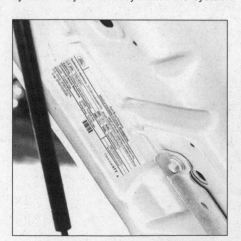

1.5 The Vehicle Emission Control Information (VECI) label contains such essential information as the types of emission control systems installed on the engine

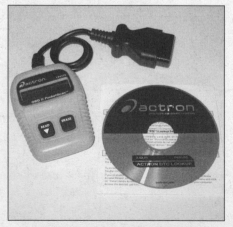

2.4a Simple code readers are an economical way to extract trouble codes when the CHECK ENGINE light comes on

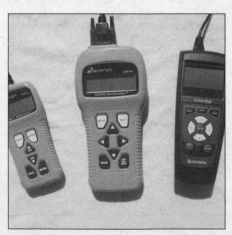

2.4b Hand-held scan tools like these can extract computer codes and also perform diagnostics

Information Sensors

Accelerator Pedal Position (APP) sensor - as you press the accelerator pedal, the APP sensor alters its voltage signal to the PCM in proportion to the angle of the pedal, and the PCM commands a motor inside the throttle body to open or close the throttle plate accordingly

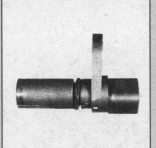

Camshaft Position (CMP) sensor - produces a signal that the PCM uses to identify the number 1 cylinder and to time the firing sequence of the fuel injectors

Crankshaft Position (CKP) sensor - produces a signal that the PCM uses to calculate engine speed and crankshaft position, which enables it to synchronize ignition timing with fuel injector timing, and to detect misfires

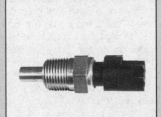

Engine Coolant Temperature (ECT) sensor - a thermistor (temperature-sensitive variable resistor) that sends a voltage signal to the PCM, which uses this data to determine the temperature of the engine coolant

Fuel tank pressure sensor - measures the fuel tank pressure and controls fuel tank pressure by signaling the EVAP system to purge the fuel tank vapors when the pressure becomes excessive

Intake Air Temperature (IAT) sensor - monitors the temperature of the air entering the engine and sends a signal to the PCM to determine injector pulse-width (the duration of each injector's on-time) and to adjust spark timing (to prevent spark knock)

Knock sensor - a piezoelectric crystal that oscillates in proportion to engine vibration which produces a voltage output that is monitored by the PCM. This retards the ignition timing when the oscillation exceeds a certain threshold

Manifold Absolute Pressure (MAP) sensor - monitors the pressure or vacuum inside the intake manifold. The PCM uses this data to determine engine load so that it can alter the ignition advance and fuel enrichment

Mass Air Flow (MAF) sensor - measures the amount of intake air drawn into the engine. It uses a hot-wire sensing element to measure the amount of air entering the engine

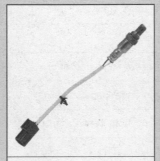

Oxygen sensors - generates a small variable voltage signal in proportion to the difference between the oxygen content in the exhaust stream and the oxygen content in the ambient air. The PCM uses this information to maintain the proper air/fuel ratio. A second oxygen sensor monitors the efficiency of the catalytic converter

Throttle Position (TP) sensor - a potentiometer that generates a voltage signal that varies in relation to the opening angle of the throttle plate inside the throttle body. Works with the PCM and other sensors to calculate injector pulse width (the duration of each injector's on-time)

Photos courtesy of Wells Manufacturing, except APP and MAF sensors.

Accessing the DTCs

6   On these models, all of which are equipped with On-Board Diagnostic II (OBD-II) systems, the Diagnostic Trouble Codes (DTCs) can only be accessed with a code reader or a scan tool (see illustrations 2.4a and 2.4b). Simply plug the connector of the tool into the Data Link Connector (DLC) or diagnostic connector (see illustration), which is located under the lower edge of the dash, just to the right of the steering column. Then follow the instructions included with the scan tool to extract the DTCs.

7   Once you have viewed all of the stored DTCs, look them up on the accompanying DTC chart.

8   After troubleshooting the source of each DTC, make any necessary repairs or replace the defective component(s).

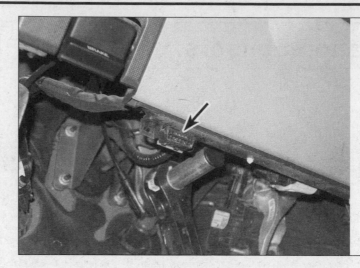

2.6 The 16-pin Data Link Connector (DLC), also referred to as the diagnostic connector, is located under the left part of the dash

Clearing the DTCs

9   Clear the DTCs with the scan tool in accordance with the instructions provided by the scan tool's manufacturer.

Trouble codes - gasoline engines

OBD-II trouble codes

Note: *Not all trouble codes apply to all models.*

Code	Probable cause
P0010	Intake camshaft position actuator, circuit open
P0011	Intake camshaft position actuator, over advanced
P0012	Intake camshaft position timing, over retarded
P0016	Crankshaft position correlation, bank 1, sensor A
P0020	Intake camshaft position actuator, circuit open, Bank 2
P0021	Intake camshaft position actuator, over advanced, Bank 2
P0022	Intake camshaft position actuator, over retarded, Bank 2
P0030	Upstream oxygen sensor, bank 1, heater circuit
P0040	Upstream oxygen sensors swapped (crossed wiring harnesses)
P0041	Downstream oxygen sensors swapped (crossed wiring harnesses)
P0050	Oxygen sensor heater control circuit open or short, Bank 2, sensor 1
P0054	Oxygen sensor heater control circuit resistance, Bank 1, sensor 2
P0055	Oxygen sensor heater control circuit resistance, Bank 1, sensor 3
P0059	Oxygen sensor heater control circuit resistance, Bank 2, sensor 1
P0060	Downstream oxygen sensor heater circuit open or short, Bank 2, sensor 2
P060A	Internal Powertrain Control Module (PCM) error
P060B	Internal Powertrain Control Module (PCM) analog to digital processing error

Code	Probable cause
P060C	Internal Powertrain Control Module (PCM) main processor error
P061B	Internal Powertrain Control Module (PCM) torque calculation error
P061C	Internal Powertrain Control Module (PCM) engine rpm calculation error
P061D	Internal Powertrain Control Module (PCM) air mass error
P061F	Internal Powertrain Control Module (PCM) Throttle Actuator Controller (TAC) error
P0068	Throttle Position (TP) sensor inconsistent with Mass Air Flow sensor
P0097	Intake Air Temperature sensor circuit, sensor 2, circuit low
P0098	Intake Air Temperature sensor circuit, open or short
P0102	Mass Air Flow (MAF) sensor circuit, low input
P0103	Mass Air Flow (MAF) sensor circuit, high input
P0104	Mass Air Flow (MAF) sensor circuit, intermittent failure
P0106	Barometric (BARO) pressure sensor circuit, performance problem
P0107	Barometric (BARO) pressure sensor/MAP sensor circuit, low voltage
P0108	Barometric (BARO) pressure sensor/MAP sensor circuit, high voltage
P0109	BARO/MAP sensor circuit intermittent
P0111	Intake Air Temperature (IAT) sensor 1 circuit, range/performance problem
P0112	Intake Air Temperature (IAT) sensor 1 circuit, low input
P0113	Intake Air Temperature (IAT) sensor circuit, high input
P0114	Intake Air Temperature (IAT) sensor circuit, intermittent failure
P0117	Engine Coolant Temperature (ECT) sensor circuit, low input
P0118	Engine Coolant Temperature (ECT) sensor circuit, high input
P0121	Throttle Position (TP) circuit out of range or performance problem
P0122	Throttle pedal position sensor, A circuit low
P0123	Throttle Position (TP) sensor circuit, A high input
P0127	IAT2 sensor, problem with Charge Air Cooler
P012B	Turbocharger inlet pressure sensor, circuit range/performance
P012C	Turbocharger inlet pressure sensor, circuit low
P012D	Turbocharger inlet pressure sensor, circuit high
P012E	Turbocharger inlet pressure sensor, circuit erratic
P0130	Oxygen sensor heater control circuit resistance, Bank 1, sensor 1
P0131	Upstream oxygen sensor circuit problem (right cylinder bank)

OBD-II trouble codes (continued)

Note: *Not all trouble codes apply to all models.*

Code	Probable cause
P0132	Upstream oxygen sensor circuit, high voltage (right cylinder bank)
P0134	Oxygen sensor circuit, open or damaged sensor
P0136	Downstream oxygen sensor circuit problem (right cylinder bank)
P0138	Downstream oxygen sensor circuit, high voltage (right cylinder bank)
P0139	Upstream oxygen sensor heater circuit, slow response, Bank 1, sensor 2
P0141	Downstream oxygen sensor heater circuit problem, Bank 1, sensor 2
P0144	Oxygen sensor, Bank 1, sensor 3, high voltage
P0147	Oxygen sensor, Bank 1, sensor 3, circuit open or short
P0148	Fuel delivery error
P0150	Oxygen sensor heater control circuit resistance, Bank 2, sensor 1
P0151	Upstream oxygen sensor circuit, low voltage (left cylinder bank)
P0152	Upstream oxygen sensor circuit, high voltage (left cylinder bank)
P0153	Heated oxygen sensor circuit, slow response (left cylinder bank)
P0154	Upstream oxygen sensor heater circuit problem, Bank 2, sensor 1
P0155	Oxygen sensor, Bank 2, sensor 2, circuit open or short
P0156	Downstream oxygen sensor circuit problem (left cylinder bank)
P0158	Downstream oxygen sensor circuit, high voltage (left cylinder bank)
P0159	Downstream oxygen sensor circuit, slow response, Bank 2, sensor 2
P0161	Downstream oxygen sensor heater circuit problem, Bank 2, sensor 2
P0171	System too lean (right cylinder bank)
P0172	System too rich (right cylinder bank)
P0174	System too lean (left cylinder bank)
P0175	System too rich (left cylinder bank)
P0176	Flexible Fuel (FF) sensor circuit malfunction
P0180	Fuel Rail Temperature (FRT) sensor A circuit, open or short
P0181	Fuel Rail Temperature (FRT) sensor A circuit, range/performance
P0182	Fuel Rail Temperature (FRT) sensor circuit, low input
P0183	Fuel Rail Temperature (FRT) sensor circuit, high input
P0190	Fuel Rail Pressure (FRP) sensor A circuit, reference voltage

Code	Probable cause
P0191	Fuel Rail Pressure (FRP) sensor circuit, range/performance
P0192	Fuel Rail Pressure (FRP) sensor circuit, low input
P0193	Fuel Rail Pressure (FRP) sensor circuit, high input
P0196	Engine Oil Temperature sensor circuit range/performance problem
P0197	Engine Oil Temperature sensor circuit, low input
P0198	Engine Oil Temperature sensor circuit, high input
P0201	Injector no. 1 circuit malfunction
P0202	Injector no. 2 circuit malfunction
P0203	Injector no. 3 circuit malfunction
P0204	Injector no. 4 circuit malfunction
P0205	Injector no. 5 circuit malfunction
P0206	Injector no. 6 circuit malfunction
P0207	Injector no. 7 circuit malfunction
P0208	Injector no. 8 circuit malfunction
P0217	Engine coolant over-temperature
P0218	Transmission fluid over-temperature
P0219	Engine over speed condition
P0221	Throttle Position (TP) sensor B circuit range/performance problem
P0222	Throttle Position (TP) sensor B circuit, low input
P0223	Throttle Position (TP) sensor B circuit, high input
P0230	Fuel pump primary circuit malfunction
P0231	Fuel pump secondary circuit low
P0232	Fuel pump secondary circuit high
P0261	Cylinder 1 injector circuit low
P0262	Cylinder 1 injector circuit high
P0264	Cylinder 2 injector circuit low
P0265	Cylinder 2 injector circuit high
P0267	Cylinder 3 injector circuit low
P0268	Cylinder 3 injector circuit high
P0270	Cylinder 4 injector circuit low
P0271	Cylinder 4 injector circuit high

OBD-II trouble codes (continued)

Note: *Not all trouble codes apply to all models.*

Code	Probable cause
P0273	Cylinder 5 injector circuit low
P0274	Cylinder 5 injector circuit high
P0276	Cylinder 6 injector circuit low
P0277	Cylinder 6 injector circuit high
P0297	Vehicle overspeed condition
P0298	Engine oil over temperature condition
P0300	Random misfire detected
P0301	Cylinder no. 1 misfire detected
P0302	Cylinder no. 2 misfire detected
P0303	Cylinder no. 3 misfire detected
P0304	Cylinder no. 4 misfire detected
P0305	Cylinder no. 5 misfire detected
P0306	Cylinder no. 6 misfire detected
P0307	Cylinder no. 7 misfire detected
P0308	Cylinder no. 8 misfire detected
P0315	PCM unable to learn crankshaft pulse wheel tooth spacing
P0316	Misfire occurred during first 1000 engine revolutions
P0320	Ignition engine speed input circuit malfunction
P0325	Knock sensor 1 circuit malfunction (right cylinder head)
P0326	Knock sensor 1 circuit range/performance (right cylinder bank)
P0330	Knock sensor 2 circuit malfunction (left cylinder bank)
P0331	Knock sensor 2 circuit range/performance (left cylinder bank)
P0340	Camshaft Position (CMP) sensor circuit malfunction (right cylinder head)
P0341	Camshaft position sensor, A circuit, Bank 1, range or performance
P0344	Camshaft position sensor, A circuit, intermittent
P0345	Camshaft Position (CMP) sensor circuit malfunction (left cylinder head)
P0346	Camshaft position sensor, A circuit, range or performance
P0349	Camshaft position sensor, A circuit, Bank 2, intermittent
P0350	Ignition coil primary or secondary circuit malfunction

Code	Probable cause
P0351-P0358	Ignition coil primary or secondary circuit malfunction, coils 1 through 8
P0400	EGR flow failure (outside the minimum or maximum limits)
P0401	Exhaust Gas Recirculation (EGR) valve, insufficient flow detected
P0402	Exhaust Gas Recirculation (EGR) valve, excessive flow detected
P0403	EEGR electric motor windings or circuits to PCM shorted or open (6.8L engine)
P0405	EGR sensor, A circuit, low
P0406	EGR sensor, A circuit, high
P0410	Secondary Air Injection (AIR) system, low flow
P0411	Secondary Air Injection (AIR) system, upstream flow
P0412	Secondary Air Injection (AIR) system, circuit malfunction
P0420	Catalyst system efficiency below threshold (right cylinder bank)
P0430	Catalyst system efficiency below threshold (left cylinder bank)
P0442	EVAP control system, small leak detected
P0443	EVAP control system, canister purge valve circuit malfunction
P0446	EVAP control system canister vent solenoid circuit malfunction
P0451	Fuel tank pressure sensor circuit out of range or performance problem
P0452	Fuel tank pressure sensor circuit, low input
P0453	Fuel tank pressure sensor circuit, high input
P0454	Fuel tank pressure sensor circuit, noisy
P0455	EVAP control system, big leak detected
P0456	EVAP control system, very small leak detected
P0457	EVAP control system, leak detected (fuel filler neck cap loose or off)
P0460	Fuel level sensor circuit malfunction
P0461	Fuel level sensor circuit range or performance problem
P0462	Fuel level sensor circuit, low input
P0463	Fuel level sensor circuit, high input
P0480	Fan 1, control primary circuit malfunction
P0481	Fan 2, control circuit, open or short
P0482	Medium Fan Control (MFC) primary circuit failure
P0483	Cooling fan, binding or mechanical failure
P0491	Secondary Air Injection (AIR) system, Bank 1, low flow

OBD-II trouble codes (continued)

Note: *Not all trouble codes apply to all models.*

Code	Probable cause
P0500	Vehicle Speed Sensor (VSS), circuit malfunction
P0501	Vehicle Speed Sensor (VSS) range/performance problem
P0503	Vehicle Speed Sensor (VSS), intermittent malfunction
P0504	Brake switch circuit, correlation to brake position switch
P0505 - P0507	Idle Air Control (IAC) system malfunction - on models w/o IAC, code indicates restricted air intake or damaged throttle body preventing proper idle
P0511	Idle Air Control (IAC) circuit malfunction
P0512	Starter relay circuit, short
P0528	Visctronic Drive Fan (VDF) speed sensor circuit malfunction
P052A	Cold start cam position timing, Bank 1, over-advanced
P052B	Cold start cam position timing, Bank 1, over-retarded
P052C	Cold start cam position timing, Bank 2, over-advanced
P052D	Cold start cam position timing, Bank 2, over-retarded
P0532	Air conditioning pressure sensor circuit, low voltage
P0533	Air conditioning pressure sensor circuit, high voltage
P0534	Low air conditioning cycling period
P0537	Air conditioning evaporator temperature circuit, low input
P0538	Air conditioning evaporator temperature circuit, high input
P053A	PCV heater system, Bank 1, control circuit open
P0552	Power Steering Pressure (PSP) sensor circuit malfunction
P0553	Power Steering Pressure (PSP) sensor circuit malfunction
P0562	System voltage low
P0563	System voltage high
P0571	Brake switch, A circuit
P0572	Brake switch, A circuit, low voltage
P0573	Brake switch, A circuit, high voltage
P0579	Cruise control multifunction input A circuit range or performance problem
P0581	Cruise control multifunction input A circuit, high
P0600	Serial communication link (PCM) error
P0602	Control module programming error

Code	Probable cause
P0603	Powertrain Control Module (PCM) Keep-Alive-Memory (KAM) test error
P0604	Powertrain Control Module (PCM) Random (RAM memory corrupted
P0605	Powertrain Control Module (PCM) Read-Only-Memory (ROM) error
P0606	Powertrain Control Module (PCM), internal communication error
P0607	Powertrain Control Module (PCM), PCM needs reprogramming
P060A to P060D	Powertrain Control Module (PCM), processor performance
P0610	Powertrain Control Module (PCM), vehicle options error
P0611	Fuel injector control module performance
P0620	Generator control circuit failure
P0622	Generator field terminal, circuit failure
P0625	Generator field terminal, circuit low
P0626	Generator field terminal, circuit high
P0627	Fuel pump, A circuit, control/open
P062F	Powertrain Control Module (PCM) EEPROM error
P0634	PCM/ECM/TCM internal temperature too high
P0641	Sensor reference voltage A circuit or open
P0642	Reference voltage circuit, low voltage
P0643	Reference voltage circuit, high voltage
P0645	Air conditioning clutch relay (wide open throttle a/c cutoff) primary circuit malfunction
P0657	Transmission solenoid actuator, supply voltage circuit open
P0685	Powertrain control module, power relay control circuit open
P0689	Ignition switch or PATS circuits, open or short
P0690	Powertrain control module, power relay control circuit, high
P0703	Brake Pedal Position (BPP) switch circuit input malfunction
P0704	Clutch pedal position switch malfunction
P0705	Transmission range sensor, A circuit input
P0707	Transmission range sensor, A circuit, low
P0708	Transmission range sensor, A circuit, high
P0720	Insufficient input from Output Shaft Speed (OSS) sensor
P0721	Noise interference on Output Shaft Speed (OSS) sensor signal
P0722	No signal from Output Shaft Speed (OSS) sensor

OBD-II trouble codes (continued)

Note: *Not all trouble codes apply to all models.*

Code	Probable cause
P0723	Output Shaft Speed (OSS) sensor circuit, intermittent failure
P0812	Reverse Switch (RS) input circuit malfunction
P0815	Automatic transmission, upshift switch circuit
P0830	Clutch pedal switch A circuit
P0833	Clutch pedal switch, B circuit
P0840	Transmission fluid temp/pressure sensor, switch A circuit

3   Powertrain Control Module (PCM) - removal and installation

Warning: *The models covered by this manual are equipped with a Supplemental Restraint System (SRS), more commonly known as airbags. Always disable the airbag system before working in the vicinity of any airbag system components to avoid the possibility of accidental deployment of the airbag, which could cause personal injury (see Chapter 12).*

Caution: *To avoid electrostatic discharge damage to the PCM, handle the PCM only by its case. Do not touch the electrical terminals during removal and installation. If available, ground yourself to the vehicle with an anti-static ground strap, available at computer supply stores.*

Note: *The replacement of the PCM requires the original programming to be retrieved via a scan tool before PCM removal. The programming would then be downloaded to the new PCM and the original instrument cluster. This is a procedure best performed at a dealership.*

1   The Powertrain Control Module (PCM) is located in the cowl, accessible from the engine compartment (see illustration).

2   Disconnect the cable(s) from the negative battery terminal(s) (see Chapter 5, Section 3).

3   Disconnect the electrical connectors from the PCM (see illustration).

4   Remove the two nuts and pull the PCM from the cowl.

5   Installation is the reverse of removal.

4   Throttle Position Sensor (TPS) - replacement

1   The Throttle Position Sensor (TPS) is part of the throttle body and is serviced as one unit. See Chapter 4A, Section 13.

3.1 The PCM is located on the right side of the cowl

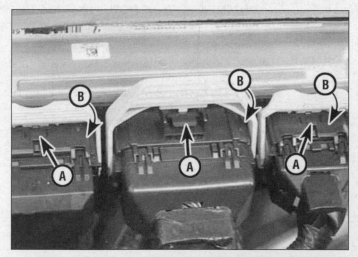

3.3 To disconnect the electrical connectors from the PCM, depress the lock tabs (A) and swing open the levers (B)

5   Mass Airflow (MAF) sensor - replacement

Note: *A drop-in style MAF sensor is used and is located in the air inlet duct.*

1   Disconnect the cable from the negative battery terminal.
2   Disconnect the electrical connector from the MAF sensor (see illustration).
3   Remove the MAF sensor mounting screws and remove the MAF sensor from the air inlet duct.
4   Installation is the reverse of removal.

6   Intake Air Temperature (IAT) sensor - replacement

1   The Intake Air Temperature (IAT) sensor is part of the Mass Airflow (MAF) and is replaced along with the MAF. See Section 5.

7   Cylinder Head Temperature (CHT) sensor - replacement

Note: *The cylinder head temperature sensor is a major component of the Fail Safe Cooling system. This sensor varies the value of its voltage output in accordance with temperature changes. The change in the resistance values will directly affect the voltage signal from the CHT sensor. As the sensor temperature decreases, the resistance values will increase (voltage increases). If the cylinder head temperature exceeds 265-degrees F, the PCM disables four fuel injectors at a time. The cylinders that do not receive fuel act as cooling air pumps for the other cylinders. If the temperature exceeds 330-degrees F, the PCM disables all the fuel injectors.*

1   To remove the cylinder head temperature sensor, the intake manifold must be removed from the engine (see Chapter 2A).

2   Locate and unplug the electrical connector (see illustration), then carefully unscrew the CHT sensor.
Caution: *Handle the CHT sensor with care. Damage to this sensor will affect the operation of the entire fuel injection system.*
3   Install the sensor and tighten it to the torque listed in this Chapter's Specifications.
4   Remainder of installation is the reverse of removal.

8   Crankshaft Position (CKP) sensor - replacement

Note: *The crankshaft position sensor (CKP) determines the timing on each cylinder for the fuel injectors and ignition system. The crankshaft sensor is mounted in the crankshaft rear main oil seal retainer plate. A problem in the crankshaft sensor circuit will set a diagnostic trouble code.*
Note: *After installing the CKP sensor, use a scan tool to perform the Misfire Monitor Neutral Profile Correction procedure.*
1   Remove the intake manifold (see Chapter 2A, Section 11).
2   Locate and disconnect the crankshaft position sensor electrical connector.
3   Remove the bolt and pull the sensor out of the crankshaft seal retainer plate.
Note: *The CKP is very long.*
Caution: *The Crankshaft Position (CKP) sensor must be flush against the boss on the engine block before the bolt is installed. If the sensor is installed incorrectly, it can be damaged.*
4   Installation is the reverse of removal.

9   Camshaft Position (CMP) sensor - replacement

Note: *The Camshaft Position (CMP) sensor determines the position of the cylinder for ig-*

nition start-up signals and for sequential fuel injection signals to each cylinder. The engine is equipped with two CMPs, one for each camshaft. The CMPs are mounted on the inner flanks of the cylinder heads at the rear of the engine.
1   Locate and disconnect the electrical connector from the camshaft position sensor.
2   Remove the bolt and pull the sensor out of the cylinder head.
3   Lubricate the O-ring with clean engine oil prior to installation.
4   Installation is the reverse of removal.

10   Variable Camshaft Timing (VCT) variable force solenoid - removal and installation

Note: *The Variable Camshaft Timing (VCT) variable force solenoid supplies force to actuate the VCT oil control valve. The engine is equipped with two VCTs, one under each valve cover for each camshaft. The PCM sends a signal based on engine speed and load, and the solenoid moves the VCT oil control valve to advance or retard or hold position. The camshaft is then re-positioned in relation to crankshaft timing to allow for optimum engine performance with lower emissions and reduced fuel consumption.*
1   Disconnect the VCT sensor connector.
2   Remove the valve cover for the sensor to removed (see Chapter 2A, Section 4).
3   Unscrew the bolts and remove the VCT from the cylinder head.
Note: *The bolts are part of the VCT and cannot be removed.*
Caution: *The VCT variable force solenoid pins must be fully depressed to avoid interference with the VCT valve tips when installing the solenoids. Failure to follow these instructions can result in damage to the engine.*
4   Installation is reverse of removal. Tighten the bolts to the torque listed in this Chapter's Specifications.

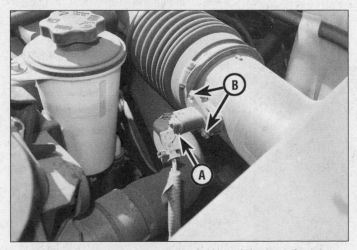

5.2 Slide out the red connector lock (A) and depress the tab to disconnect the electrical connector, then remove the MAF sensor mounting screws (B)

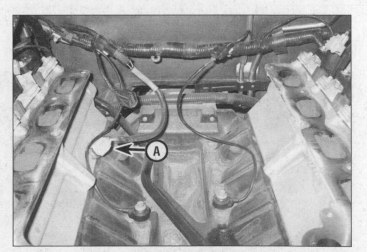

7.2 CHT sensor (A)

11 Oxygen sensors - general information and replacement

General information

1   All gasoline models covered by this manual have On-Board Diagnostics II (OBD-II) engine management systems, which means that they have the ability to verify the accuracy of the basic feedback loop between the oxygen sensor and the PCM. They accomplish this by using an oxygen sensor in front of the catalytic converter and an oxygen sensor behind the catalytic converter. By sampling the exhaust gas before and after the catalytic converter, the PCM can determine the efficiency of the converter and can even predict when it will fail.

2   The primary (upstream) oxygen sensor is located in the exhaust manifold and the secondary (downstream) oxygen sensor is located behind the catalytic converter. The upstream and downstream oxygen sensors on all models is a heated oxygen sensor. The PCM uses a supply wire and ground wire to control the power to the O2 sensor heater during warm-up.

Note: *The downstream oxygen sensors may also be referred to as Catalyst Monitor Sensor (CMS).*

3   Special care must be taken whenever a sensor is serviced.

a) *Oxygen sensors have a permanently attached pigtail and an electrical connector which should not be removed from the sensor. Damage or removal of the pigtail or electrical connector can adversely affect operation of the sensor.*

b) *Grease, dirt and other contaminants should be kept away from the electrical connector and the louvered end of the sensor.*

c) *Do not use cleaning solvents of any kind on an oxygen sensor.*

d) *Do not drop or roughly handle an oxygen sensor.*

e) *The silicone boot must be installed in the correct position to prevent the boot from being melted and to allow the sensor to operate properly.*

Replacement

Note: *Because it is installed in the exhaust manifold or pipe, which contracts when cool, the oxygen sensor may be very difficult to loosen when the engine is cold. Rather than risk damage to the sensor, assuming you are planning to reuse it in another manifold or pipe, start and run the engine for a minute or two, then shut it off. Be careful not to burn yourself during the following procedure.*

4   Raise the vehicle and secure it on jackstands. Access the oxygen sensor harness and unplug the electrical connector.

5   Unscrew the sensor from the exhaust manifold or exhaust pipe (see illustration).

Note: *The best tool for removing an oxygen sensor is a special slotted socket, especially if you're planning to reuse a sensor. If you don't have this tool, and you plan to reuse the sensor, be extremely careful when unscrewing the sensor.*

6   Apply anti-seize compound to the threads of the sensor to facilitate future removal. The threads of new sensors should already be coated with this compound, but if you're planning to reuse an old sensor, recoat the threads. Install the sensor and tighten it securely.

7   Reconnect the electrical connector of the pigtail lead to the main wiring harness.

8   Lower the vehicle, test drive the car and verify that no trouble codes have been set.

12 Knock sensor - replacement

Note: *The knock sensor are located under the intake manifold. The knock control system is designed to reduce spark knock during periods of heavy detonation. This allows the engine to use optimal spark advance to improve driveability. The knock sensor detects abnormal vibration in the engine and produces a voltage output which increases with the severity of the knock. The voltage signal is monitored by the PCM, which retards ignition timing until the detonation ceases. A problem in the knock sensor circuit will set a diagnostic trouble code.*

1   Remove the intake manifold (see Chapter 2A).

2   Disconnect the electrical connector and remove the bolt securing the knock sensor to the lifter valley and remove the knock sensor.

3   Coat the threads of the knock sensor bolt with thread sealant. New sensor bolts are pre-coated with thread sealant; do not apply any additional sealant or the operation of the sensor may be affected.

4   Install the knock sensor and tighten the bolt to the torque listed in this Chapter's Specifications.

Caution: *Do not tighten the knock sensor bolt without a torque wrench. The knock sensor readings may be inaccurate or damage may occur to the knock sensor or engine block.*

5   Remainder of installation is the reverse of removal.

13 Fuel tank pressure (FTP) sensor - replacement

Warning: *Gasoline is extremely flammable, so take extra precautions when you work on any part of the fuel system. Don't smoke or allow open flames or bare light bulbs near the work area, and don't work in a garage where a gas-type appliance (such as a water heater or clothes dryer) is present. Since gasoline is carcinogenic, wear fuel-resistant gloves when there's a possibility of being exposed to fuel, and, if you spill any fuel on your skin, rinse it off immediately with soap and water. Mop up any spills immediately and do not store fuel-soaked rags where they could ignite. When you perform any kind of work on the fuel system, wear safety glasses and have a Class B type fire extinguisher on hand. The fuel system is under pressure, so if any lines must be disconnected, the pressure in the system must be relieved first (see Chapter 4A for more information).*

1   The fuel tank pressure (FTP) sensor is used to monitor the fuel tank pressure or vacuum during the OBD-II test portion for emissions integrity. This test scans various sensors and output actuators to detect abnormal amounts of fuel vapors that may not be purging into the canister or the intake system for recycling. The FTP sensor helps the PCM monitor this pressure differential (pressure vs. vacuum) inside the fuel tank. A problem in the fuel tank pressure sensor circuit will set a diagnostic trouble code.

Note: *The Fuel Tank Pressure (FTP) sensor is part of the vapor tube assembly and serviced as one unit.*

2   Remove the fuel tank (see Chapter 4A, Section 6).

3   Disconnect the vapor tube from the quick-disconnect fittings and remove the vapor tube and FTP sensor.

4   Installation is the reverse of removal.

5   Turn the ignition on and check for leaks.

11.5a Use a slotted socket to remove the oxygen sensor

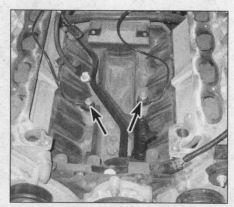

12.2 The knock sensors are located in the valley between the cylinder banks

14 Accelerator pedal module - replacement

1   To replace the accelerator pedal module, see Chapter 4A, Section 14.

15 Positive Crankcase Ventilation (PCV) system

1   The Positive Crankcase Ventilation (PCV) system reduces hydrocarbon emissions by scavenging crankcase vapors. It does this by circulating fresh air from the air cleaner through the crankcase, where it mixes with blow-by gases and is then rerouted through a PCV valve to the intake manifold (on gasoline engines) or through a breather housing on the left valve cover (on diesel engines).

2   The main components of the PCV system are the PCV valve, a blow-by filter and the vacuum hoses connecting these two components with the engine. The PCV valve is located on the rear of the passenger valve cover.

3   To maintain idle quality, the PCV valve restricts the flow when the intake manifold vacuum is high. If abnormal operating conditions (such as piston ring problems) arise, the system is designed to allow excessive amounts of blow-by gases to flow back through the crankcase vent tube into the air cleaner to be consumed by normal combustion.

4   Checking and replacement of the PCV valve is covered in Chapter 1.

16 Evaporative emissions control (EVAP) system

Warning: *Gasoline is extremely flammable, so take extra precautions when you work on any part of the fuel system. Don't smoke or allow open flames or bare light bulbs near the work area, and don't work in a garage where a gas-type appliance (such as a water heater or clothes dryer) is present. Since gasoline is carcinogenic, wear fuel-resistant gloves when there's a possibility of being exposed to fuel, and, if you spill any fuel on your skin, rinse it off immediately with soap and water. Mop up any spills immediately and do not store fuel-soaked rags where they could ignite. When you perform any kind of work on the fuel system, wear safety glasses and have a Class B type fire extinguisher on hand. The fuel system is under pressure, so if any lines must be disconnected, the pressure in the system must be relieved first (see Chapter 4A for more information).*

General description

1   This system is designed to trap and store fuel vapors that evaporate from the fuel tank, throttle body and intake manifold during non-operation or idling, store them in the charcoal canister and then route them into the combustion chamber to be burned during engine operation.

2   The Evaporative Emission Control System (EVAP) consists of a charcoal-filled canister and the lines connecting the canister to the fuel tank, a fuel vapor management valve (VMV), a fuel tank pressure sensor, fuel filler cap, a canister vent solenoid, a fuel vapor vent valve, ported vacuum and intake manifold vacuum.

3   Fuel vapors are transferred from the fuel tank, throttle body and intake manifold to a canister where they are stored when the engine is not operating. When the engine is running, the fuel vapors are purged from the canister by a vapor management valve (VMV), which is PCM controlled and consumed in the normal combustion process. The fuel tank pressure sensor detects internal fuel tank pressure and relays the information to the PCM which in turn regulates the EVAP system purge controls.

Replacement

Charcoal canister

4   Disconnect the negative battery cable (see Chapter 5, Section 3).
5   Remove the spare tire.
6   Remove the fuel filler cap to relieve the pressure inside the fuel tank.
7   Disconnect the charcoal canister quick-disconnect fittings and hose (see illustration).
8   Disconnect the charcoal canister electrical connector.
9   Remove the charcoal canister bracket mounting bolts.
10   Remove the charcoal canister and bracket from the vehicle. Remove the canister from the bracket.
11   Installation is the reverse of removal.

Canister vent solenoid

Note: *The canister vent solenoid is located on the charcoal canister and is part of the lid it is connected to.*

12   Remove the charcoal canister from the vehicle.
13   Remove the canister vent solenoid tube.

14   Spread the four tabs and pull the canister vent solenoid off of the canister.
15   Installation is the reverse of removal. Ensure a click is heard when installing the solenoid to the canister.

Canister purge valve

Note: *The canister purge valve is located on the left (driver's) side of the intake manifold, just behind the throttle body.*

16   Remove the air intake duct and expansion resonator (see Chapter 4A Section 12).
17   Disconnect the electrical connector and quick-connect fitting from the valve.
18   Remove the two bolts and disconnect the valve from the intake manifold.
19   Installation is the reverse of removal. Tighten the purge valve bolts to the torque listed in this Chapter's Specifications.

17 Catalytic converter

Note: *Because of a Federally mandated extended warranty which covers emissions-related components such as the catalytic converter, check with a dealer service department before replacing the converter at your own expense.*

General description

1   The catalytic converter is an emission control device added to the exhaust system to reduce pollutants from the exhaust gas stream. The three-way catalyst lowers the levels of oxides of nitrogen (NOx) as well as hydrocarbons (HC) and carbon monoxide (CO).

Check

2   The test equipment for a catalytic converter is expensive and highly sophisticated. If you suspect that the converter on your vehicle is malfunctioning, take it to a dealer or authorized emissions inspection facility for diagnosis and repair.
3   Whenever the vehicle is raised for servicing of underbody components, check the converter for leaks, corrosion, dents and other

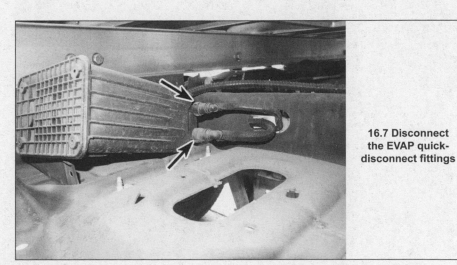

16.7 Disconnect the EVAP quick-disconnect fittings

damage. Check the welds/flange bolts that attach the front and rear ends of the converter to the exhaust system. If damage is discovered, the converter should be replaced.

4   Although catalytic converters don't break too often, they can become plugged. The easiest way to check for a restricted converter is to use a vacuum gauge to diagnose the effect of a blocked exhaust on intake vacuum.

a) *Connect a vacuum gauge to an intake manifold vacuum source (see Chapter 2C).*

b) *Warm the engine to operating temperature, place the transmission in Park (automatic) or Neutral (manual) and apply the parking brake.*

c) *Note and record the vacuum reading at idle.*

d) *Quickly open the throttle to near full throttle and release it shut. Note and record the vacuum reading.*

e) *Perform the test three more times, recording the reading after each test.*

f) *If the reading after the fourth test is more than one in-Hg lower than the reading recorded at idle, the exhaust system may be restricted (the catalytic converter could be plugged or an exhaust pipe or muffler could be restricted).*

Replacement

5   Be sure to spray the nuts on the exhaust flange studs before removing them from the catalytic converter(s).

6   Remove the nuts and separate the catalytic converter from the exhaust system.

7   Installation is the reverse of removal.

Chapter 6 Part B
Emissions and engine control systems - diesel engine

Contents

Specifications

Torque specifications

Ft-lbs (unless otherwise indicated)

Note: *One foot-pound (ft-lb) of torque is equivalent to 12 inch-pounds (in-lbs) of torque. Torque values below approximately 15 foot-pounds are expressed in inch-pounds, because most foot-pound torque wrenches are not accurate at these smaller values.*

CKP sensor bolt	89 in-lbs
CMP sensor bolt	89 in-lbs
IAC valve mounting bolts	
Step 1	89 in-lbs
Step 2	Tighten additional 90 degrees
EGR cooler bolts	89 in-lbs
EGR mounting bolts (in sequence - see illustration 19.15)	
Step 1 - Bolts 1 through 12	106 in-lbs
Step 2 - Bolts 1 through 12	142 in-lbs
Step 3 - Bolts 1 through 4	142 in-lbs
Step 4 - Bolt 13	106 in-lbs
Engine Coolant Temperature (ECT) sensor	
Primary	20
Secondary (2014 and earlier models)	159 in-lbs
Engine oil pressure switch	177 in-lbs
Engine oil temperature sensor	177 in-lbs

1   General Information

1   To prevent pollution of the atmosphere from incompletely burned and evaporating gases, and to maintain good driveability and fuel economy, a number of emission control systems are incorporated. They include the:

a) *On-Board Diagnostic (OBD) II system*
b) *Electronic Fuel Injection (EFI) system*
c) *Exhaust Gas Recirculation (EGR) system*
d) *Evaporative Emissions Control (EVAP) system*
e) *Positive Crankcase Ventilation (PCV) system*
f) *Catalytic converter*

2   Before assuming that an emissions control system is malfunctioning, check the fuel and ignition systems carefully. The diagnosis of some emission control devices requires specialized tools, equipment and training. If checking and servicing become too difficult or if a procedure is beyond your ability, consult a dealer service department or other repair shop. Remember, the most frequent cause of emissions problems is simply a loose or broken wire or vacuum hose, so always check the hose and wiring connections first.

3   This doesn't mean, however, that emissions control systems are particularly difficult to maintain and repair. You can quickly and easily perform many checks and do most of the regular maintenance at home with common tune-up and hand tools.

Note: *Because of a Federally mandated warranty which covers the emissions control system components, check with your dealer about warranty coverage before working on any emissions-related systems. Once the warranty has expired, you may wish to perform some of the component checks and/or replacement procedures in this Chapter to save money.*

4   Pay close attention to any special precautions outlined in this Chapter. It should be noted that the illustrations of the various systems may not exactly match the system installed on your vehicle because of changes made by the manufacturer during production or from year-to-year.

5   A Vehicle Emissions Control Information (VECI) label is attached to the underside of the hood (see illustration). This label contains important emissions specifications and adjustment information. Part of this label, the Vacuum Hose Routing Diagram, provides a vacuum hose schematic with emissions components identified. When servicing the engine or emissions systems, the VECI label and the vacuum hose routing diagram in your particular vehicle should always be checked for up-to-date information.

2   On Board Diagnosis (OBD) system and diagnostic trouble codes

General description

1   All models are equipped with the second generation OBD-II system. This system consists of an on-board computer known as the Powertrain Control Module (PCM), and information sensors, which monitor various functions of the engine and send data to the PCM. This system incorporates a series of diagnostic monitors that detect and identify fuel injection and emissions control system faults and store the information in the computer memory. This system also tests sensors and output actuators, diagnoses drive cycles, freezes data and clears codes.

2   The PCM is the brain of the electronically controlled fuel and emissions system. It receives data from a number of sensors and other electronic components (switches, relays, etc.). Based on the information it receives, the PCM generates output signals to control various relays, solenoids (fuel injectors) and other actuators. The PCM is specifically calibrated to optimize the emissions, fuel economy and driveability of the vehicle.

3   It isn't a good idea to attempt diagnosis or replacement of the PCM or emission control components at home while the vehicle is under warranty. Because of a federally-mandated warranty which covers the emissions system components and because any owner-induced damage to the PCM, the sensors and/or the control devices may void this warranty, take the vehicle to a dealer service department if the PCM or a system component malfunctions.

Scan tool information

4   Because extracting the Diagnostic Trouble Codes (DTCs) from an engine management system is now the first step in troubleshooting many computer-controlled systems and components, a code reader, at the very least, will be required (see illustration). More powerful scan tools can also perform many of the diagnostics once associated with expensive factory scan tools (see illustration). If you're planning to obtain a generic scan tool for your vehicle, make sure that it's compatible with OBD-II systems. If you don't plan to purchase a code reader or scan tool and don't have access to one, you can have the codes extracted by a dealer service department or an independent repair shop.

Note: *Some auto parts stores even provide this service.*

Obtaining and clearing Diagnostic Trouble Codes (DTCs)

5   Before outputting any DTCs stored in the PCM, thoroughly inspect ALL electrical connectors and hoses. Make sure that all electrical connections are tight, clean and free of corrosion. Make sure that all hoses are correctly connected, fit tightly and are in good condition (no cracks or tears). Also, make sure that the engine is tuned up. A poorly running engine is probably one of the biggest causes of emission-related malfunctions. Often, simply giving the engine a good tune-up will correct the problem.

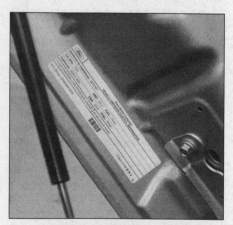

1.5 The Vehicle Emission Control Information (VECI) label contains such essential information as the types of emission control systems installed on the engine

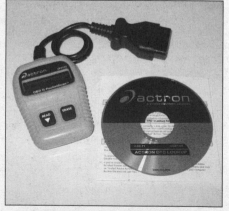

2.4a Simple code readers are an economical way to extract trouble codes when the CHECK ENGINE light comes on

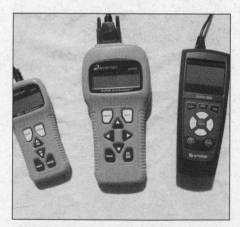

2.4b Hand-held scan tools like these can extract computer codes and also perform diagnostics

Information Sensors

Accelerator Pedal Position (APP) sensor - as you press the accelerator pedal, the APP sensor alters its voltage signal to the PCM in proportion to the angle of the pedal, and the PCM commands a motor inside the throttle body to open or close the throttle plate accordingly

Camshaft Position (CMP) sensor - produces a signal that the PCM uses to identify the number 1 cylinder and to time the firing sequence of the fuel injectors

Crankshaft Position (CKP) sensor - produces a signal that the PCM uses to calculate engine speed and crankshaft position, which enables it to synchronize ignition timing with fuel injector timing, and to detect misfires

Engine Coolant Temperature (ECT) sensor - a thermistor (temperature-sensitive variable resistor) that sends a voltage signal to the PCM, which uses this data to determine the temperature of the engine coolant

Fuel tank pressure sensor - measures the fuel tank pressure and controls fuel tank pressure by signaling the EVAP system to purge the fuel tank vapors when the pressure becomes excessive

Intake Air Temperature (IAT) sensor - monitors the temperature of the air entering the engine and sends a signal to the PCM to determine injector pulse-width (the duration of each injector's on-time) and to adjust spark timing (to prevent spark knock)

Knock sensor - a piezoelectric crystal that oscillates in proportion to engine vibration which produces a voltage output that is monitored by the PCM. This retards the ignition timing when the oscillation exceeds a certain threshold

Manifold Absolute Pressure (MAP) sensor - monitors the pressure or vacuum inside the intake manifold. The PCM uses this data to determine engine load so that it can alter the ignition advance and fuel enrichment

Mass Air Flow (MAF) sensor - measures the amount of intake air drawn into the engine. It uses a hot-wire sensing element to measure the amount of air entering the engine

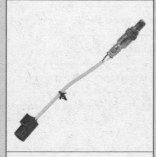

Oxygen sensors - generates a small variable voltage signal in proportion to the difference between the oxygen content in the exhaust stream and the oxygen content in the ambient air. The PCM uses this information to maintain the proper air/fuel ratio. A second oxygen sensor monitors the efficiency of the catalytic converter

Throttle Position (TP) sensor - a potentiometer that generates a voltage signal that varies in relation to the opening angle of the throttle plate inside the throttle body. Works with the PCM and other sensors to calculate injector pulse width (the duration of each injector's on-time)

Photos courtesy of Wells Manufacturing, except APP and MAF sensors.

Accessing the DTCs

6   On these models, all of which are equipped with On-Board Diagnostic II (OBD-II) systems, the Diagnostic Trouble Codes (DTCs) can only be accessed with a code reader or a scan tool (see illustrations 2.4a and 2.4b). Simply plug the connector of the tool into the Data Link Connector (DLC) or diagnostic connector (see illustration), which is located under the lower edge of the dash, just to the right of the steering column. Then follow the instructions included with the scan tool to extract the DTCs.

7   Once you have viewed all of the stored DTCs, look them up on the accompanying DTC chart.

8   After troubleshooting the source of each DTC, make any necessary repairs or replace the defective component(s).

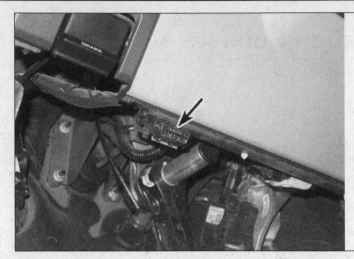

2.6 The 16-pin Data Link Connector (DLC), also referred to as the diagnostic connector, is located under the left part of the dash

Clearing the DTCs

9   Clear the DTCs with the scan tool in accordance with the instructions provided by the scan tool's manufacturer.

Trouble codes - diesel engine

OBD-II trouble codes

Note: *Not all trouble codes apply to all models.*

Code	Probable cause
P0AXX	DC - AC converter malfunction
P00B7	Engine coolant flow/temperature performance problem
P0001	Fuel Volume Regulator Control Valve (FVCV)
P0003	Fuel Volume Regulator Control Valve (FVCV) driver circuit - low voltage
P0004	Fuel Volume Regulator Control Valve (FVCV) driver circuit - high voltage
P000E	Fuel Volume Regulator Control Valve (FVCV) driver circuit - exceeding learn values
P0046	Turbocharger boost control solenoid - performance problem
P0069	Manifold Absolute Pressure (MAP) sensor correlation problem
P006B	Manifold Absolute Pressure (MAP) sensor/Exhaust Pressure correlation problem
P0087	Fuel rail system pressure - low
P0088	Fuel rail system pressure - high
P008C	Fuel cooling pump circuit - open
P008D	Fuel cooling pump circuit - low voltage
P008E	Fuel cooling pump circuit - high voltage
P008F	Engine Coolant Temperature (ECT)/Fuel temperature correlation problem

Code	Probable cause
P0090	Fuel Pressure Control Valve (FPCV)
P0091	Fuel Pressure Control Valve (FPCV) circuit - low voltage
P0092	Fuel Pressure Control Valve (FPCV) - high voltage
P0096	Intake Air Temperature (IAT) sensor circuit number 2 - performance problem
P0097	Intake Air Temperature (IAT) sensor circuit number 2 - low voltage
P0098	Intake Air Temperature (IAT) sensor circuit number 2 - high voltage
P0101	Mass Airflow (MAF) sensor circuit - performance problem
P0102	Mass Airflow (MAF) sensor circuit - low voltage
P0103	Mass Airflow (MAF) sensor circuit - high voltage
P0104	Mass Airflow (MAF) sensor circuit - intermittent, erratic
P0106	Barometric Pressure (BARO) sensor circuit - performance problem
P0107	Barometric Pressure (BARO) sensor circuit - low voltage
P0108	Barometric Pressure (BARO) sensor circuit - high voltage
P0112	Intake Air Temperature (IAT) sensor circuit) - low voltage
P0113	Intake Air Temperature (IAT) sensor circuit) - high voltage
P0114	Intake Air Temperature (IAT) sensor circuit) - voltage erratic
P0117	Engine Coolant Temperature (ECT) sensor circuit - low voltage
P0118	Engine Coolant Temperature (ECT) sensor circuit - high voltage
P0122	Accelerator Pedal Position (APP) sensor circuit - low voltage
P0123	Accelerator Pedal Position (APP) sensor circuit - high voltage
P0128	Coolant below regulating temperature for thermostat - possible stuck open
P012F	Engine coolant temperature/fuel temperature - correlation problem
P0148	Fuel delivery error
P0149	Fuel timing error
P0168	Fuel temperature excessive - high
P0181	Fuel temperature sensor circuit - performance problem
P0182	Fuel temperature sensor circuit - low voltage
P0183	Fuel temperature sensor circuit - high voltage
P0191	Fuel rail pressure sensor circuit - performance
P0192	Fuel rail pressure sensor circuit - low voltage
P0193	Fuel rail pressure sensor circuit - high voltage

Trouble codes - diesel engine
OBD-II trouble codes
Note: Not all trouble codes apply to all models.

Code	Probable cause
P0194	Fuel rail pressure sensor circuit - erratic
P0196	Engine Oil Temperature (EOT) sensor circuit performance problem
P0197	Engine Oil Temperature (EOT) sensor circuit - low voltage
P0198	Engine Oil Temperature (EOT) sensor circuit - high voltage
P0201-208	Cylinder X (X = cylinder no.) - injector circuit open
P0216	Injection/injector timing control circuit
P0219	Engine overspeed condition
P0220	Throttle Switch B circuit malfunction
P0221	Throttle Switch B circuit performance problem
P0230	Fuel pump relay driver circuit
P0231	Fuel pump relay secondary circuit - low voltage
P0232	Fuel pump relay secondary circuit - high voltage
P0234	Turbocharger overboost condition
P0236	Turbocharger boost sensor A - performance problem
P0237	Turbocharger boost sensor A - low voltage
P0238	Turbocharger boost sensor A - high voltage
P0261	Cylinder 1 - injector circuit low voltage
P0262	Cylinder 1 - injector circuit high voltage
P0263	Cylinder 1 balance system - possible rough idle detected
P0264	Cylinder 2 - injector circuit low voltage
P0265	Cylinder 2 - injector circuit high voltage
P0266	Cylinder 2 balance system - possible rough idle detected
P0267	Cylinder 3 - injector circuit low voltage
P0268	Cylinder 3 - injector circuit high voltage
P0269	Cylinder 3 balance system - possible rough idle detected
P0270	Cylinder 4 - injector circuit low voltage
P0271	Cylinder 4 - injector circuit high voltage
P0272	Cylinder 4 balance system - possible rough idle detected
P0273	Cylinder 5 - injector circuit low voltage

Code	Probable cause
P0274	Cylinder 5 - injector circuit high voltage
P0275	Cylinder 5 balance system - possible rough idle detected
P0276	Cylinder 6 - injector circuit low voltage
P0277	Cylinder 6 - injector circuit high voltage
P0278	Cylinder 6 balance system - possible rough idle detected
P0279	Cylinder 7 - injector circuit low voltage
P0280	Cylinder 7 - injector circuit high voltage
P0281	Cylinder 7 balance system - possible rough idle detected
P0282	Cylinder 8 - injector circuit low voltage
P0283	Cylinder 8 - injector circuit high voltage
P0284	Cylinder 8 balance system - possible rough idle detected
P0297	Vehicle overspeed condition
P0298	Engine oil over-temperature condition
P0299	Turbocharger underboost condition
P02CC	Fuel injector number 1 offset learning at minimal limit
P02CD	Fuel injector number 1 offset learning at maximum limit
P02CE	Fuel injector number 2 offset learning at minimal limit
P02CF	Fuel injector number 2 offset learning at maximum limit
P02D0	Fuel injector number 3 offset learning at minimal limit
P02D1	Fuel injector number 3 offset learning at maximum limit
P02D2	Fuel injector number 4 offset learning at minimal limit
P02D3	Fuel injector number 4 offset learning at maximum limit
P02D4	Fuel injector number 5 offset learning at minimal limit
P02D5	Fuel injector number 5 offset learning at maximum limit
P02D6	Fuel injector number 6 offset learning at minimal limit
P02D7	Fuel injector number 6 offset learning at maximum limit
P02D8	Fuel injector number 7 offset learning at minimal limit
P02D9	Fuel injector number 7 offset learning at maximum limit
P02DA	Fuel injector number 8 offset learning at minimal limit
P02DB	Fuel injector number 8 offset learning at maximum limit
P0300	Random misfire detected

Trouble codes - diesel engine
OBD-II trouble codes

Note: *Not all trouble codes apply to all models.*

Code	Probable cause
P030X	Cylinder misfire detected (X = cylinder number)
P0335	Crankshaft Position (CKP) sensor circuit - malfunction
P0336	Crankshaft Position (CKP) sensor circuit - performance problem
P0337	Crankshaft Position (CKP) sensor circuit - low voltage
P0340	Camshaft Position (CMP) sensor circuit - malfunction
P0341	Camshaft Position (CMP) sensor circuit - performance problem
P0344	Camshaft Position (CMP) sensor circuit - intermittent failure
P0380	Glow plug circuit - performance problem
P0381	Glow plug light indicator circuit - malfunction
P0401	Exhaust Gas Recirculation (EGR) flow - insufficient
P0402	Exhaust Gas Recirculation (EGR) flow - excessive
P0403	Exhaust Gas Recirculation (EGR) flow - malfunction
P0404	Exhaust Gas Recirculation (EGR) flow - performance problem
P0405	Exhaust Gas Recirculation (EGR) flow - low voltage
P0406	Exhaust Gas Recirculation (EGR) flow - high voltage
P040B	Exhaust Gas Recirculation Temperature (EGRT) sensor A circuit - performance problem
P040C	Exhaust Gas Recirculation Temperature (EGRT) sensor A circuit - low voltage
P040D	Exhaust Gas Recirculation Temperature (EGRT) sensor A circuit - high voltage
P041B	Exhaust Gas Recirculation Temperature (EGRT) sensor B circuit - performance problem
P041C	Exhaust Gas Recirculation Temperature (EGRT) sensor B circuit - low voltage
P041D	Exhaust Gas Recirculation Temperature (EGRT) sensor B circuit - high voltage
P0420	Catalytic converter efficiency - below threshold (Bank 1)
P042E	Exhaust Gas Recirculation (EGR) control - stuck open
P042F	Exhaust Gas Recirculation (EGR) control - stuck closed
P0460	Fuel tank level sensor circuit - performance problem
P0462	Fuel tank level sensor circuit - low voltage
P0463	Fuel tank level sensor circuit - high voltage
P0470	Exhaust backpressure sensor circuit - malfunction
P0471	Exhaust backpressure sensor circuit - performance problem

Code	Probable cause
P0472	Exhaust backpressure sensor circuit - low voltage
P0473	Exhaust backpressure sensor circuit - high voltage
P0475	Exhaust pressure control valve - malfunction
P0476	Exhaust pressure control valve - performance problem
P0478	Exhaust pressure control valve - high voltage
P0480	Cooling fan control circuit
P0488	Throttle control circuit - performance problem
P0494	Cooling fan speed - low
P0495	Cooling fan speed - high
P0500	Vehicle Speed Sensor (VSS) circuit
P0503	Vehicle Speed Sensor (VSS) circuit - interference
P0512	Starter request circuit - shorted
P0528	Cooling fan sensor circuit - no signal
P0529	Cooling fan sensor circuit - intermittent fault
P0541	Manifold intake air heater - voltage high
P0542	Manifold intake air heater - connection
P0544	Exhaust gas temperature sensor circuit (Bank 1 Sensor 1) - malfunction
P0545	Exhaust gas temperature sensor circuit (Bank 1 Sensor 1) - low voltage
P0546	Exhaust gas temperature sensor circuit (Bank 1 Sensor 1) - high voltage
P0560	System voltage - diagnostic monitor system problem
P0562	System voltage low
P0563	System voltage high
P0565	Cruise control ON circuit - not detecting
P0566	Cruise control OFF circuit - not detecting
P0567	Cruise control RESUME circuit - not detecting
P0568	Cruise control SET circuit - not detecting
P0569	Cruise control COAST circuit - not detecting
P0571	Cruise control Brake Switch circuit - not detecting
P0600	PCM communication link
P0602	PCM programming error
P0603	PCM long-term memory reset

Trouble codes - diesel engine
OBD-II trouble codes
Note: *Not all trouble codes apply to all models.*

Code	Probable cause
P0604	PCM Random Access Memory (RAM) problem
P0605	PCM Read-Only Memory (ROM)
P0606	PCM malfunction
P06XX	PCM internal control module - malfunction(s)
P0611	Fuel Injector Control Module (FICM) - performance problem
P0620	Alternator control circuit - malfunction
P0623	Alternator light control circuit - malfunction
P0625	Alternator field circuit - low
P0626	Alternator field circuit - high
P0627	Fuel pump control circuit - open
P0628	Fuel pump control circuit - low voltage
P0629	Fuel pump control circuit - high voltage
P062X	Fuel injector driver circuit (X = cylinder number) - performance problem
P0640	Manifold intake air heater - voltage low
P0642	Sensor reference A voltage - low voltage
P0643	Sensor reference A voltage - high voltage
P0645	A/C clutch relay control circuit - malfunction
P0646	A/C clutch relay control circuit - low voltage
P0647	A/C clutch relay control circuit - high voltage
P0649	Cruise control light circuit - malfunction
P0652	Sensor reference B voltage - low voltage
P0653	Sensor reference B voltage - high voltage
P0657	Transmission actuator supply voltage circuit - open
P0670	Glow plug module control circuit malfunction
P067X	Glow plug circuit failure (X = cylinder number)
P0683	Glow plug diagnostic signal communication fault
P0684	Glow plug control module-to-PCM communication fault
P0691	Cooling fan control circuit - low voltage
P0692	Cooling fan control circuit - high voltage

Code	Probable cause
P0700	Transmission control system - malfunction
P0703	Brake On Off (BOO) switch - malfunction
P0704	Transmission clutch switch circuit
P0705*	Transmission Range (TR) sensor circuit - malfunction
P0707*	Transmission Range (TR) sensor circuit - low voltage
P0708*	Transmission Range (TR) sensor circuit - high voltage
P0712*	Transmission Fluid Temperature (TFT) sensor - low voltage
P0713*	Transmission Fluid Temperature (TFT) sensor - high voltage
P0715*	Transmission Shift Solenoid (TSS) sensor circuit - malfunction
P0717*	Transmission Shift Solenoid (TSS) sensor circuit - intermittent failure
P0718*	Transmission Shift Solenoid (TSS) - noisy
P0720*	Shift Solenoid (OSS) sensor circuit - malfunction
P0721*	Shift Solenoid (OSS) sensor circuit - noisy
P0732	Transmission gear 2 ratio error
P0733	Transmission gear 3 ratio error
P0741	Torque Converter Clutch (TCC) circuit - performance problem
P0743*	Torque Converter Clutch (TCC) system - electrical failure
P0750	Shift solenoid 1 - malfunction
P0755*	Shift solenoid 2 - malfunction
P0781*	1 - 2 shift malfunction
P0782*	2 - 3 shift malfunction
P0783*	3 - 4 shift malfunction
P0830	Clutch pedal switch A circuit - malfunction
P0833	Clutch pedal switch B circuit - malfunction
P1000	OBD-II monitor checks incomplete, require another drive cycle
P1001	Key On Engine Running (KOER) test aborted
P1102	Mass Airflow (MAF) sensor circuit - low voltage
P1103	Mass Airflow (MAF) sensor values - higher than normal
P1105	Dual alternator monitor circuit - fault
P1106	Dual alternator control circuit - fault

*Transmission Control Indicator Light (TCIL) will flash when fault code present

Trouble codes - diesel engine
OBD-II trouble codes
Note: *Not all trouble codes apply to all models.*

Code	Probable cause
P1107	Dual alternator control circuit - malfunction
P1108	Dual alternator BATT light circuit - malfunction
P1118	Manifold Air Temperature sensor circuit - low voltage
P1119	Manifold Air Temperature sensor circuit - high voltage
P1139	Water in Fuel indicator circuit - malfunction
P1140	Water in Fuel condition
P1148	Alternator number 2 control circuit - malfunction
P1149	Alternator number 2 control circuit - high voltage
P115A	Low fuel level indication - limited power
P117B	Exhaust gas temperature sensor - correlation problem
P1184	Engine oil temperature sensor circuit - performance problem
P120F	Fuel pressure regulator - excessive variation
P1209	Injection control system pressure peak fault
P1210	Injection control pressure above expected level
P1211*	Injection control pressure not controllable - pressure above/below normal
P1212*	Injection control pressure voltage - not at normal values
P1218	Cylinder identification (CID) values - stuck high
P1219	Cylinder identification (CID) values - stuck low
P123C	Cold start turbocharger protection - block heater inoperative
P1247	Turbocharger boost pressure - low
P1248	Turbocharger boost pressure - not detected
P1249	Turbocharger wastegate Fail Steady State test
P1250	Electronic passive anti-theft system failure
P1260	Electronic anti-theft system - vehicle immobiized
P126X	Injector high to low side circuit short (X = cylinder no.)
P127X	Injector high to low side circuit open (X = cylinder no.)
P127A	Fuel pressure failure - aborted KOER test
P1280	Injection control pressure out of range - low

Transmission Control Indicator Light (TCIL) will flash when fault code present

Code	Probable cause
P1281	Injection control pressure out of range - high
P1282	Injection control pressure not controllable - excessive
P1283	Injection Pressure Regulator (IPR) circuit - failure
P1284	Injection control pressure circuit - testing failure
P1291	Injector high side (number 1) circuit - short to ground or battery
P1292	Injector high side (number 2) circuit - short to ground or battery
P1293	Injector high side (Bank 1) circuit - open
P1294	Injector high side (Bank 2) circuit - open
P1295*	Injector Bank 1 circuit - multiple faults
P1296*	Injector Bank 2 circuit - multiple faults
P1297	Injector high side circuits - shorted together
P1298	Injector Driver Module (IDM) - failure
P1316	Injector circuit - Injection Driver Module (IDM) codes detected
P132X	Turbocharger boost control - malfunction
P1335	Exhaust Gas Recirculation (EGR) sensor - performance problem
P1336	Crankshaft (CKP)/Camshaft (CMP) sensor information - erratic
P1378	Fuel Injector Control Module (FICM) supply voltage circuit - low voltage
P1379	Fuel Injector Control Module (FICM) supply voltage circuit - high voltage
P138D	Turbocharger boost control - high
P1397	Glow plug system voltage - out of self test range
P1408	Exhaust Gas Recirculation (EGR) flow - out of self test range
P1464	A/C on during KOER test procedure
P1501	Vehicle moved during testing procedure
P1502	Invalid testing procedure - APCM functioning
P1531	Invalid test - accelerator pedal movement during test procedure
P1536	Parking brake applied during testing - circuit failure
P1551-1558	Cylinder injector circuit - performance problem
P1561	Brake line pressure sensor circuit - malfunction
P1586	Electronic throttle control error
P1610	Interactive reprogramming code - diagnose PCM

Transmission Control Indicator Light (TCIL) will flash when fault code present

Trouble codes - diesel engine

OBD-II trouble codes

Note: *Not all trouble codes apply to all models.*

Code	Probable cause
P1611	Interactive reprogramming code - diagnose PCM
P1615	Interactive reprogramming code - erase flash error
P1616	Interactive reprogramming code - flash error, low voltage
P1617	Interactive reprogramming code - block programming error
P1618	Interactive reprogramming code - block programming error
P162E	PTO internal control module malfunction
P1633	Keep Alive power voltage - low
P1635	Tire/axle out of acceptable range
P1639	Vehicle ID block corrupted
P1662	Injector Driver Module (IDM) EN circuit - failure
P1663	Fuel Demand Command Signal (FDCS) circuit - failure
P1667	Cylinder Identification (CID) circuit failure
P1668	Injector Driver Module (IDM)/PCM circuit - failure
P1670	Electronic Feedback signal not detected
P1690	Turbocharger wastegate control valve malfunction
P1702*	Digital Transmission Range (TR) sensor - intermittent circuit failure
P1703	Brake switch out of self test range
P1704	Digital Transmission Range (TR) sensor - failed to transition state
P1705	Digital Transmission Range (TR) sensor - out of self test range
P1711	Transmission Fluid Temperature (TFT) sensor - out of self test range
P1713*	Transmission Fluid Temperature (TFT) sensor - failure below 50 degrees F
P1714	Shift solenoid A inductive signature malfunction
P1715	Shift solenoid B inductive signature malfunction
P1718*	Transmission Fluid Temperature (TFT) sensor - failure above 250 degrees F
P1725	Insufficient engine speed during self test
P1726	Excessive engine speed during self test
P1728*	Torque Converter Clutch (TCC) transmission slip error
P1729*	4WD Low switch error

Transmission Control Indicator Light (TCIL) will flash when fault code present

Code	Probable cause
P1744	Torque Converter Clutch (TCC) system performance
P1746	Exhaust Pressure Control (EPC) solenoid - open circuit
P1747	Exhaust Pressure Control (EPC) solenoid - short circuit
P1754	Coast Clutch Solenoid (CCS) circuit - malfunction
P1760*	Exhaust Pressure Control (EPC) circuit - intermittent failure
P1780	TCS circuit out of self test range
P1781	4WD Low circuit out of self test range
P1783*	Transmission over-temperature condition
P179A	Controller Area Network (CAN)/PCM/turbocharger communication error

Transmission Control Indicator Light (TCIL) will flash when fault code present

3 Powertrain Control Module (PCM) - removal and installation

Note: *The replacement of the PCM requires the original programming to be retrieved via a scan tool before PCM removal. The programming would then be downloaded to the new PCM and the original instrument cluster. This is a procedure best performed at a dealership.*

1   The Powertrain Control Module (PCM) removal and installation procedure is the same as for gasoline engine models. See Chapter 6A, Section 3.

4 Manifold Absolute Pressure (MAP) sensor - replacement

1   The MAP sensor is mounted on the intake manifold (see illustration). The Manifold Absolute Pressure (MAP) sensor monitors the intake manifold pressure changes resulting from changes in engine load and speed and converts the information into a voltage output. The PCM uses the MAP sensor to control fuel delivery and ignition timing. A problem in any of the MAP sensor circuits will set a diagnostic trouble code.

2   Make sure the ignition key is in the Off position.

3   Disconnect the electrical connector from the MAP sensor.

4   Unscrew the mounting screw and remove the MAP sensor from the intake manifold.

5   Installation is the reverse of removal.

5 Mass Airflow (MAF) sensor - replacement

1   Diesel engines are equipped with a drop-in style MAF sensor similar to those used on gasoline engines. The sensor is mounted to the top of the air intake tube on the air filter housing lid (see illustration).

2   Disconnect the electrical connector, then

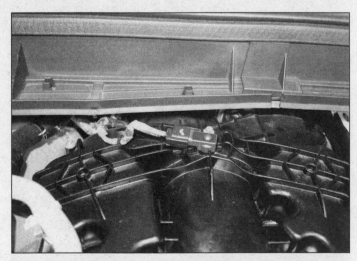

4.1 The MAP sensor is located on top of the upper intake manifold

5.1 MAF sensor details

1   *Connector lock (slide outward)*
2   *Release tab (depress and and pull off connector after sliding out connector lock)*
3   *Mounting screws*

remove the two screws securing the MAF sensor. Remove the sensor from the housing.
3   Installation is reverse of removal.

6   Intake Air Temperature (IAT) sensor - replacement

1   The Intake Air Temperature (IAT) sensor is part of the Mass Airflow (MAF) sensor and is replaced along with the MAF sensor. See Section 5.

7   Engine Coolant Temperature (ECT) sensor - replacement

Warning: *Wait until the engine has cooled completely before beginning this procedure.*
1   The engine coolant temperature (ECT) sensor is a thermistor. The thermistor is a resistor which varies the value of its resistance in accordance with temperature changes. The

7.2 Location of the primary cooling system ECT sensor

change in the resistance values will directly affect the voltage signal from the sensor to the PCM. As the sensor temperature decreases, the resistance values will increase. As the sensor temperature increases, the resistance values will decrease. 6.7L diesel engines are equipped with primary and secondary cooling systems, requiring separate sensors for each. A problem in any of the ECT sensor circuits will set a diagnostic trouble code.
Caution: *Handle the coolant sensor with care. Damage to this sensor will affect the operation of the fuel injection system.*

Primary cooling system
2   The primary cooling system ECT sensor is mounted near the thermostat housing (see illustration).
3   Drain the cooling system to a level below that of the sensor (see Chapter 1). Remove the duct between the turbocharger and the intercooler (see Chapter 4B, Section 20).
4   Disconnect the electrical connector and unscrew the sensor.
5   Installation is the reverse of removal. Tighten the sensor to the torque listed in this Chapter's Specifications.
6   Refill the cooling system (see Chapter 1).

Secondary cooling system
Note: *On 2014 and earlier models, the secondary cooling system ECT sensor is located on the EGR cooler (above the passenger valve cover) near the inlet and outlet hoses. On 2015 and later models, the secondary cooling system ECT sensor is located in the EGR cooler hose, near the right side of the radiator.*
7   Drain the secondary cooling system to a point lower than that of the ECT sensor (see Chapter 1).
8   On 2015 and later models, remove the air filter housing (see Chapter 4B, Section 22).

9   On 2014 and earlier models, disconnect the electrical connector and unscrew the sensor (see illustration).
10   On 2015 and later models, pull out the retaining clip and remove the sensor from the hose.
11   Installation is the reverse of removal. On 2014 and earlier models, tighten the sensor to the torque listed in this Chapter's Specifications. On 2015 and later models, make sure the sensor retaining clip is securely seated.
12   Refill the secondary cooling system (see Chapter 1).

8   Engine oil temperature sensor/ pressure switch - replacement

Warning: *Wait until the engine has cooled completely before beginning this procedure.*
Caution: *Handle the oil temperature sensor with care. Damage to this sensor will affect the operation of the entire diesel fuel injection system.*
1   The oil temperature sensor and pressure switch are located on the oil filter housing (see illustration). The oil temperature sensor points to the front of the vehicle and the oil pressure switch points toward the rear of the vehicle. The engine oil temperature sensor is a thermistor which varies the value of its resistance in accordance with temperature changes. The change in the resistance values will directly affect the voltage signal from the sensor to the PCM. The PCM uses this signal to calculate the fuel quantity, injection timing, glow plug operation and exhaust back pressure. Low oil temperatures signal the PCM to increase idle for complete warm-up operation. A problem in any of the oil temperature sensor circuits will set a diagnostic trouble code.
2   Disconnect the electrical connector and carefully unscrew the sensor or switch.
3   Installation is the reverse of removal.

7.9 Location of the secondary cooling system ECT sensor (2014 and earlier models)

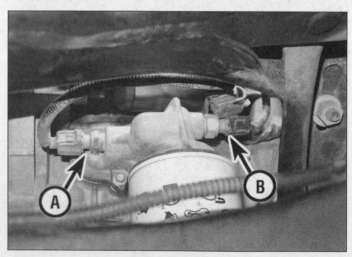

8.1 Location of the engine oil temperature sensor (A) and oil pressure switch (B)

9   Crankshaft Position (CKP) sensor - replacement

1   The crankshaft position sensor (CKP) determines the timing on each cylinder for the fuel injectors and ignition system. The crankshaft sensor is mounted near the transmission bellhousing. A problem in the crankshaft sensor circuit will set a diagnostic trouble code (see Section 2).
2   Raise the vehicle and support it securely on jackstands.
3   Working under the vehicle, remove the rubber plug on the driver's side rear of the engine block for access to the sensor (see illustration).
4   Disconnect the crankshaft position sensor electrical connector.
5   Remove the bolt and detach the sensor.
6   Installation is the reverse of removal. Tighten the sensor to the torque listed in this Chapter's Specifications. Make sure the rubber access plug seats securely.

10   Camshaft Position (CMP) sensor - replacement

1   The camshaft position (CMP) sensor determines the position of the cylinder for ignition start-up signals and for sequential fuel injection signals to each cylinder. The camshaft position sensor is mounted near the top of the crankshaft pulley.
2   Disconnect the electrical connector from the camshaft position sensor (see illustration).
3   Remove the bolt and detach the sensor.
4   Installation is the reverse of removal. Tighten the sensor mounting bolt to the torque listed in this Chapter's Specifications.

11   Charge Air Cooler Temperature (CACT) sensor - replacement

Note: *The Charge Air Cooler Temperature (CACT) sensor is located in the air intake tube between the air filter assembly and the intercooler, near the intercooler.*
Note: *The Charge Air Cooler (CAC) is also known as the intercooler.*
1   Remove the throttle body air inlet tube from between the throttle body and intercooler (see Chapter 4B, Section 17).
2   Locate the CACT sensor in the air intake tube at the intercooler and disconnect the electrical connector (see illustration).
3   Remove the CACT from its grommet in the intake tube.
4   Installation is reverse of removal.

12   Fuel tank pressure (FTP) sensor - replacement

1   See Chapter 6A, Section 13 for this procedure.

13   Accelerator Pedal Position (APP) sensor - replacement

1   To replace the APP sensor, see Chapter 4A, Section 14.

14   Exhaust Gas Temperature (EGT) sensor - replacement

Caution: *If the EGT sensor bends during installation, do not attempt to straighten it - install a new sensor.*

1   Four Exhaust Gas Temperature (EGT) sensors are installed in the catalyst and diesel particulate filter. The EGT sensors provide temperature information to the PCM to ensure the catalyst and diesel particulate filter is running at the designed temperature. If there is a problem with the EGT sensors, a DTC will set (see Section 2).
Note: *If necessary, warm the exhaust slightly and apply a penetrating oil to the sensor before attempting removal.*
2   Disconnect the EGT sensor electrical connector and any harness clips.
3   Unscrew the EGT sensor from the catalyst/diesel particulate filter (see illustration).
4   Apply anti-seize lubricant to the sensor threads before installing.
5   Installation is the reverse of removal.

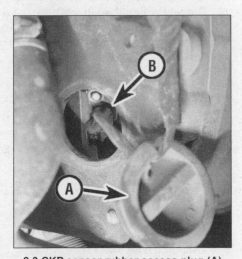

9.3 CKP sensor rubber access plug (A) and the CKP sensor (B)

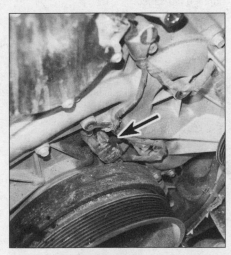

10.2 Locating the CMP sensor behind the crankshaft pulley

11.2 Locating the charge air cooler temperature sensor

14.3 Identifying an EGT sensor

15   Exhaust pressure sensor - replacement

Note: *The exhaust pressure sensor is located on the passenger's rear of the engine, threaded into a tube connected to the right-side EGR pipe.*
Caution: *Ensure the engine is cool prior to replacing the exhaust pressure sensor.*

1   Remove the upper intake manifold (see Chapter 2B).
2   Locate the exhaust pressure sensor and disconnect the electrical connector (see illustration).
3   Unscrew the sensor from the pressure sensor tube.
4   Installation is reverse of removal.

16   Diesel Particulate Filter (DPF) pressure sensor - replacement

Caution: *The hose must not be kinked after installation.*
1   The Diesel Particulate Filter (DPF) pressure sensor is installed on the diesel particulate filter. The DPF provides information to the PCM for the proper operation of the diesel particulate filter. A problem with the DPF pressure sensor circuit will set a trouble code (see Section 2).
2   Disconnect the sensor harness connector.
3   Unbolt the DPF sensor from the diesel particulate filter.
4   Install the hose onto the exhaust pressure tube on the particulate filter.
5   Installation is the reverse of removal.

17   Reductant system

Overview

1   The selective reduction catalyst improves the exhaust emissions and fuel efficiency by injecting a reductant into the exhaust system. The reductant (Diesel Exhaust Fluid - DEF), is a mixture of urea in deionized water. The DEF is injected at the inlet of the catalyst. When the DEF enters the system, it atomizes and mixes evenly with exhaust gases. During this time, the heat of the exhaust gases causes the urea in the reductant to split into carbon dioxide (CO_2) and ammonia. As the ammonia and oxides of nitrogen (NOx) pass over the catalyst, a reduction reaction takes place and the ammonia and NOx are converted to nitrogen and water.

Replacement

2   Raise and support the vehicle on jackstands.

Reductant tank

Note: *The vehicle may be equipped with an inboard or outboard frame mounted tank.*
3   Locate the inboard or outboard reductant tank and disconnect the filler and pump hoses.
4   Disconnect the electrical connectors for the pump and heater.
5   On inboard tanks, remove the bolts attaching the reductant tank bracket to the frame and remove the tank and bracket as an assembly.
6   On outboard tanks, remove the tank strap bolts and straps and remove the tank.
7   Installation is reverse of removal.

Pump

8   Remove the reductant tank from the vehicle.
9   Remove the three bolts and pull the pump off of the reductant heater and sender assembly.
10   Inspect the O-ring for the reductant pump. Replace if it is damaged.
11   Installation is reverse of removal.

Heater and sender assembly

12   Remove the reductant tank from the vehicle.
13   Remove the reductant pump.
14   Unscrew the lock ring securing the heater and sender assembly to the tank and remove the heater and sender assembly.
15   Inspect the seal for the heater and sender assembly. Replace if it is damaged.
16   Installation is reverse of removal.

Injector and pressure line

Note: *The reductant injector is installed in the top of the catalyst/particulate filter assembly, near the rear of the transfer case or transmission output shaft.*
17   Locate the injector and pressure sensor and slide the heat shieldoff of the injector line (see illustration).
18   Disconnect the electrical connector from the injector.
19   Gently push the pressure line connector toward the injector and hold. Press the pressure line connector tabs and pull the pressure line connector straight off the injector nipple.
20   Remove the nuts securing the injector and pressure sensor to the articulate filter and remove the injector and pressure sensor.
21   Installation is reverse of removal. Be sure to use a new gasket and tighten the nuts to the torque listed in this Chapter's Specifications.

18   Nitrogen oxides (NOx) sensor/ module - replacement

1   The NOx sensor is similar in construction to an oxygen sensor. The NOx sensor is installed at the end of the catalyst/diesel particulate filter before the exhaust pipe. The NOx sensor module is attached to the frame, near the NOx sensor location. The PCM uses the NOx sensor module input to control injection operation. If a problem occurs with the NOx sensor or module, a DTC will set (see Section 2).
Caution: *DO NOT use silicone based spray or lubricant around the NOx sensor as it will damage the sensor.*

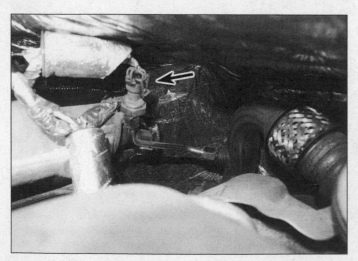

15.2 Location of the exhaust pressure sensor

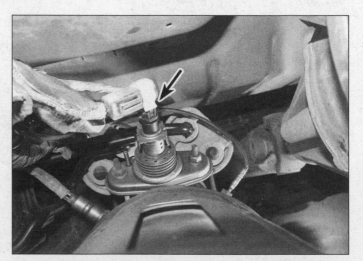

17.17 Location of the reductant injector and pressure line

Sensor

Note: *Because it is installed in the exhaust manifold or pipe, which contracts when cool, the NOx sensor may be very difficult to loosen when the engine is cold. Rather than risk damage to the sensor, assuming you are planning to reuse it in another manifold or pipe, start and run the engine for a minute or two, then shut it off. Be careful not to burn yourself during the following procedure.*

2   The procedure to replace the NOx sensor is similar to that of an oxygen sensor. See Chapter 6A, Section 11.

Module

3   Locate the NOx sensor module and disconnect the electrical connectors (see illustration).
4   Remove the bolts and the NOx sensor module from the vehicle.
5   Installation is reverse of removal. If a new module is installed, perform the following:

   a) *Using a scan tool, reset/clear the specified function for the nitrogen oxide.*
   b) *Clear continuous PCM DTCs and reset the emission monitors.*

19   Exhaust Gas Recirculation (EGR) system

General description

1   The EGR system reduces oxides of nitrogen (NOx) by recirculating exhaust gases from the exhaust ports through the EGR valve and back into the intake manifold for recirculation into the engine which lowers the peak flame temperature during combustion.The EGR system reduces peak combustion temperatures and NOx emissions by 90%. The EGR comes from the right exhaust manifold.

2   The EGR system has a bypass feature that skips the EGR cooler to allow the engine to warm up faster. The EGR system uses two EGR coolers, but introduces a hot-side valve at the front of the first cooler that controls the volume of air allowed into the system instead of using a conventional cool-side valve behind the second cooler.

Replacement
EGR temperature (EGRT) sensor

Note: *The EGR temperature sensor is located in the EGR by-pass pipe attached to the EGR cooler.*

3   Remove the air intake tube between the upper intake manifold and the air filter housing lid (see Chapter 4B, Section 22).
4   Disconnect the EGRT sensor connector.
5   Unscrew and remove the EGRT sensor from the by-pass pipe (see illustration).
6   Coat the threads of the EGRT sensor with anti-seize before installing.
7   Installation is reverse of removal.

EGR valve

Warning: *The engine must be completely cool before beginning this procedure.*
Note: *The EGR valve is located on the front of the EGR cooler on the passenger side of the engine compartment.*

8   Drain the secondary cooling system (see Chapter 1, Section 25).
9   Remove the air filter housing (see Chapter 4A).
10   Remove the coolant vent hose and disconnect the coolant hose retainer (see illustration).
11   Disconnect the EGR valve electrical connector.
12   Remove the bolts and the EGR pipe from the exhaust manifold and the EGR valve.
13   Remove the 13 bolts and remove the EGR valve from the vehicle (see illustration 19.15).
14   Remove the EGR gasket, clean the valve and cooler of any remaining gasket material and clean the EGR valve gasket surface if valve is to be reused.

18.3 The NOx sensor module is located on the right-side frame rail

19.5 Locating the EGR temperature sensor

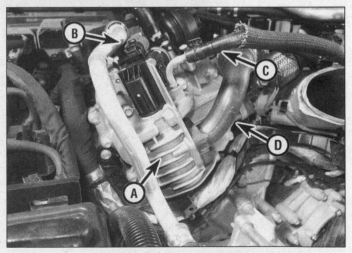

19.10 EGR component details

A EGR valve	C EGR coolant vent hose
B EGR valve electrical connector	D EGR pipe

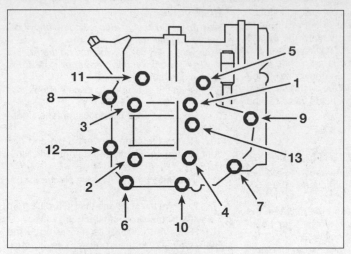

19.15 EGR valve bolt tightening sequence

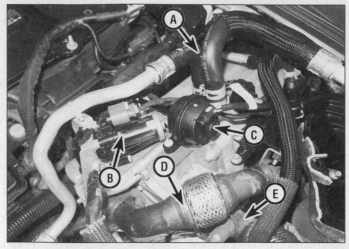

19.18 EGR cooler components

A	Coolant hoses	D	EGR inlet tube
B	EGR valve	E	EGR by-pass outlet tube
C	EGR actuator		

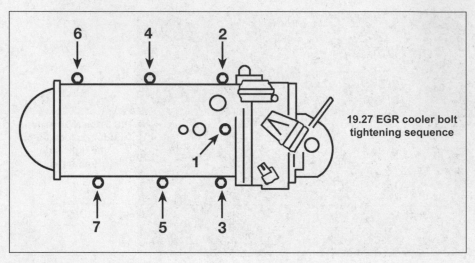

19.27 EGR cooler bolt tightening sequence

15   Installation is the reverse of removal. Tighten the EGR valve bolts, in sequence, (see illustration) to the torque listed in this Chapter's Specifications.

16   Refill the cooling system (see Chapter 1).

EGR cooler

Warning: *The engine must be completely cool before beginning this procedure.*

17   Drain the primary and secondary cooling systems (see Chapter 1, Section 25).

18   Disconnect the coolant hoses from the top of the cooler (see illustration).

19   Remove the coolant hose retainer.

20   Disconnect the electrical connector from the EGR valve and cooler and remove the harness bracket bolts and harness bracket from the front of the EGR.

21   Disconnect the bypass solenoid electrical connector and vacuum lines. Remove the bolts and the bypass solenoid together with the bracket.

22   Remove the EGR inlet tube from the exhaust manifold and the EGR valve.

23   Disconnect the EGR temperature sensor and remove the EGR by-pass outlet tube.

24   Remove the exhaust pressure sensor tube bracket at the rear of the EGR cooler.

25   Remove bolts and pull the EGR cooler assembly straight up and off of the mount. Note the locations of the bolts, as they are different lengths.

26   Use NEW O-rings when installing the cooler.

27   Installation is reverse of removal. Tighten the bolts, in the proper sequence (see illustration), to the torque listed in this Chapter's Specifications.

28   Refill the cooling systems (see Chapter 1).

20   Crankcase vent oil separator

General description

1   The Closed Crankcase Ventilation (CCV) system uses intake manifold vacuum and crankcase pressure to vent blow-by gases from the crankcase and return the gases to the intake for combustion. The separator also removes oil from the blow-by gas and returns it to the engine.

Replacement

2   Relieve fuel pressure (see Chapter 4B, Section 3).

3   Remove the retainers and remove the driver's side sound shield from the engine.

4   Remove the intake tube between the intercooler and the throttle body (see Chapter 4B, Section 22).

5   Disconnect the CCV hose (see illustration) from the intake manifold (above the fuel injector lines on the driver's side of the engine) one of two ways:

  a) *Disconnect the electrical connector and disconnect the hose by rotating the release ring counterclockwise. This CCV can be reinstalled.*

  b) *Cut the rubber sleeve from the hose fitting and break the fitting tabs to remove. The CCV must be replaced if the hose is disconnected on this type of system.*

6   Disconnect the secondary fuel filter hoses (see Chapter 4B, Section 8). Remove the filter-to-fuel injection pump hose.

7   Remove the bolts attaching the driver's side fuel supply line to the valve cover.

8   Detach the vacuum hose retainer behind the secondary fuel filter and remove the fuel injection pump supply and return line bracket bolt from the filter base.

9   Remove the secondary (engine-mounted) fuel filter from the engine (see Chapter 1).

10   Detach the harness clip from the CCV bolt stud.

11   Remove the two bolts at the rear of the left-side valve cover attaching the CCV to the valve cover.

12   Remove the single stud bolt and regular bolt at the center of the valve cover and pull the CCV upwards to disengage from the valve cover.

13   On models where the CCV can be reused, replace the O-rings on the CCV. DO NOT lubricate the O-rings before installation.

14   Installation is the reverse of removal.

21   Catalyst and particulate filter

Note: *Because of a Federally mandated extended warranty which covers emissions-related components such as the catalytic converter, check with a dealer service department before replacing the converter at your own expense.*

20.5a CCV hose (reusable type shown)

General description

1   The oxy catalyst (OC) and particulate filter assembly oxidizes hydrocarbons in the exhaust and generates heat for DPF regeneration. As soot collects it begins to restrict the filter. The filter needs to be periodically cleaned of soot and can be cleaned in two different ways: passive regeneration and active regeneration. Both methods occur automatically.

Passive regeneration

2   Passive regeneration occurs as a result of normal engine operation. During passive regeneration, exhaust constituents/temperature are at an appropriate level where some soot can be burned, resulting in cleaning of the filter.

Active regeneration

3   Active regeneration (controlled by the PCM) occurs when not enough passive regeneration occurs. In active regeneration, the DPF is cleaned by raising the exhaust temperature to a point where the soot is burned away. After soot is burned off, the exhaust temperature and back pressure return to normal levels.

Replacement

4   Be sure to spray the nuts on the exhaust flange studs before removing them from the catalytic converter.

5   Remove the nuts and separate the catalytic converter/particulate filter from the exhaust system.

6   Installation is the reverse of removal.

Notes

Chapter 7 Part A
Automatic transmission

Contents

Specifications

Torque specifications

Ft-lbs (unless otherwise indicated)

Note: *One foot-pound (ft-lb) of torque is equivalent to 12 inch-pounds (in-lbs) of torque. Torque values below approximately 15 foot-pounds are expressed in inch-pounds, because most foot-pound torque wrenches are not accurate at these smaller values.*

Engine mount	
Left nut	148
Right	
Nut	85
Stud	59
Shift lever bolt	159 in-lbs
Torque converter-to-driveplate nuts	35
Transmission-to-engine bolts	35
Transmission cooler bolt (TorqShift6)	18
Transmission crossmember-to-frame support	81
Transmission mount-to-crossmember	
Nut	85
Stud	55
Transmission mount-to-transmission bolts	76

1   General Information

1   The TorqShift transmission is a fully automatic, electronic-shift five-speed with a lock-up torque converter, known as a torque converter clutch, or TCC (the TCC provides a direct connection between the engine and the drive wheels for improved efficiency and fuel economy) and has a selectable Tow/Haul feature. The TorqShift6 is a six-speed transmission.

2   Because of the complexity of the clutches and the electronic and hydraulic control systems, and because of the special tools and expertise needed to overhaul an automatic transmission, diagnosis and repair of this transmission must be handled by a dealer service department or a transmission repair shop. The procedures in this Chapter are limited to general diagnosis, routine maintenance and adjustment: replacing the shift lever, replacing and adjusting the shift cable, and similar jobs. Serious repair work, however, must be done by a transmission specialist. But if the transmission must be rebuilt or replaced, you can save money by removing and installing it yourself, so instructions for that procedure are included as well.

2   Diagnosis - general

Note: *Automatic transmission malfunctions may be caused by five general conditions: poor engine performance, improper adjustments, hydraulic malfunctions, mechanical malfunctions or malfunctions in the computer or its signal network. Diagnosis of these problems should always begin with a check of the easily repaired items: fluid level and condition (see Chapter 1), and shift cable adjustment. Next, perform a road test to determine if the problem has been corrected or if more diagnosis is necessary. If the problem persists after the preliminary tests and corrections are completed, additional diagnosis should be done by a dealer service department or transmission repair shop. Refer to the* Troubleshooting *Section at the front of this manual for information on symptoms of transmission problems.*

Note: *All of the vehicles covered in this manual require battery power to be available at all times to maintain strategy parameters which are stored in the keep alive memory (KAM). Therefore, whenever the battery is to be disconnected, first note the following to ensure that there are no unforeseen consequences of this action:*

a) The KAM will lose the information stored in its memory when the battery is disconnected.

b) This is a temporary condition. Whenever the battery is disconnected, the information relating to operating values will have to be re-programmed into the unit's memory. The PCM does this by itself, but until then, there may be a generally inferior level of performance. Once the PCM relearns these values it will return to normal operating condition.

Preliminary checks

1   Drive the vehicle to warm the transmission to normal operating temperature.

2   Check the fluid level as described in Chapter 1:

a) If the fluid level is unusually low, add enough fluid to bring the level within the designated area of the dipstick, then check for external leaks (see below).

b) If the fluid level is abnormally high, drain off the excess, then check the drained fluid for contamination by coolant. The presence of engine coolant in the automatic transmission fluid indicates that a failure has occurred in the internal radiator walls that separate the coolant from the transmission fluid (see Chapter 3).

c) If the fluid is foaming, drain it and refill the transmission, then check for coolant in the fluid or a high fluid level.

3   Check the engine idle speed.

Note: *If the engine is malfunctioning, do not proceed with the preliminary checks until it has been repaired and runs normally.*

4   Inspect the shift cable (see Section 4). Make sure it's properly adjusted and operates smoothly.

Fluid leak diagnosis

5   Most fluid leaks are easy to locate visually. Repair usually consists of replacing a seal or gasket. If a leak is difficult to find, the following procedure may help.

6   Identify the fluid. Make sure it's transmission fluid and not engine oil or brake fluid (automatic transmission fluid is a deep red color).

7   Try to pinpoint the source of the leak. Drive the vehicle several miles, then park it over a large sheet of cardboard. After a minute or two, you should be able to locate the leak by determining the source of the fluid dripping onto the cardboard.

8   Make a careful visual inspection of the suspected component and the area immediately around it. Pay particular attention to gasket mating surfaces. A mirror is often helpful for finding leaks in areas that are hard to see.

9   If the leak still cannot be found, clean the suspected area thoroughly with a degreaser or solvent, then dry it.

10   Drive the vehicle for several miles at normal operating temperature and varying speeds. After driving the vehicle, visually inspect the suspected component again.

11   Once the leak has been located, the cause must be determined before it can be properly repaired. If a gasket is replaced but the sealing flange is bent, the new gasket will not stop the leak. The bent flange must be straightened.

12   Before attempting to repair a leak, check to make sure the following conditions are corrected or they may cause another leak.

Note: *Some of the following conditions cannot be fixed without highly specialized tools and expertise. Such problems must be referred to a transmission repair shop or a dealer service department.*

Gasket leaks

13   Check the pan periodically. Make sure the bolts are tight, no bolts are missing, the gasket is in good condition and the pan is flat (dents in the pan may indicate damage to the valve body inside).

14   If the pan gasket is leaking, the fluid level or the fluid pressure may be too high, the vent may be plugged, the pan bolts may be too tight, the pan sealing flange may be warped, the sealing surface of the transmission housing may be damaged, the gasket may be damaged or the transmission casting may be cracked or porous. If sealant instead of gasket material has been used to form a seal between the pan and the transmission housing, it may be the wrong sealant.

Seal leaks

15   If a transmission seal is leaking, the fluid level or pressure may be too high, the vent may be plugged, the seal bore may be damaged, the seal itself may be damaged or improperly installed, the surface of the shaft protruding through the seal may be damaged or a loose bearing may be causing excessive shaft movement.

16   Make sure the dipstick tube seal is in good condition and the tube is properly seated. Periodically check the area around the speedometer gear or sensor for leakage. If transmission fluid is evident, check the O-ring for damage.

Case leaks

17   If the case itself appears to be leaking, the casting is porous and will have to be repaired or replaced.

18   Make sure the oil cooler hose fittings are tight and in good condition.

Fluid comes out vent pipe or fill tube

19   If this condition occurs, the transmission is overfilled, there is coolant in the fluid, the case is porous, the dipstick is incorrect, the vent is plugged or the drain back holes are plugged.

3   Shift lever - removal and installation

Note: *The Transmission Control Switch (TCS) contains the tow/haul switch and upshift/downshift buttons. The TCS is part of the shift lever and not serviceable separately.*

Shift lever

1   Remove the steering column covers (see Chapter 11).

2   Unplug the electrical connector for the TCS (see illustration).

3   Remove the shift lever bolt (see illustration).

4   Remove the shift lever.

5   Installation is the reverse of removal. Be sure to replace the shift lever bolt; do NOT use the old bolt.

3.2 Disconnect the TCS connector

3.3 Remove the shift lever bolt

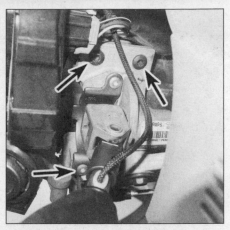

3.9 Remove the three bolts to remove the shift lever mechanism

Shift lever mechanism

6   Remove the shift lever.
7   Disconnect the shift cable from the shift lever mechanism (see Section 4).
8   Disconnect the BTSI solenoid connector.
9   Remove the three bolts and the shift lever mechanism from the column (see illustration).
10   Installation is the reverse of removal.

4   Shift cable - removal, installation and adjustment

Removal and installation

1   Remove the steering column covers (see Chapter 11, Section 35).
2   Shift the transmission to Drive position.
3   Detach the shift cable from the steering column shift tube lever and lift the locking tab and detach the shift cable from the steering column bracket (see illustration).
4   Detach the cable retainer on the floor.
5   Push the rubber grommet and shift cable through the floor.
Note: *Pull the carpet back to access the cable as necessary.*
6   Raise the vehicle and place it securely on jackstands.
7   Detach the shift cable from the manual lever and cable bracket (see illustration).
8   Installation is the reverse of removal. Be sure to adjust the cable before reattaching it to the manual lever (see below).

Adjustment

9   Working inside the vehicle, put the shift lever in the Drive position.
10   Raise the vehicle and support it securely on jackstands.
11   With the shift cable detached from the manual lever, unlock the lock tab on the shift cable.
12   Move the manual lever to the Drive position.

4.3 Detach the shift cable from the lever (A) and bracket (B)

13   Reattach the shift cable to the manual lever.
14   Lock the shift cable locking tab.
15   Remove the jackstands and lower the vehicle.
16   Move the shift lever through all gear positions and verify that the indicated positions correspond with the actual gear positions at the manual lever. Also verify that the engine will start only in Park and Neutral, and that the back-up lights come on when the shift lever is placed in Reverse. If necessary, readjust the cable until these conditions are met. It may also be necessary to adjust the transmission range sensor (see Section 5).

5   Transmission Range (TR) sensor - description, adjustment and replacement

Note: *The TorqShift transmission uses a transmission range sensor that is inside the*

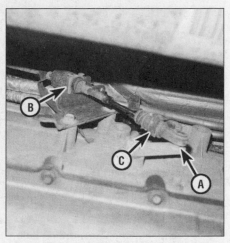

4.7 Detach the shift cable from the shift lever (A) and bracket (B). (C) is the adjuster lock tab

transmission *and requires a special tool for service. Replacement should be performed by a qualified repair facility.*

Description

1   The Transmission Range (TR) sensor, which is located inside the transmission at the manual lever, is an information sensor for the Powertrain Control Module (PCM). Among its functions are those normally handled by a conventional Park/Neutral switch: it prevents the engine from starting in any gear other than Park or Neutral, and closes the circuit for the back-up lights when the shift lever is moved to Reverse.

Adjustment

2   If the engine starts in any position other than Park or Neutral, the shift cable is either out of adjustment or the TR sensor is defective. The TR sensor is non-adjustable; adjust the shift cable to adjust the TR sensor operation (see Section 4).

6.6 Move the brake shift interlock solenoid spindle outward

6.12 Identifying the BTSI solenoid (A) and park position switch (B)

7.1 Locating the TCM under the cab

8.14 Location of the transmission fluid cooler (diesel models)

6 Shift interlock system - description, check and actuator replacement

Description

1 The shift interlock system prevents the shift lever from being moved out of the Park position unless the brake pedal is depressed. The system consists of a Brake Transmission Shift Interlock (BTSI) solenoid mounted on the steering column. When the ignition key is turned to the Run position, the actuator is energized. When the brake pedal is depressed, the shift lever can be moved from Park.

Check

2 Check the following if the BTSI solenoid isnt working correctly:

   a) Fuses F18 (10A) and F28 (15A) in the Body Control Module (BCM).

   b) Brake lights operate when pedal is depressed.

Override

3 Set the parking brake and turn the ignition off.
4 Tilt the steering wheel down and telescope fully out.
5 Detach the shift lever boot from the column covers by pressing rearward from the front of the boot.
6 Use a suitable tool to move the brake shift interlock solenoid spindle outward (see illustration).
7 With the spindle held outward, press the brake pedal and move the shift lever to Neutral (N).

Actuator replacement

8 Remove the steering column covers (see Chapter 11, Section 35).
9 Remove the shift lever mechanism (see Section 3).
10 Remove the shift lock actuator bolt.
11 Carefully pry out the BTSI solenoid clip and pull the BTSI solenoid from the shift lever mechanism.
12 Press the spring loaded pin for the park position switch and rotate the switch clockwise to remove (see illustration).
13 Remove the BTSI solenoid from the shift lever mechanism.
14 Installation is the reverse of removal.

7 Transmission Control Module (TCM) - replacement

Note: *The TCM is located on a bracket attached to the left frame rail under the cab.*
1 Locate the TCM under the vehicle (see illustration).
2 Slide the connector lock tab away from the frame to disengage.
3 Pull the connector straight off of the TCM or damage to the terminals may occur.
4 Remove the two nuts attaching the TCM to the bracket.

5 Installation is reverse of removal.
Note: *If a new module is installed, module configuration must be restored and solenoid body strategy must be entered by a qualified repair center.*

8 Auxiliary cooler - removal and installation

Gasoline models

1 Remove the grille (refer to Chapter 11).
2 Remove the upper cooler mounting bolts.
3 Remove the plastic push-style retainers from the lower splash shields.
4 Remove the four lower splash shield retainers.
5 Working under the front of the vehicle, remove the lower bumper filler by first removing the six retainers.
6 Unbolt the power steering cooler (if so equipped) and secure it out of the way.
7 Remove the two lower splash shield push-style retainers for access to the condenser bracket.
8 Disconnect the transmission cooler hoses.
9 Remove both lower air conditioning condenser mounting brackets.
10 Remove the transmission cooler bolts. Pull the air conditioning condenser forward for clearance and then lift the transmission cooler up and forward as you remove it.
11 Installation is the reverse of removal.

Diesel models

Warning: *Wait until the engine is completely cool before beginning this prodedure.*
12 Raise and support the front of the vehicle on jackstands.
13 Drain the diesel secondary cooling system (see Chapter 1, Section 25).
14 Locate the transmission cooler on the passenger frame rail, above the stabilizer bar (see illustration).

9.20 Remove the bolt and disconnect the cooler lines (TorqShift6 model shown)

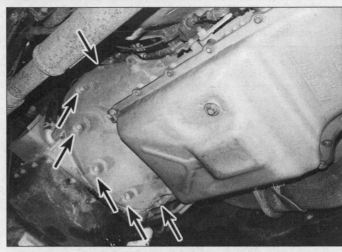

9.26 Remove the transmission-to-bellhousing bolts (TorqShift6 model shown)

15   Disconnect the harness clip from the cooler.
16   Disconnect the coolant and fluid hoses from the cooler.
17   Remove the three bolts and remove the cooler.
18   Installation is reverse of removal. Top off the transmission fluid and refill the secondary cooling system.

9   Automatic transmission - removal and installation

Removal

1   Disconnect the cable(s) from the negative battery terminal(s) (see Chapter 5, Section 3).
2   Put the shift lever in the Neutral position
3   Raise the vehicle and support it securely on jackstands.
4   Remove the transmission dipstick.
5   If necessary, drain the transmission fluid by removing the drain plug.
6   On gasoline models, remove the exhaust Y-pipe (see Chapter 4A).
7   Remove the driveshaft(s) (see Chapter 8).
8   On 4WD models, remove the skid plate, if equipped, and transfer case (see Chapter 7B).
9   On manual shift transfer case models, remove the shift lever bracket from the transmission.
10   Place a floor jack under the engine. Place a wood block between the jack head and the engine oil pan.
11   Place a transmission jack or a floor jack under the transmission and secure the transmission to the jack with safety chains.
12   Remove the left and right engine mount nuts at the crossmember. If the studs come out of the right engine mount, replace them

with new ones and tighten them to the torque listed in this Chapter's Specifications.
Caution: *Use only hand tools to remove the nuts to prevent damage to the mounts.*
13   Remove the transmission crossmember. If the studs come out of the transmission mount, replace them and tighten them to the torque listed in this Chapter's Specifications.
14   Remove the transmission mount from the transmission.
15   Disconnect the shift cable from the transmission and bracket and secure out of the way (see Section 4).
16   Disconnect all connectors and harnesses from the transmission and secure out of the way.
17   Remove the driveplate inspection bolts and cover.
18   Remove the oil filler tube mounting nut and remove the tube.
19   On TorqShift models, while holding the transmission case fitting with one wrench, use a second wrench to disconnect the transmission oil cooler lines.
20   On TorqShift6 models, remove the bolt and disconnect the cooler lines at the case (see illustration).
21   Remove the starter motor (see Chapter 5) and starter motor dust shield.
22   On diesel models, remove the engine oil filter adapter (see Chapter 2B). Discard the O-ring.
23   Disconnect the steering damper to allow access to the crankshaft pulley.
24   On gasoline models, remove the plastic or rubber plug from the access hole on the driver's side of the engine block to access the torque converter nuts.
Note: *On diesel models the nuts are accessed through the starter opening.*
25   Mark the relationship of the torque converter to the driveplate, then remove the four torque converter retaining nuts. Rotate the crankshaft to bring each nut within reach

through the inspection cover hole.
26   Remove the transmission-to-engine bellhousing bolts (see illustration).
27   Make a final check that all wires have been disconnected from the transmission, then move the transmission and jack toward the rear of the vehicle until the torque converter is separated from the driveplate. Secure the torque converter to the transmission so it won't fall out during removal.
28   Remove the dipstick tube and discard the O-ring.

Installation
29   Installation is reverse of removal, noting the following points:
a)   *Use new nuts for the engine and transmission mounts and install using hand tools.*
b)   *Tighten all fasteners to the torque listed in this Chapter's Specifications.*
c)   *Use a new O-ring on the dipstick tube.*
d)   *On TorqShift6, install new oil cooler O-rings.*
e)   *On models so equipped, flush the PTO prior to installation.*
f)   *Flush the transmission cooler prior to installation.*
g)   *If the transmission had a mechanical failure, replace the auxiliary oil cooler (see Section 8).*
h)   *Lubricate the torque converter pilot hub (where the torque converter enters the crankshaft).*
i)   *Adjust the shift cable (see Section 4).*
j)   *After filling the transmission with new fluid, check the transmission fluid level at operating temperature (150 to 170 degrees) and top off as necessary. Use a scan tool to verify transmission fluid temperature.*
k)   *If a new valve body has been installed, the solenoid body strategy must be updated by a qualified technician.*

10 Automatic transmission overhaul - general information

1   In the event of a fault occurring, it will be necessary to establish whether the fault is electrical, mechanical or hydraulic in nature, before repair work can be contemplated.

Diagnosis requires detailed knowledge of the transmission's operation and construction, as well as access to specialized test equipment, and so is deemed to be beyond the scope of this manual. It is therefore essential that problems with the automatic transmission are referred to a dealer service department or other qualified repair facility for assessment.

2   Note that a faulty transmission should not be removed before the vehicle has been assessed by a knowledgeable technician equipped with the proper tools, as troubleshooting must be performed with the transmission installed in the vehicle.

Chapter 7 Part B
Transfer case

Contents

Specifications

Fluid specifications
Transfer case fluid capacity .... See Chapter 1

Torque specifications   Ft-lbs (unless otherwise indicated)

Note: *One foot-pound (ft-lb) of torque is equivalent to 12 inch-pounds (in-lbs) of torque. Torque values below approximately 15 foot-pounds are expressed in inch-pounds, because most foot-pound torque wrenches are not accurate at these smaller values.*

Transfer case-to-transmission bolts	40
Rear output shaft-to-flange nut	186
Front output shaft-to-flange nut	164
Drain plug	20
Fill plug	20

2.3 Shift motor electrical connectors

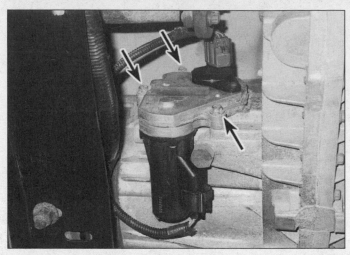

2.4 Shift motor mounting bolts

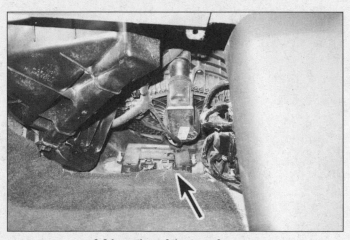

3.2 Location of the transfer case
control module

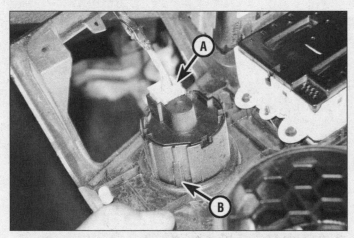

4.2 Disconnect the electrical connector (A), then depress the
retaining tabs (B, one of three shown) and push the switch
through the front of the panel

1   General Information

1   New Venture Gear transfer cases are either manual or electric shift.

Mechanical shift

2   The mechanical shift system on 4WD vehicles allows the driver to manually select one of three ranges: 2WD High, 4WD Low or 4WD High. The hub locks are engaged manually. The driver can shift from 4WD High to 2WD High at speeds up to 55 mph. 4WD Low can only be engaged or disengaged with the brake pedal depressed, the transmission in Neutral, and the vehicle at a complete stop.

Electronic shift-on-the-fly (ESOF) system

3   The electronic shift-on-the-fly (ESOF) system on 4WD vehicles allows the driver to engage 4WD High while the vehicle is moving. The hub locks are engaged automatically, and an electronic shift is initiated. A switch on the dash allows a selection of 2WD, 4WD

High or 4WD Low. The vehicle must be at a complete stop, with the transmission in Neutral and the brake applied before 4WD Low can be engaged.

2   Electric shift motor - replacement

Note: *The electric shift motor must be replaced as an assembly.*
1   Place the shift range selector switch into the 4X4 High position.
2   Raise the vehicle and place it securely on jackstands (if necessary).
3   Unplug the shift motor electrical connectors (see illustration).
4   Remove the three electric shift motor bolts (see illustration).
5   Remove the electric shift motor.
6   Clean the grease from the motor adapter and apply a new coat of multi-purpose grease.
7   The remainder of installation is the reverse of removal.

3   Transfer Case Control Module (TCCM) - replacement

Note: *The TCCM is located below the passenger side of the instrument panel, near the A pillar, attached to the floor.*
1   Disconnect the negative battery cable (see Chapter 5, Section 3).
2   Locate the TCCM and disconnect the electrical connectors (see illustration).
3   Remove the two nuts and remove the module from the vehicle.
4   Installation is reverse of removal.

4   Shift range selector switch (electric-shift models) - replacement

1   Remove the center finish panel from the instrument panel (see Chapter 11).
2   Unplug the electrical connector from the backside of the switch, and carefully pry out the switch through the front (see illustration).
3   Installation is the reverse of removal.

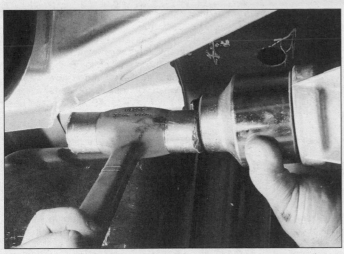

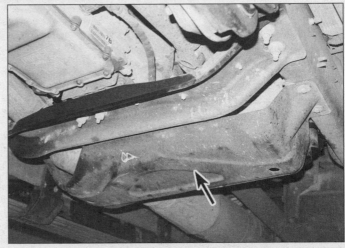

7.7 Drive the seal into place with a large socket or similar

8.5 Remove the transfer case skid plate

5 Manual shift lever - removal and installation

1   Shift the transfer case shift lever in the 4H position.
2   Remove the decorative trim bezel and boot and slide the boot up the shifter.
3   Remove the fasteners that attach the boot and bracket to the floor and slide the boot up the shifter.
4   Remove the shifter retaining bolt.
5   Remove the shifter.
6   Disconnect the linkage from the lower shift arm.
7   Remove the pivot and control bracket bolts, noting their positions, and remove the transfer case shift control assembly.
8   Installation is the reverse of removal. Verify the transfer case shift properly between all modes.

6 Mode indicator switch - replacement

Note: *The mode indicator switch is screwed into the top of the transfer case.*
1   Manually shifted transfer case models are equipped with a mode indicator switch to identify the mode the transfer case is in.
2   Locate the switch and disconnect the electrical connector.
3   Unscrew the switch and remove from the transfer case.
4   Installation is reverse of removal.

7 Oil seal - replacement

1   Raise the vehicle and support it securely on jackstands.
2   Remove the driveshaft(s) depending on which seal is being replaced (see Chapter 8).
3   Remove the flange mounting nut.

4   Remove the companion flange; a small puller may be required for removal.
5   Pry out the seal with a screwdriver or a seal removal tool. Don't damage the seal bore.
6   Lubricate the new seal lips with petroleum jelly.
7   Drive the seal into place with a large socket or section of tubing (see illustration). The outside diameter of the socket should be slightly smaller than the outside diameter of the seal.
8   Install the flange, slinger and retaining nut onto the output shaft. Tighten the nut to the torque listed in this Chapter's Specifications.
9   Install the driveshaft (see Chapter 8).
10   The remainder of installation is the reverse of removal.

8 Transfer case - removal and installation

Removal

1   Make sure the transfer case is shifted to 2WD High.
2   Disconnect the cable(s) from the negative battery terminal(s) (see Chapter 5, Section 3).
3   Raise the vehicle and support it securely on jackstands.
4   Shift the transmission to Neutral.
5   If equipped, remove the transfer case skid plate bolts, then remove the skid plate (see illustration).
6   If necessary, drain the transfer case lubricant (see Chapter 1).
7   Remove the front and rear driveshafts (see Chapter 8).
8   On manual-shift vehicles, remove the shifter (see Section 5).
9   On manual-shift vehicles, unplug the mode indicator switch electrical connector on the top of the transfer case.

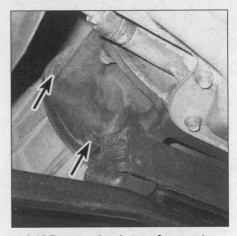

8.15 Remove the six transfer case-to-extension housing bolts

10   On electric-shift vehicles, unplug the electric shift motor connectors (see Section 2).
11   Disconnect the vent hose.
12   Support the transfer case with a jack - preferably a special jack made for this purpose. Safety chains or tie-downs will help steady the transfer case on the jack.
13   On models equipped with the TorqShift6 transmission, remove the transmission crossmember from the vehicle.
14   Disconnect the transfer case vent hose.
15   Remove the transfer case bolts that connect to the transmission extension housing (see illustration).
Note: *The TorqShift6 transmission has 11 bolts attaching the transfer case.*
16   Make a final check that all wires and hoses have been disconnected from the transfer case, then move the transfer case and jack toward the rear of the vehicle until the transfer case is clear of the transmission. Keep the transfer case level as this is done. Once the input shaft is clear, lower the transfer case and remove it from under the vehicle.

Installation

17   Installation is the reverse of removal. Be sure to install a new mounting gasket and tighten the transmission-to-transfer case bolts to the torque listed in this Chapter's Specifications.

18   Fill the transfer case with the specified fluid (see Chapter 1), run the engine and check for fluid leaks.

9   Transfer case overhaul - general information

1   Overhauling a transfer case is a difficult job for the do-it-yourselfer. It involves the disassembly and reassembly of many small parts. Numerous clearances must be precisely measured and, if necessary, changed with select-fit spacers and snap-rings. As a result, if transfer case problems arise, it can be removed and installed by a competent do-it-yourselfer, but overhaul should be left to a transmission repair shop. Rebuilt transfer cases may be available - check with your dealer parts department and auto parts stores. At any rate, the time and money involved in an overhaul is almost sure to exceed the cost of a rebuilt unit.

2   Nevertheless, it's not impossible for an inexperienced mechanic to rebuild a transfer case if the special tools are available and the job is done in a deliberate step-by-step manner so nothing is overlooked.

3   The tools necessary for an overhaul include internal and external snap-ring pliers, a bearing puller, a slide hammer, a set of pin punches, a dial indicator and possibly a hydraulic press. In addition, a large, sturdy workbench and a vise or transfer case stand will be required.

4   During disassembly of the transfer case, make careful notes of how each piece comes off, where it fits in relation to other pieces and what holds it in place. Noting how they are installed when you remove the parts will make it much easier to get the transfer case back together.

5   Before taking the transfer case apart for repair, it will help if you have some idea what area of the transfer case is malfunctioning. Certain problems can be closely tied to specific areas in the transfer case, which can make component examination and replacement easier. Refer to the *Troubleshooting* section at the front of this manual for information regarding possible sources of trouble.

Chapter 8
Driveline

Contents

Specifications

Torque specifications

Ft-lbs (unless otherwise indicated)

Note: *One foot-pound (ft-lb) of torque is equivalent to 12 inch-pounds (in-lbs) of torque. Torque values below approximately 15 foot-pounds are expressed in inch-pounds, because most foot-pound torque wrenches are not accurate at these smaller values.*

Hub lock screws	53 in-lbs
ABS sensor-to-rear differential bolt	30
Driveshaft center bearing mounting bolts	52
Driveshaft flange bolts (front/rear)	76
Driveshaft u-joint bearing strap bolts	
Front	26
Rear	46
Reverse slip driveshaft flange-to-driveshaft nut	425
Hub bearing-to-steering knuckle nuts (4WD)	133
Pinion nut	
Front	Bearing preload must be 5 in-lbs greater than before removal.
Rear	
Dana 80	470
Ford 10.50 inch	See Section 20
Rear axleshaft flange bolts	
Dana	98
Ford 10.50 inch	80
Front wheel hub extender nuts (dual rear wheel models)	130
Front axle trailing arm bolts	221
Rear axle u-bolt nuts	
Dana 80	
Step 1	37
Step 2	74
Step 3	111
Step 4	148
Ford 10.5 inch	
Step 1	48
Step 2	96
Step 3	148
Step 4	195

1   General information

1   The information in this Chapter deals with the components from the rear of the engine to the rear wheels, except for the transmission (and transfer case, if equipped), which is dealt with in the previous Chapter. For the purposes of this Chapter, these components are grouped into two categories: driveshaft and axles. Separate Sections within this Chapter offer general descriptions and checking procedures for components in each of the two groups.

2   Since nearly all the procedures covered in this Chapter involve working under the vehicle, make sure it's securely supported on sturdy jackstands or on a hoist where the vehicle can be easily raised and lowered.

2   Driveshaft(s) - general information

1   The driveshaft is of tubular construction and may be of a one- or two-section type according to the wheelbase of the vehicle. The front driveshaft on 4WD models is connected by two strap-type clamps at the front axle pinion and bolted to a flange at the transfer case. The attachment of the rear driveshaft to the rear axle pinion flange may be connected by bolted flange or two strap-type clamps, while attachment to the transmission or transfer case may include strap-type clamps, a bolted flange or a splined sliding sleeve (slip-yoke - one-piece driveshafts) connecting it to the output shaft. The type of connection used depends on the wheelbase and transmission type. Where a two-section shaft is used, the shaft is supported near its forward end on a ball bearing which is flexibly mounted in a bracket attached to the frame crossmember.

2   The driveshaft is finely balanced during manufacture and it is recommended that care be used when universal joints are replaced

to help maintain this balance. It is sometimes better to have the universal joints replaced by a dealership or shop specializing in this type of work. If you replace the joints yourself, mark each individual yoke in relation to the one opposite in order to maintain the balance. Do not drop the assembly during servicing operations.

3   Driveshaft(s) - removal and installation

Note: *The manufacturer recommends replacing driveshaft fasteners with new ones when installing the driveshaft.*

Removal

Rear

Note: *Where a two-piece driveshaft is involved, the rear shaft must be removed before the front shaft.*

1   Raise the vehicle and support it securely on jackstands.

2   Use chalk or a scribe to "index" the relationship of the driveshaft to the differential mating flange. This ensures correct alignment when the driveshaft is reinstalled (see illustration).

3   Remove the bolts securing the driveshaft flange or universal joint clamps to the differential pinion flange (see illustration).

4   If the vehicle is equipped with a two-piece driveshaft, remove the center support bearing mounting bolts (see illustration).

5   On 4WD models, "index" the relationship of the driveshaft to the transfer case flange. Remove the mounting bolts.

6   Some automatic transmissions have a splined slip yoke. Mark the relationship of the output shaft to slip yoke and simply slide the yoke out of the transmission.

7   Pry the universal joint away from its mating flange and remove the shaft from the flange. Be careful not to let the caps fall off of the universal joint (which would cause con-

tamination and loss of the needle bearings).

Front (4WD models)

8   Using a scribe, white paint or a hammer and punch, place marks on the driveshaft and differential flanges in line with each other. This is to make sure the driveshaft is reinstalled in the same position to preserve the balance.

9   Make alignment marks on the driveshaft and transfer case flanges.

10   Unbolt the flange that secures the driveshaft double-cardan joint to the transfer case.

11   Remove the bolts and straps that secure the front end of the driveshaft to the differential yoke.

12   Wrap tape around the universal joint bearings at the axle end of the driveshaft so they won't fall off.

Installation

13   Installation is the reverse of removal. If the shaft cannot be lined up due to the components of the differential or transmission having been rotated, put the vehicle in Neutral or rotate one wheel to allow the original alignment to be achieved. Make sure the universal joint caps are properly placed in the flange seat. Tighten the fasteners to the torque listed in this Chapter's Specifications.

4   Driveshaft center support bearing - check and replacement

Note: *Driveshaft center bearing replacement requires the use of a hydraulic press. Remove the driveshaft and have a repair shop or automotive machine shop remove the bearing.*

Check

1   The center support bearing can be checked in a similar manner as the universal joints are examined (see Section 5). Check for looseness or deterioration of the flexible rubber mounting.

3.2 Mark the relationship of the driveshaft U-joint to the differential pinion flange

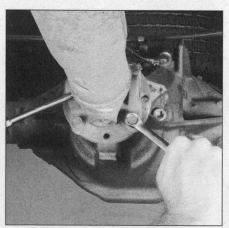

3.3 Insert a screwdriver through the U-joint to prevent the driveshaft from turning as you break loose the four U-joint-to-pinion flange bolts

3.4 Remove the two bolts for the center support bearing

2   Further examination of the center bearing can be made by running the vehicle in gear with the rear wheels raised in the air. However, this should be done with the vehicle supported on a lift and by a dealer service department or other qualified repair facility who can perform the tests safely.

Replacement

3   Raise the vehicle and support it securely on jackstands.
4   Remove the driveshaft assembly (see Section 3).
5   With the driveshaft removed from the vehicle and the shaft sections placed on a suitable work bench, hold the driveshaft flange with a chain wrench to prevent the driveshaft from rotating, then remove the flange retaining nut.
6   Remove the strap which retains the rubber cushion to the bearing support bracket.
7   Separate the cushion, bracket and remove the rubber insulator.
8   Have the old bearing and dust slinger pressed off.
9   Install a new slinger and bearing.
10   Pack the space between the inner dust slinger and the bearing with lithium-base grease.
11   Carefully tap the bearing and slinger assembly onto the driveshaft journal until the components are tight against the shoulder on the shaft. Use a suitable piece of tubing to do this, taking care not to damage the shaft splines.
12   Lubricate the shaft splines with lithium-base grease.
13   Install the bearing rubber cushion, bracket and strap. The center bearing bracket should be installed with the deep flange rearward.
14   Install the flange and retaining nut. Tighten the nut to the torque listed in this Chapter's Specifications.
15   The remainder of installation is the reverse of removal.

5   Universal joints - general information, lubrication and check

1   Universal joints are mechanical couplings which connect two rotating components that meet each other at different angles.
2   These joints are composed of a yoke on each side connected by a crosspiece called a trunnion. Cups at each end of the trunnion contain needle bearings which provide smooth transfer of the torque load. Snap-rings, either inside or outside of the bearing cups, hold the assembly together.
3   Refer to Chapter 1 for details on universal joint lubrication. Also see the routine maintenance schedule at the beginning of Chapter 1.
4   Wear in the needle roller bearings is characterized by vibration in the driveline, noise during acceleration, and in extreme cases of lack of lubrication, metallic squeaking and ultimately grating and shrieking sounds as the bearings disintegrate.
5   It is easy to check if the needle bearings are worn with the driveshaft in position, by trying to turn the shaft with one hand, the other hand holding the rear axle flange when the rear universal joint is being checked, and the front half coupling when the front universal joint is being checked. Any movement between the driveshaft and the front half couplings, and around the rear half couplings, is indicative of considerable wear. Another method of checking for universal joint wear is to use a pry bar inserted into the gap between the universal joint and the driveshaft or flange. Leave the vehicle in gear and try to pry the joint both radially and axially. Any looseness should be apparent with this method. A final test for wear is to attempt to lift the shaft and note any movement between the yokes of the joints.
6   If any of the above conditions exist, replace the universal joints with new ones.

6   Universal joints - replacement

Single-cardan U-joints

Note: *A press or large vise will be required for this procedure. It may be advisable to take the driveshaft to a local dealer service department, service station or machine shop where the universal joints can be replaced for you, normally at a reasonable charge.*

1   Remove the driveshaft as outlined in Section 3.
2   On U-joints with external snap-rings, use a small pair of pliers to remove the snap-rings from the spider (see illustration).
3   Supporting the driveshaft, place it in position on a workbench equipped with a vise.
4   Place a piece of pipe or a large socket, having an inside diameter slightly larger than the outside diameter of the bearing caps, over one of the bearing caps. Position a socket with an outside diameter slightly smaller than that of the opposite bearing cap against the cap (see illustration) and use the vise or press to force the bearing cap out (inside the pipe or large socket). Use the vise or large pliers to work the bearing cap the rest of the way out.
5   Transfer the sockets to the other side and press the opposite bearing cap out in the same manner.
6   Pack the new universal joint bearings with grease. Ordinarily, specific instructions for lubrication will be included with the universal joint servicing kit and should be followed carefully.
7   Position the spider in the yoke and partially install one bearing cap in the yoke.
8   Start the spider into the bearing cap and then partially install the other cap. Align the spider and press the bearing caps into position, being careful not to damage the dust seals.
9   Install the snap-rings. If difficulty is encountered in seating the snap-rings, strike the driveshaft yoke sharply with a hammer. This will spring the yoke ears slightly and

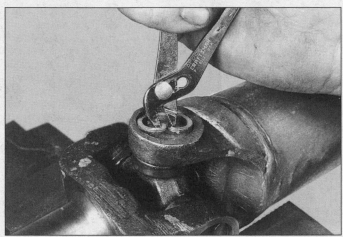

6.2 A pair of needle-nose pliers can be used to remove the universal joint snap-rings

6.4 To press the universal joint out of the driveshaft yoke, set it up in a vise with the small socket pushing the joint and bearing cap into the large socket

allow the snap-rings to seat in the groove.

10   Install the grease fitting and fill the joint with grease. Be careful not to overfill the joint, as this could blow out the grease seals.

11   Install the driveshaft (see Section 3).

Double-cardan U-joints

12   Use the above procedure, but note that it will have to be repeated because the double-cardan joint is made up of two single-cardan joints. Also pay attention to how the spring, centering ball and bearing are arranged.

Note: *Some of the following conditions cannot be fixed without highly specialized tools and expertise. Such problems must be referred to a transmission repair shop or a dealer service department.*

---

7   Rear axle - general information

1   The rear axle assembly consists of a straight, hollow housing enclosing a differen-

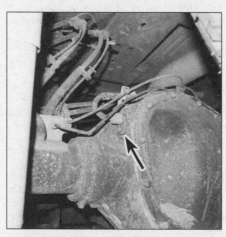

8.4a Disconnect the ABS sensor (Dana 80 axle)

tial assembly and axleshafts. These assemblies support the vehicle's 'sprung' weight components through leaf springs attached between the axle housings and the vehicle's frame rails.

2   The rear axle assemblies employed on vehicles covered by this manual are heavy duty full-floating axleshafts. Full-floating axleshafts do not themselves bear any of the vehicle's weight, can be removed independent of the tapered roller wheel bearings and are designed only to transfer power to the rear wheels.

3   Due to the need for special tools and equipment, it is recommended that operations on these models be limited to those described in this Chapter. Where repair or overhaul is required, remove the axle assembly and take it to a rebuilder, or exchange it for a new or reconditioned unit. Always make sure that an axle unit is exchanged for one of identical type and gear ratio.

---

8   Rear axle assembly - removal and installation

Removal

1   Raise the rear of the vehicle and support it with jackstands placed under the frame rails.

2   Remove the rear wheels.

3   Disconnect the driveshaft from the rear axle yoke (see Section 3).

4   On Dana 80 models, disconnect the ABS sensor. On Ford 10.5" ring gear models, remove the sensors from the backing plates (see illustrations).

5   On Ford 10.5" ring gear models, disconnect the Electronic Locking Differential (ELD) connector if equipped.

6   Disconnect the parking brake cable from the parking brake lever (see Chapter 9).

7   Disconnect the axle vent hose and

remove the brake hose junction block.

8   Detach the stabilizer bar from the axle housing.

9   Disconnect the brake lines from the axle housing. Remove the rear brake calipers (see Chapter 9).

Caution: *Tie the calipers up with wire to keep any strain off the flexible brake lines.*

10   Support the rear axle with a jack or other suitable device. It may take two jacks to do this safely as the offset of the rear differential causes this assembly to be heavily weighted to one side.

11   Remove the lower mounting bolts securing the rear shocks to the axle.

12   With the jack(s) supporting the axle, remove the U-bolts and spring plates securing the axle to the springs (see illustration).

13   Lower the axle assembly and remove it from under the vehicle.

Installation

14   Installation is the reverse of the removal procedure.

15   Tighten the U-bolt nuts to the torque listed in the Chapter 10 Specifications. If necessary, check and fill the axle with the specified lubricant (see Chapter 1).

---

9   Rear axleshaft - removal and installation

Removal

1   Remove the bolts which attach the axleshaft flange to the hub (see illustration). There is no need to remove the wheel or jack up the vehicle.

2   Tap the flange with a soft-faced hammer to loosen the shaft, then grip the rib of the face of the flange with a pair of locking pliers; twist the shaft slightly in both directions and withdraw it from the axle tube (see illustration).

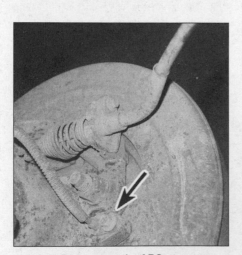

8.4b Disconnect the ABS sensors (Ford 10.5 inch axle)

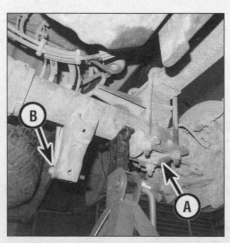

8.12 Remove the U-bolts and spring plates (A) and remove the lower bolts for the shocks (B)

9.1 Remove the axleshaft flange bolts

Installation

3   Inspect the O-ring seal for cracks or wear and if necessary, replace it with a new one, coated with clean differential lubricant (see illustration). Installation is the reverse of removal, but hold the axleshaft level in order to engage the splines at its inner end with those in the differential side gear. Use new bolts, lock washers and/or a thread locking compound and tighten the bolts to the torque listed in this Chapter's Specifications.
Note: *Final tightening of the axleshaft retaining bolts should be done after the wheel lug nuts have been tightened (if the wheel was removed).*
4   If a loss of fluid is observed, check the differential lubricant level and re-fill it if required (see Chapter 1).
5   Test drive the vehicle and check for leaks.

10   Rear wheel hub bearings and grease seal - removal, inspection and installation

Removal

1   Remove the axleshafts (see Section 9).
2   Raise the rear of the vehicle and place it securely on jackstands.
3   Remove the rear wheels.

Dana axle

4   Remove the disc brake caliper (see Chapter 9).
5   Install a special hub wrench so that the drive tangs of the tool engage the four slots in the hub nut and remove the nut (see illustration).
6   Remove the outer bearing from the hub and pull the hub and brake disc straight off the spindle.
7   Separate the hub from the disc by removing the bolts on the inside of the disc.

8   Use a large screwdriver or pry bar to pry the seal from the back of the wheel hub. Remove the inner bearing.

Ford 10.50-inch axle

9   Remove the disc brake caliper (see Chapter 9).
10   Install a special hub wrench so that the drive tangs of the tool engage the four slots in the hub nut and remove the nut (see illustration 17.5).
Note: *Push inward on the socket to disengage the ratchet mechanism to allow the locknut to be loosened.*
Note: *The left side nut has a left-hand thread (turn the nut clockwise to loosen it).*
11   Using a step plate installed in the axle tube and a 3-jaw puller, loosen the rear hub to the point of removal (see illustration).
12   Remove the rear hub. Take care not to drop the outer hub bearing while removing the hub.
13   Use a large screwdriver or prybar to pry the seal from the back of the wheel hub. Remove the inner bearing.

Inspection

14   Clean the hub nut and bearings with cleaning solvent; allow the parts to air dry.
15   Clean the inside of the wheel hub to remove all axle lubricant and grease. Clean the axle spindle.
16   Inspect the bearing assemblies for signs of wear, pitting, galling and other damage. Replace the bearings if any of these conditions exist. Inspect the bearing races for signs of erratic wear, galling and other damage.
17   If the bearing races need replacement, drive out the bearing races from the wheel hub with a brass drift. Install the new races with a bearing cup replacement tool or other suitable tool designed for this purpose. Never use a drift or punch for this operation as these races must be driven squarely, seated correctly and can be damaged easily.

Installation

Dana axle

18   Prior to installation, pack the inner and outer wheel bearings with high temperature grease (see Chapter 1) or equivalent. If you do not have access to a bearing packer, pack each bearing carefully by hand and make sure the entire bearing is filled with grease.
19   Install the newly packed inner wheel bearing into the hub. Install a new hub inner seal with a suitable drive tool (tubular drift, large socket or special tool) being careful not to damage the seal.
20   Install the rear disc onto the hub (see Chapter 9).
21   Prior to installing the wheel hub, coat the inner seal lip with grease. Also, cover the spindle with a light coat of grease. Pour one ounce of differential lubricant in the center of the hub.
22   Carefully slide the hub and disc assembly over the spindle, being very careful to keep it straight so as not to contact the spindle with the seal (which would damage it).

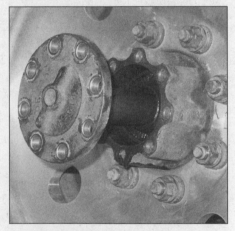

9.2 Pull the axleshaft straight out of the axle housing

9.3 Inspect this O-ring seal for cracks or wear

10.5 The drive tangs of the tool engage with the four slots in the hub nut

10.11 Use a puller to loosen the hub

23　Install the newly packed outer wheel bearing over the spindle and into the wheel hub.

24　Install the hub nut, making sure the tab is aligned in the keyway slot.

25　Start the nut while applying inward pressure to the socket, then tighten the nut to 70 ft-lbs. Rotate the hub occasionally while tightening the locknut.

26　Back the locknut off 90-degrees (1/4 turn counterclockwise), then tighten it to 15 to 20 ft-lbs.

27　After the final torque, the endplay should be zero and the maximum torque to rotate the hub should be no greater than 20 in-lbs.

28　The remainder of installation is the reverse of removal. Be sure to tighten the fasteners securely. Check the axle lubricant level and adjust if necessary (see Chapter 1).

Ford 10.50-inch axle

29　Prior to installation, pack the inner and outer wheel bearings with high temperature grease (see Chapter 1) or equivalent. If you do not have access to a bearing packer, pack each bearing carefully by hand and make sure the entire bearing is penetrated with grease.

30　Install the newly packed inner wheel bearing into the hub. Install a new hub inner seal with a suitable drive tool (tubular drift, large socket or special tool) being careful not to damage the seal.

31　On dual rear wheel axles, install the rear disc onto the hub (see Chapter 9).

32　Prior to installing the wheel hub, coat the inner seal lip with grease. Also, cover the spindle with a light coat of grease. Pour one ounce of differential lubricant in the center of the hub.

33　Carefully slide the hub assembly over the spindle, being very careful to keep it straight so as not to contact the spindle with the seal (which would damage it).

34　Install the newly packed outer wheel bearing over the spindle and into the wheel hub.

35　Install the hub nut, making sure the tab is aligned in the keyway. Using the locknut wrench, tighten the locknut clockwise for right-hand threads and counterclockwise for left-hand threads. Rotate the hub occasionally while tightening the locknut. The hub nut should ratchet as torque is applied.

36　Tighten the hub nut to 60 ft-lbs.

37　Back the locknut off 180-degrees (1/2 turn counterclockwise), then tighten it to 15 ft-lbs.

Note: *Push inward on the socket to disengage the ratchet mechanism to allow the locknut to be loosened.*

38　The remainder of installation is the reverse of removal. Be sure to tighten all fasteners securely. Check the axle lubricant level and adjust if necessary (see Chapter 1).

---

11 Hub bearing assembly and front axleshaft (4WD models) - removal and installation

Note: *For 2WD model front hub service, see Chapter 9, Section 6.*

Removal

1　Loosen the wheel lug nuts. Raise the vehicle and support it securely on jackstands. Remove the wheel.

2　Remove the brake caliper and support it out of the way with wire. Remove the caliper mounting bracket and the brake disc (see Chapter 9).

3　Remove the hub lock (see Section 14).

4　Remove the axleshaft snap-ring and thrust washers (see illustration). Note the order of the thrust washers for installation.

5　Disconnect the ABS wheel sensor harness connector and routing clips.

6　Remove the hub-to-knuckle nuts (see illustration).

7　Remove the hub bearing from the steering knuckle and axleshaft.

Note: *The hub bearing assembly may become rusted and seized to the steering knuckle depending on the severity of driving conditions*

(snow, rain, salt, etc.). In this event, have the hub bearing assembly removed by an automotive repair shop.

8　Remove the brake dust shield and ABS sensor from the hub bearing.

9　Remove and discard the O-ring on the hub.

10　Using a brass drift, drive the axle seal out of the knuckle.

11　Carefully pull the axleshaft and seal from the axle housing. If the U-joint is worn out, it can be replaced using the procedure described in Section 6.

Installation

12　Clean the axleshaft of all dirt and grease.

13　Place the axleshaft in a padded vise so the vise grips the shaft's second step.

14　Place the new main seal onto the axle shaft. Using a hammer and seal installer, seat the main seal onto the axle shaft.

15　Apply a thin film of wheel bearing grease to the shaft splines, seal contact surface and hub bore. Install the axleshaft, engaging the splines with the differential side gears. Be very careful not to damage the axleshaft oil seals.

16　Using a hammer and seal installer, seat the main seal into the steering knuckle.

17　Install the ABS sensor and dust shield.

18　Install a new O-ring seal onto the wheel hub. Apply a small coat of high-temperature grease to the O-ring and wheel hub contact area.

19　Install the wheel hub bearing. Install the hub bearing-to-steering knuckle nuts and tighten them to the torque listed in this Chapter's Specifications.

20　Install the three thrust washers and snap-ring.

Caution: *Be sure to place the non-metallic washer between the two metallic washers. Incorrect installation will lead to severe wear to the washer, resulting in damage to the wheel hub, bearing, the axle shaft and end seal.*

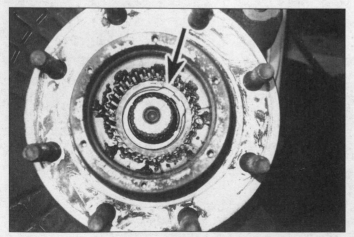

11.4 Use a pair of snap-ring pliers to remove this snap-ring

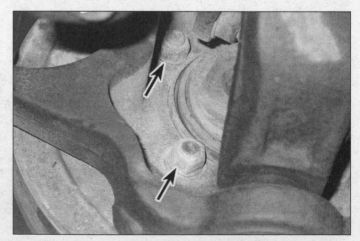

11.6 Remove the four hub-to-steering knuckle fasteners (two not shown in picture)

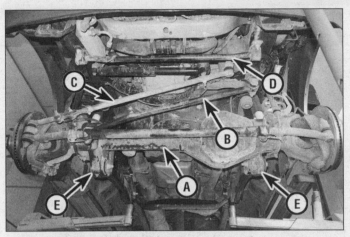

12.1 Front axle assembly components (4WD models)

A   Axle assembly
B   Track bar
C   Drag link
D   Stabilizer bar
E   Trailing arms

12.13 Unbolt the lower ends of the shock absorbers from the axle

21   Install the brake disc and caliper (see Chapter 9).
22   The remainder of installation is the reverse of removal.

12 Front axle assembly (4WD models) - removal and installation

Removal

1   Raise the front of the vehicle and support it on jackstands placed under the frame rails (see illustration).
2   Remove the wheels.
3   Disconnect the driveshaft from the front axle yoke (see Section 3).
4   Disconnect the brake hose brackets and position aside.
5   Remove the front brake calipers (see Chapter 9).
Note: *Tie the calipers up with wire to keep any strain off the flexible brake lines.*
6   Disconnect the ABS sensors and detach the harnesses from the trailing arms. Disconnect the vacuum hose for the hub locks and bracket.
7   Detach the drag link end from the steering knuckle and position aside (see Chapter 10).
8   Unbolt the stabilizer bar clamps from the axle and the stabilizer bar link nuts (see Chapter 10).
9   Disconnect the vent tube and be sure to plug the fitting to prevent anything from falling into the vent.
10   Support the front axle with a pair of floor jacks. Two jacks should be used, as the offset of the front differential causes this assembly to be heavily weighted to one side.
11   Raise the axle enough to relieve tension on the track bar. Unbolt the track bar at the axle. Disconnect the track bar and relieve the load on the suspension.
12   Move the hub-lock vacuum hoses aside.

13   Unbolt the shocks from the axle assembly (see illustration).
14   Carefully lower the axle assembly and then remove the coil springs (see Chapter 10). Unbolt the trailing arms from the front axle (see illustration).
15   Lower the axle assembly completely and remove it from the vehicle.

Installation

16   Installation is the reverse of the removal procedure.
17   Tighten all fasteners securely. If necessary, check and fill the axle with the specified lubricant (see Chapter 1).

13 Differential pinion seal - replacement

1   Raise the vehicle and place it securely on jackstands.
2   Remove the driveshaft (see Section 3).

Ford 10.50 inch rear axle and Dana model 50/60 front axles

Caution: *This procedure disturbs the pinion bearing preload adjustment. Follow the procedure very carefully to reset the pinion bearing preload during reassembly.*
3   Depending on which axle pinion seal is being replaced, remove the rear or front wheels and brake calipers.
Note: *The removal of the wheels and brake calipers is advisable to eliminate the added pinion shaft rotation resistance that otherwise might contribute to a false pinion shaft rotation preload torque value.*
4   Using an inch-pound torque wrench (scale from approximately 0 to 40 inch-pounds) on the drive pinion nut, measure and record the torque necessary to rotate the drive pinion in a load-free state (see illustration).
Note: *Measured bearing preload:* _____
5   Count the number of threads visible

12.14 The trailing arms are mounted to the front axle with two bolts (arrows)

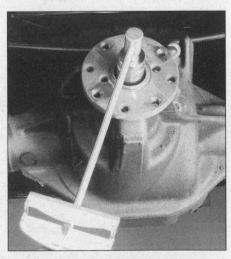

13.4 Use an inch-pound torque wrench to check the torque necessary to rotate the pinion shaft

13.6 Mark the relative positions of the pinion and flange before removing the nut

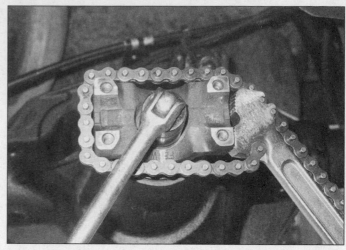

13.7 Use a chain wrench to hold the flange while loosening the pinion flange locknut

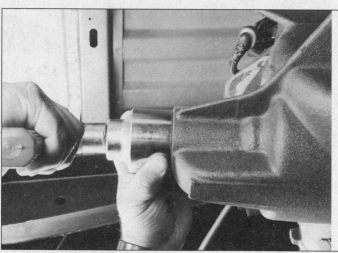

13.11 Lubricate the lips of the new pinion seal and seat it squarely in the bore, then drive it into the carrier with a seal driver or a large socket

between the end of the nut and the end of the pinion shaft and record it for use later.

Note: *Visible threads:* _____

6   Mark the drive pinion-to-companion flange orientation for proper location of flange to pinion upon reassembly (see illustration).

7   Using a flange holding tool (available at some auto parts stores), or a chain wrench to prevent the flange from turning while removing the nut, unscrew the pinion flange locknut (see illustration). Discard the nut - a new one must be used when reassembling.

8   Using a two-jaw puller, remove the companion flange from the drive pinion shaft.

Note: *Some fluid loss may occur.*

9   Avoiding contact with the pinion shaft/ threads, tap out an edge of the seal using a chisel or dull screwdriver. Using vise-grips or other similar clamping pliers, grip the exposed edge and strike the side of the pliers until the seal is removed.

10   Prior to installing the new seal, clean the seal mating surfaces.

11   Lubricate the lips of the new seal with high-temperature grease and tap it evenly

into position with a seal installation tool or a large socket. Make sure it enters the housing squarely and is tapped in to its full depth (see illustration).

12   Align the mating marks made before disassembly and install the companion flange and a new nut. If necessary, tighten the pinion nut to draw the flange into place. Do not try to hammer the flange into position.

13   Using a suitable holding tool, secure the companion flange while tightening the nut carefully until the original number of threads are exposed.

14   Measure the torque required to rotate the pinion and tighten the nut in small increments until it matches the figure recorded in Step 4. In order to compensate for the drag of the new oil seal, the nut should be tightened more until the rotational torque of the pinion exceeds earlier recording by no more than 5 in-lbs.

15   The remainder of installation is the reverse of removal. Be sure to check the differential lubricant level and add if required (see Chapter 1).

Dana model 80 rear axle

16   Mark the relationship of the companion flange to the pinion shaft.

17   Remove the pinion nut. A special flange holding tool, available at some auto parts stores, or a chain wrench can be used to keep the companion flange from moving while the self-locking pinion nut is loosened (see illustration 13.4).

18   Withdraw the companion flange. It may be necessary to use a two or three-jaw puller engaged behind the flange to draw it out. Do not attempt to pry behind the flange or hammer on the end of the pinion shaft.

19   Pry out the old seal and discard it.

20   Lubricate the lips of the new seal with high-temperature grease and tap it evenly into position with a seal installation tool or a large socket. Make sure it enters the housing squarely and is tapped in to its full depth (see illustration 13.11).

21   Align the mating marks made before disassembly and install the companion flange. If necessary, tighten the pinion nut to draw the flange into place. Do not try to hammer the flange into position.

22   Tighten the nut to the torque listed in this Chapter's Specifications.

23   Install the driveshaft (see Section 3).

24   The remainder of installation is the reverse of removal.

25   Check the differential lubricant level and add if required (see Chapter 1).

14   Hub lock - removal and installation

Removal

1   On dual rear wheel models, remove the nuts and the hub extender.

2   On all models, remove the three screws and carefully pry outward on the hub lock to remove (see illustrations).

14.2a Remove the three screws . . .

14.2b . . . carefully pry the hub lock away from the hub . . .

Note: *Use care while prying to prevent damage to the hub lock while removing it.*

3   Remove and discard the O-ring seal and/or gasket.

Installation

4   Install a new O-ring and/or gasket. Clean the contact areas for the O-ring and apply a light coat of high temperature grease to the O-ring before installing it.

5   Rotate the hub lock dial clockwise to the Lock position. Apply grease to the inner and outer splines.

6   Insert the hub lock into the wheel hub while slowly turning the wheel hub to align the splines. Lift and jiggle the axle U-joint while pushing the hub lock into the wheel hub until it is seated completely.

7   Install the screws and tighten them to the torque listed in this Chapter's Specifications.

8   The remainder of installation is reverse of removal. Ensure that the hub rotates freely between the two positions.

15   Hub lock vacuum solenoid - replacement

Note: *On gasoline-engine models, the vacuum hub lock solenoid is attached to the hub lock vacuum reservoir, on the right side of the fan shroud. on diesel engine models, the solenoid is located on the rear of the primary cooling system expansion tank.*

1   Disconnect the vacuum hose and harness connector from the solenoid.

2   Remove the hub lock solenoid from the reservoir (gasoline models) or expansion tank (diesel models).

3   Remove the mounting bolts to remove the reservoir.

4   Installation is the reverse of the removal steps.

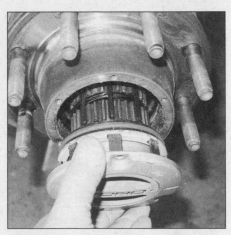

14.2c . . . and remove the hub lock from the hub

Notes

Chapter 9   Brakes

Contents

Specifications

General

Brake fluid type	See Chapter 1
Power brake booster pushrod protrusion (vacuum boosters only)	0.980 to 0.995 inch
Brake pad minimum thickness	See Chapter 1
Disc lateral runout limit	
Front	
2WD	
With single rear wheels	0.0020 inch
With dual rear wheels	0.0028 inch
4WD	0.0015 inch
Rear	0.0015 inch
Disc minimum (discard) thickness	Cast into disc

Torque specifications   **Ft-lbs** (unless otherwise indicated)

Note: *One foot-pound (ft-lb) of torque is equivalent to 12 inch-pounds (in-lbs) of torque. Torque values below approximately 15 foot-pounds are expressed in inch-pounds, because most foot-pound torque wrenches are not accurate at these smaller values.*

ABS wheel speed sensor bolt	
Front sensor	
2WD models	18
4WD models	159 in-lbs
Rear sensor	177 in-lbs
Brake hose-to-caliper fitting bolt	
Front	35
Rear	
2012 and earlier models	26
2013 and later models	35
Brake lines to master cylinder	18
Caliper mounting bolts	
Front	55
Rear	
2012 and earlier models	
Single rear wheel	26
Dual rear wheel	55
2013 and later models	55
Caliper mounting bracket bolts	
Front	166
Rear	203
Disc-to-hub flange bolts (models with dual rear wheels)	89
Rear caliper disc shield/support bracket	101
Master cylinder mounting nuts	
Hydro-boost models*	22
Vacuum booster models*	17
Power brake booster mounting nuts	18
Wheel lug nuts	See Chapter 1

* Manufacturer recommends new fasteners. Do not reuse the old fasteners.

---

1   General information

General

The vehicles covered by this manual are equipped with hydraulically operated front and rear brake systems. Both front and rear brakes are disc-type and are self-adjusting.

Hydraulic system

The hydraulic system consists of two separate circuits. The master cylinder has separate reservoirs for the two circuits, and, in the event of a leak or failure in one hydraulic circuit, the other circuit will remain operative and a warning indicator will light up on the instrument panel when a substantial amount of brake fluid is lost, showing that a failure has occurred.

Power brake booster

The power brake booster uses engine manifold vacuum to provide assistance to the brakes. It is mounted on the firewall in the engine compartment, directly behind the master cylinder.

Parking brake

Control cables are routed to the rear axle, where they operate small drum brake shoes that apply pressure to the inner diameter of the rear brake discs.

Service

After completing any operation involving disassembly of any part of the brake system, always test drive the vehicle to check for proper braking performance before resuming normal driving. When testing the brakes, perform the tests on a clean, dry, flat surface. Conditions other than these can lead to inaccurate test results.

Test the brakes at various speeds with both light and heavy pedal pressure. The vehicle should stop evenly without pulling to one side or the other. Under hard braking, the ABS system may engage, resulting in brake pedal pulsation. This is considered normal operation.

Tires, vehicle load and wheel alignment are factors which also affect braking performance.

Precautions

There are some general cautions and warnings involving the brake system on this vehicle:

a) *Use only brake fluid conforming to DOT 3 specifications.*

b) *The brake pads and linings contain fibers which are hazardous to your health if inhaled. Whenever you work on brake system components, clean all parts with brake system cleaner. Do not allow the fine dust to become airborne. Also, wear an approved filtering mask.*

c) *Safety should be paramount whenever any servicing of the brake components is performed. Do not use parts or fasteners which are not in perfect condition, and be sure that all clearances and torque specifications are adhered to. If you are at all unsure about a certain procedure, seek professional advice. Upon completion of any brake system work, test the brakes carefully in a controlled area before putting the vehicle into normal service. If a problem is suspected in the brake system, don't drive the vehicle until it's fixed.*

d) *Used brake fluid is considered a hazardous waste and it must be disposed of in accordance with federal, state and local laws. **DO NOT pour it down the sink, into septic tanks or storm drains, or on the ground.***

e) *Clean up any spilled brake fluid immediately and then wash the area with large amounts of water. This is especially true for any finished or painted surfaces.*

2   Troubleshooting

PROBABLE CAUSE	CORRECTIVE ACTION

No brakes - pedal travels to floor

1 Low fluid level 2 Air in system	1 and 2 Low fluid level and air in the system are symptoms of another problem - a leak somewhere in the hydraulic system. Locate and repair the leak
3 Defective seals in master cylinder	3 Replace master cylinder
4 Fluid overheated and vaporized due to heavy braking	4 Bleed hydraulic system (temporary fix). Replace brake fluid (proper fix)

Brake pedal slowly travels to floor under braking or at a stop

1 Defective seals in master cylinder	1 Replace master cylinder
2 Leak in a hose, line, caliper or wheel cylinder	2 Locate and repair leak
3 Air in hydraulic system	3 Bleed the system, inspect system for a leak

Brake pedal feels spongy when depressed

1 Air in hydraulic system	1 Bleed the system, inspect system for a leak
2 Master cylinder or power booster loose	2 Tighten fasteners
3 Brake fluid overheated (beginning to boil)	3 Bleed the system (temporary fix). Replace the brake fluid (proper fix)
4 Deteriorated brake hoses (ballooning under pressure)	4 Inspect hoses, replace as necessary (it's a good idea to replace all of them if one hose shows signs of deterioration)

Brake pedal feels hard when depressed and/or excessive effort required to stop vehicle

1 Power booster faulty	1 Replace booster
2 Engine not producing sufficient vacuum, or hose to booster clogged, collapsed or cracked	2 Check vacuum to booster with a vacuum gauge. Replace hose if cracked or clogged, repair engine if vacuum is extremely low
3 Brake linings contaminated by grease or brake fluid	3 Locate and repair source of contamination, replace brake pads or shoes
4 Brake linings glazed	4 Replace brake pads or shoes, check discs and drums for glazing, service as necessary
5 Caliper piston(s) or wheel cylinder(s) binding or frozen	5 Replace calipers or wheel cylinders
6 Brakes wet	6 Apply pedal to boil-off water (this should only be a momentary problem)
7 Kinked, clogged or internally split brake hose or line	7 Inspect lines and hoses, replace as necessary

Excessive brake pedal travel (but will pump up)

1 Drum brakes out of adjustment	1 Adjust brakes
2 Air in hydraulic system	2 Bleed system, inspect system for a leak

Excessive brake pedal travel (but will not pump up)

1 Master cylinder pushrod misadjusted	1 Adjust pushrod
2 Master cylinder seals defective	2 Replace master cylinder
3 Brake linings worn out	3 Inspect brakes, replace pads and/or shoes
4 Hydraulic system leak	4 Locate and repair leak

Brake pedal doesn't return

1 Brake pedal binding	1 Inspect pivot bushing and pushrod, repair or lubricate
2 Defective master cylinder	2 Replace master cylinder

Troubleshooting (continued)

PROBABLE CAUSE	CORRECTIVE ACTION

Brake pedal pulsates during brake application

PROBABLE CAUSE	CORRECTIVE ACTION
1 Brake drums out-of-round	1 Have drums machined by an automotive machine shop
2 Excessive brake disc runout or disc surfaces out-of-parallel	2 Have discs machined by an automotive machine shop
3 Loose or worn wheel bearings	3 Adjust or replace wheel bearings
4 Loose lug nuts	4 Tighten lug nuts

Brakes slow to release

PROBABLE CAUSE	CORRECTIVE ACTION
1 Malfunctioning power booster	1 Replace booster
2 Pedal linkage binding	2 Inspect pedal pivot bushing and pushrod, repair/lubricate
3 Malfunctioning proportioning valve	3 Replace proportioning valve
4 Sticking caliper or wheel cylinder	4 Repair or replace calipers or wheel cylinders
5 Kinked or internally split brake hose	5 Locate and replace faulty brake hose

Brakes grab (one or more wheels)

PROBABLE CAUSE	CORRECTIVE ACTION
1 Grease or brake fluid on brake lining	1 Locate and repair cause of contamination, replace lining
2 Brake lining glazed	2 Replace lining, deglaze disc or drum

Vehicle pulls to one side during braking

PROBABLE CAUSE	CORRECTIVE ACTION
1 Grease or brake fluid on brake lining	1 Locate and repair cause of contamination, replace lining
2 Brake lining glazed	2 Deglaze or replace lining, deglaze disc or drum
3 Restricted brake line or hose	3 Repair line or replace hose
4 Tire pressures incorrect	4 Adjust tire pressures
5 Caliper or wheel cylinder sticking	5 Repair or replace calipers or wheel cylinders
6 Wheels out of alignment	6 Have wheels aligned
7 Weak suspension spring	7 Replace springs
8 Weak or broken shock absorber	8 Replace shock absorbers

Brakes drag (indicated by sluggish engine performance or wheels being very hot after driving)

PROBABLE CAUSE	CORRECTIVE ACTION
1 Brake pedal pushrod incorrectly adjusted	1 Adjust pushrod
2 Master cylinder pushrod (between booster and master cylinder) incorrectly adjusted	2 Adjust pushrod
3 Obstructed compensating port in master cylinder	3 Replace master cylinder
4 Master cylinder piston seized in bore	4 Replace master cylinder
5 Contaminated fluid causing swollen seals throughout system	5 Flush system, replace all hydraulic components
6 Clogged brake lines or internally split brake hose(s)	6 Flush hydraulic system, replace defective hose(s)
7 Sticking caliper(s) or wheel cylinder(s)	7 Replace calipers or wheel cylinders
8 Parking brake not releasing	8 Inspect parking brake linkage and parking brake mechanism, repair as required
9 Improper shoe-to-drum clearance	9 Adjust brake shoes
10 Faulty proportioning valve	10 Replace proportioning valve

Troubleshooting (continued)

PROBABLE CAUSE	CORRECTIVE ACTION

Brakes fade (due to excessive heat)

PROBABLE CAUSE	CORRECTIVE ACTION
1 Brake linings excessively worn or glazed	1 Deglaze or replace brake pads and/or shoes
2 Excessive use of brakes	2 Downshift into a lower gear, maintain a constant slower speed (going down hills)
3 Vehicle overloaded	3 Reduce load
4 Brake drums or discs worn too thin	4 Measure drum diameter and disc thickness, replace drums or discs as required
5 Contaminated brake fluid	5 Flush system, replace fluid
6 Brakes drag	6 Repair cause of dragging brakes
7 Driver resting left foot on brake pedal	7 Don't ride the brakes

Brakes noisy (high-pitched squeal)

PROBABLE CAUSE	CORRECTIVE ACTION
1 Glazed lining	1 Deglaze or replace lining
2 Contaminated lining (brake fluid, grease, etc.)	2 Repair source of contamination, replace linings
3 Weak or broken brake shoe hold-down or return spring	3 Replace springs
4 Rivets securing lining to shoe or backing plate loose	4 Replace shoes or pads
5 Excessive dust buildup on brake linings	5 Wash brakes off with brake system cleaner
6 Brake drums worn too thin	6 Measure diameter of drums, replace if necessary
7 Wear indicator on disc brake pads contacting disc	7 Replace brake pads
8 Anti-squeal shims missing or installed improperly	8 Install shims correctly

Note: *Other remedies for quieting squealing brakes include the application of an anti-squeal compound to the backing plates of the brake pads, and lightly chamfering the edges of the brake pads with a file. The latter method should only be performed with the brake pads thoroughly wetted with brake system cleaner, so as not to allow any brake dust to become airborne.*

Brakes noisy (scraping sound)

PROBABLE CAUSE	CORRECTIVE ACTION
1 Brake pads or shoes worn out; rivets, backing plate or brake shoe metal contacting disc or drum	1 Replace linings, have discs and/or drums machined (or replace)

Brakes chatter

PROBABLE CAUSE	CORRECTIVE ACTION
1 Worn brake lining	1 Inspect brakes, replace shoes or pads as necessary
2 Glazed or scored discs or drums	2 Deglaze discs or drums with sandpaper (if glazing is severe, machining will be required)
3 Drums or discs heat checked	3 Check discs and/or drums for hard spots, heat checking, etc. Have discs/drums machined or replace them
4 Disc runout or drum out-of-round excessive	4 Measure disc runout and/or drum out-of-round, have discs or drums machined or replace them
5 Loose or worn wheel bearings	5 Adjust or replace wheel bearings
6 Loose or bent brake backing plate (drum brakes)	6 Tighten or replace backing plate
7 Grooves worn in discs or drums	7 Have discs or drums machined, if within limits (if not, replace them)
8 Brake linings contaminated (brake fluid, grease, etc.)	8 Locate and repair source of contamination, replace pads or shoes
9 Excessive dust buildup on linings	9 Wash brakes with brake system cleaner
10 Surface finish on discs or drums too rough after machining (especially on vehicles with sliding calipers)	10 Have discs or drums properly machined
11 Brake pads or shoes glazed	11 Deglaze or replace brake pads or shoes

Troubleshooting (continued)

PROBABLE CAUSE	CORRECTIVE ACTION

Brake pads or shoes click

PROBABLE CAUSE	CORRECTIVE ACTION
1 Shoe support pads on brake backing plate grooved or excessively worn	1 Replace brake backing plate
2 Brake pads loose in caliper	2 Loose pad retainers or anti-rattle clips
3 Also see items listed under Brakes chatter	

Brakes make groaning noise at end of stop

PROBABLE CAUSE	CORRECTIVE ACTION
1 Brake pads and/or shoes worn out	1 Replace pads and/or shoes
2 Brake linings contaminated (brake fluid, grease, etc.)	2 Locate and repair cause of contamination, replace brake pads or shoes
3 Brake linings glazed	3 Deglaze or replace brake pads or shoes
4 Excessive dust buildup on linings	4 Wash brakes with brake system cleaner
5 Scored or heat-checked discs or drums	5 Inspect discs/drums, have machined if within limits (if not, replace discs or drums)
6 Broken or missing brake shoe attaching hardware	6 Inspect drum brakes, replace missing hardware

Rear brakes lock up under light brake application

PROBABLE CAUSE	CORRECTIVE ACTION
1 Tire pressures too high	1 Adjust tire pressures
2 Tires excessively worn	2 Replace tires
3 Defective proportioning valve	3 Replace proportioning valve

Brake warning light on instrument panel comes on (or stays on)

PROBABLE CAUSE	CORRECTIVE ACTION
1 Low fluid level in master cylinder reservoir (reservoirs with fluid level sensor)	1 Add fluid, inspect system for leak, check the thickness of the brake pads and shoes
2 Failure in one half of the hydraulic system	2 Inspect hydraulic system for a leak
3 Piston in pressure differential warning valve not centered	3 Center piston by bleeding one circuit or the other (close bleeder valve as soon as the light goes out)
4 Defective pressure differential valve or warning switch	4 Replace valve or switch
5 Air in the hydraulic system	5 Bleed the system, check for leaks
6 Brake pads worn out (vehicles with electric wear sensors - small probes that fit into the brake pads and ground out on the disc when the pads get thin)	6 Replace brake pads (and sensors)

Brakes do not self adjust

Disc brakes

PROBABLE CAUSE	CORRECTIVE ACTION
1 Defective caliper piston seals	1 Replace calipers. Also, possible contaminated fluid causing soft or swollen seals (flush system and fill with new fluid if in doubt)
2 Corroded caliper piston(s)	2 Same as above

Drum brakes

PROBABLE CAUSE	CORRECTIVE ACTION
1 Adjuster screw frozen	1 Remove adjuster, disassemble, clean and lubricate with high-temperature grease
2 Adjuster lever does not contact star wheel or is binding	2 Inspect drum brakes, assemble correctly or clean or replace parts as required
3 Adjusters mixed up (installed on wrong wheels after brake job)	3 Reassemble correctly
4 Adjuster cable broken or installed incorrectly (cable-type adjusters)	4 Install new cable or assemble correctly

Rapid brake lining wear

PROBABLE CAUSE	CORRECTIVE ACTION
1 Driver resting left foot on brake pedal	1 Don't ride the brakes
2 Surface finish on discs or drums too rough	2 Have discs or drums properly machined
3 Also see Brakes drag	

3 Anti-lock Brake System (ABS) - general information

Description

1   The Anti-lock Brake System is designed to maintain vehicle maneuverability, directional stability and optimum deceleration under severe braking conditions on most road surfaces. It does so by monitoring the rotational speed of the wheels and controlling the brake line pressure during braking. This prevents the wheels from locking up prematurely.

2   Two types of systems are used: One system uses only three wheel speed sensors, while the other system utilities four wheel speed sensors. Both systems use two front speed sensors mounted to each front wheel while the three channel system uses a single rear speed sensor mounted in the differential. The four channel system uses a wheel speed sensor attached to each rear wheel.

Hydraulic control unit

3   The HCU (Hydraulic Control Unit) is located in the left front corner of the engine compartment (see illustration).

Control module

4   The control module for either system is mated to the hydraulic control unit and is the "brain" for the system. The function of the control module is to accept analog voltage inputs from the speed sensors, process that data, and control hydraulic line pressure to avoid wheel lockup.

5   The control modules for both systems constantly monitor the system, even under normal driving conditions, to detect malfunctions. If a problem develops within the system, the control module illuminates an ABS warning light on the instrument cluster, and may even shut down the anti-lock system if it's a serious malfunction. A diagnostic trouble code will also be stored, which, when retrieved by a service technician, will indicate the problem area or component.

Speed sensors

6   A speed sensor produces an "analog" (continuously variable) voltage output, which is transmitted to the control module, where it's converted to digital information, compared to the control unit's program, and interpreted as wheel rotation speed (see illustrations). There are two different types of electronics used for the anti-lock systems. Known as a three channel and a four channel ABS system. One uses a single rear ABS sensor (located in the top left upper corner [front side] of the differential housing) while the other system uses two rear ABS wheel speed sensors on each side. Both systems use two front wheel ABS wheel speed sensors.

7   The three channel ABS rear speed sensor is mounted on the front left upper corner of the differential. On 4WD drive models there is an additional sensor mounted on the rear of the differential. This is the ELD field coil (Electronic Locking Differential) sensor and is not part of the ABS system. Do not confuse the ELD sensor with an ABS sensor (see illustration).

Brake light switch

8   The brake light switch is referred to as the BPP (Brake Pedal Postiion) switch (see illustration). When the brake pedal is depressed both a positive and negative signal is sent to the PCM which in turn sends that signal to the BCM and then onto the respective stop/turn signal bulbs. The SCCM (Steering Column Control module) receives a signal as to what position the turn signal lever is in and adjusts the data signal to the BCM in order to inform the BCM which brake lamps should be on. (For example, when turning right with the brake pedal applied, the right rear stop/turn bulb is flashing while the left stop/turn bulb is on solid.) The PCM also uses the brake light switch signal to determine the cruise control system's activation. The ABS systems utilizes the brake light switch data to determine its operation as well. Unlike as in older models, the CHSML (Center High Mount Stop Lamp)

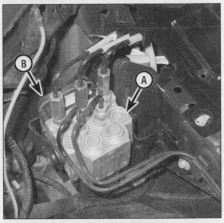

3.3 Hydraulic control unit (HCU) is also known as the ABS controller

A   ABS controller (Hydraulic)
B   ABS controller (Electrical control module)

The ABS electrical control module can be separated from the hydraulic portion without disconnecting the brake lines

3.6a Front 4WD speed sensor location

(On 2WD models the speed sensor is on the front lower portion of the steering knuckle)

3.6b Rear speed sensor location for the four channel ABS

3.7 On 4WD models equipped with the ELD system, this is the ELD sensor location. Do not confuse this sensor with the ABS sensors

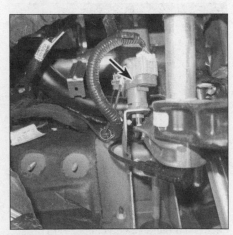

3.8 Replacing the BPP sensor requires the use of a scanner to properly calibrate the brake pedal position

is not wired directly from the brake light switch to the CHSML. On these models the CHSML runs through the BCM. The brake light switch is also responsible for some transmission functions which include various down shift points when the brake is applied as well as torque converter lock up and unlock.

Diagnosis and repair

9 If the ABS warning light on the instrument cluster comes on and stays on, make sure the parking brake is released and there's no problem with the brake hydraulic system. If neither of these is the cause, the anti-lock system is probably malfunctioning. Although special test procedures are necessary to properly diagnose the system, the home mechanic can perform a few preliminary checks before taking the vehicle to a dealer service department or independent repair facility.

a) *Make sure the brake, pads and calipers are in good condition.*
b) *Check the electrical connectors at the control unit.*
c) *Check the fuses.*
d) *Follow the wiring harness to the speed sensors and brake light on-off (BOO) switch and make sure all connections are secure and the wiring isn't damaged.*

10 If the above preliminary checks don't rectify the problem, the vehicle should be diagnosed by a dealer service department or other qualified repair shop.

Wheel speed sensor replacement

Front - 2WD models

11 Loosen the front wheel lug nuts, then jack and support the vehicle be sure to block the rear wheels and set the parking brake. Remove the wheel.
12 Disconnect the speed sensor electrical connector.
13 Detach the speed sensor bracket from the radius arm.

14 Remove the bolt securing the speed sensor to the housing.
15 With a slight twisting action, work the sensor out of the housing.
16 Installation is the reverse of removal.

Front - 4WD models

17 Loosen the front wheel lug nuts, then jack and support the vehicle and be sure to block the rear wheels and set the parking brake. Remove the wheel.
18 Remove the brake dust shield (see Section 6).
19 Detach the speed sensor electrical connector.
20 Remove the bolt securing the speed sensor bracket to the radius arm.
21 Remove the three bolts securing the speed sensor and harness.
22 With a slight twisting action, work the sensor from the housing.
23 Installation is the reverse of removal.

Rear - single sensor (Three channel ABS)

24 Block the front tires, then raise and support the rear of the vehicle.
25 Disconnect the speed sensor electrical connection.
26 The sensor is in the rear section of the upper corner on the driver's side of the differential. Remove the bolt securing the speed sensor to the differential.
27 With a slight twisting action, remove the speed sensor from the housing.
28 Installation is the reverse of removal.

Rear - two sensors (Four channel ABS)

29 Block the front tires, raise and support the rear of the vehicle.
30 Disconnect the electrical connector from the speed sensor.
31 Remove the bolt securing the speed sensor bracket to the rear axle.
32 Remove the bolt securing the speed sensor to the housing.

33 With a twisting action, remove the speed sensor.
34 Installation is the reverse of removal.

4 Disc brake pads - replacement

Warning: *Disc brake pads must be replaced on both wheels at the same time - never replace the pads on only one wheel. Also, brake system dust is hazardous to your health. DO NOT blow it out with compressed air and DO NOT inhale it. An approved filtering mask should be worn when working on the brakes. DO NOT use gasoline or petroleum-based solvents to remove the dust. Use brake system cleaner only!*

Note: *This procedure applies to the front and rear brake pads.*

1 Loosen the wheel lug nuts, raise the vehicle and support it securely on jackstands. Block the wheels at the opposite end. Remove the wheels.
2 Remove about two-thirds of the fluid from the master cylinder reservoir and discard it; as the pistons are pushed in for clearance to allow the new pads to be installed, the fluid will be forced back into the reservoir. Position a drain pan under the brake assembly and clean the caliper and surrounding area with brake system cleaner (see illustration).
3 To replace the brake pads, follow the accompanying photos, beginning with illustration 4.3a. Be sure to stay in order and read the caption under each illustration. Work on one brake assembly at a time so that you'll have something to refer to if necessary.
4 While the pads are removed, inspect the caliper for brake fluid leaks and ruptures in the piston boots. If necessary, replace the caliper (see Section 5). Also inspect the brake disc carefully (see Section 6). If machining is necessary, follow the information in that Section to remove the disc.
5 Before installing the caliper mounting bolts, clean them and check them for corrosion

4.2 Wash the disc and caliper with brake system cleaner to remove brake dust; DO NOT blow off brake dust with compressed air

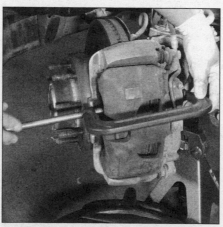

4.3a Push the pistons back into their bores with a C-clamp to provide room for the new brake pads

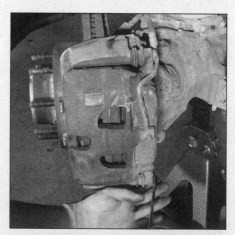

4.3b Remove the two caliper mounting bolts, check them for thread damage and replace as necessary

4.3c Pull off the caliper . . .

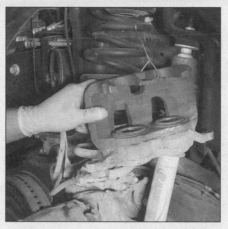

4.3d . . . and hang it with a piece of wire; DON'T allow the caliper to hang by the brake hose!

4.3e Remove the V-springs - if applicable

4.3f Remove the outer brake pad

4.3g Remove the inner brake pad

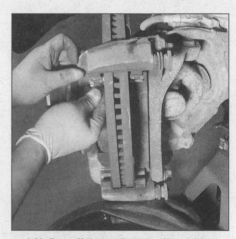

4.3h Pry off the anti-rattle clips with a small screwdriver (check them for cracks and make sure they fit tightly)

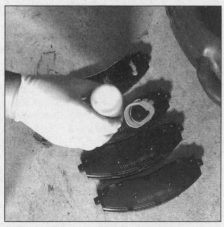

4.3i Apply some anti-squeal compound to the backing plates of the new pads (follow the instructions on the product)

4.3j Install the anti-rattle clips; make sure they're fully seated on the caliper mounting bracket

4.3k Install the new inner brake pad; make sure both ends of the pad are properly seated in the mounting bracket and anti-rattle clips

4.3l Install the new outer brake pad; make sure both ends of the pad are properly seated in the mounting bracket and anti-rattle clips

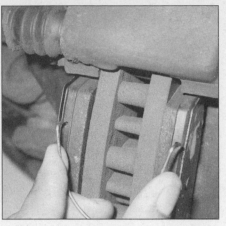

4.3m Install the ends of the V-springs into the holes in the pad backing plates - if applicable

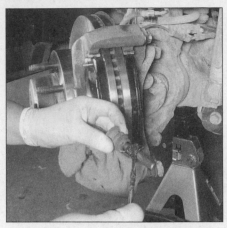

4.3n Apply high temperature grease to the caliper slide pins - install both bolts, tighten the lower bolt first

4.3o Install the caliper over the pads, install the bolts and tighten them to the torque listed in this Chapter's Specifications. Be sure to pump the pedal a few times to bring the pads into contact with the disc, and also check the brake fluid level

5.2 Remove the brake hose inlet fitting bolt

and damage. If they're significantly corroded or damaged, replace them. Be sure to tighten the caliper mounting bolts to the torque listed in this Chapter's Specifications.

6   Install the brake pads on the opposite wheel, then install the wheels and lower the vehicle. Tighten the lug nuts to the torque listed in the Chapter 1 Specifications.

7   Add brake fluid to the reservoir until it's full (see Chapter 1). Pump the brakes several times to seat the pads against the disc, then check the fluid level again.

8   Check the operation of the brakes before driving the vehicle in traffic. Try to avoid heavy brake applications until the brakes have been applied lightly several times to seat the pads.

5   Disc brake caliper - removal and installation

Warning: *The dust created by the brake system is harmful to your health. Never blow it out with compressed air and don't inhale any of it. An approved filtering mask should be worn when working on the brakes. Do not, under any circumstances, use petroleum-based solvents to clean brake parts. Use brake system cleaner only!*

Note: *If replacement is indicated (usually because of fluid leakage), it is recommended that the calipers be replaced, not overhauled. New and factory rebuilt units are available on an exchange basis. Always replace the calipers in pairs - never replace just one of them.*

Removal

1   Loosen the wheel lug nuts, raise the vehicle and place it securely on jackstands.

Block the wheels at the opposite end. Remove the wheel(s).

Note: *If you are simply removing the caliper for access to other components, leave the brake hose connected and suspend the caliper with a length of wire - don't let it hang by the hose.*

2   Remove the inlet fitting bolt and disconnect the brake hose from the caliper (see illustration). Discard the old sealing washers. Plug the brake hose immediately to keep contaminants and air out of the brake system and to prevent losing any more brake fluid than is necessary.

3   Remove the caliper mounting bolts (see illustration 4.3b) and detach the caliper from the mounting bracket.

Installation

4   Installation is the reverse of removal. Don't forget to use new sealing washers on each side of the brake hose inlet fitting and be sure to tighten the fitting bolt and the caliper mounting bolts to the torque values listed in this Chapter's Specifications.

5   Bleed the brake system (see Section 9).

Note: *If the brake hose was not disconnected, bleeding won't be required.*

Warning: *If air has found its way into the HCU (Hydraulic Control Unit), the system must be bled with the use of a scan tool. If the brake pedal feels "spongy" even after bleeding the brakes, or the ABS light on the instrument panel does not go off, or if you have any doubts whatsoever about the effectiveness of the brake system, have the vehicle towed to a dealer service department or other repair shop equipped with the necessary tools for bleeding the system.*

6   Make sure there are no leaks from the hose connections. Test the brakes carefully before returning the vehicle to normal service.

6.2 The brake pads on this vehicle were obviously neglected - they wore down to the rivets and cut deep grooves into the disc (wear this severe means the disc must be replaced)

6.3 Measure the brake disc runout with a dial indicator

6.4a The minimum (discard) thickness of the brake disc is cast into the disc

6.4b Measure the brake disc thickness at several points with a micrometer

6.6 The caliper mounting bracket is retained by two bolts (front shown, rear similar)

6   Brake disc - inspection, removal and installation

Inspection

1   Loosen the wheel lug nuts, raise the vehicle and support it securely on jackstands. Apply the parking brake. Remove the wheels.

2   Visually inspect the disc surface for score marks and other damage (see illustration). Light scratches and shallow grooves are normal after use and won't affect brake operation. Deep grooves require disc removal and refinishing by an automotive machine shop. Be sure to check both sides of the disc.

3   To check disc runout, place a dial indicator at a point about 1/2-inch from the outer edge of the disc (see illustration). If you're checking a front or rear disc on a 4WD model with single rear wheels, install the lug nuts, with the flat sides facing in, and tighten them

securely to hold the disc in place. Set the indicator to zero and turn the disc. The indicator reading should not exceed the runout limit listed in this Chapter's Specifications. If it does, the disc should be refinished by an automotive machine shop.

Caution: *The manufacturer recommends resurfacing the brake discs only in the event of pedal pulsations or hard (overheated) spots on the disc. If you elect not to have the discs resurfaced, deglaze them with sandpaper or emery cloth.*

4   The disc must not be machined to a thickness less than the specified minimum thickness, which is cast into the disc (see illustration). The disc thickness can be checked with a micrometer (see illustration).

Removal and installation

5   Remove the brake calipers (don't disconnect the brake hoses) and hang them out of the way (see Section 5).

6   Remove the caliper mounting bracket (see illustration).

7   On 2WD model front discs, remove the grease cap, wheel bearing retainer nut, spindle nut, outer bearing retainer washer and outer wheel bearing, then remove the disc/hub (see Front wheel bearing check, repack and adjustment in Chapter 1).

8   If you're removing a front disc on a 4WD model with single rear wheels, remove the brake pads, caliper, caliper mounting bracket and then pull the disc off the hub. If you're removing a front disc on a truck with dual rear wheels, remove the hub extender bolts and detach the hub extender as well, then remove the disc as outlined.

9   If you're removing a rear disc on a dual rear wheel axle, remove the hub assembly from the rear axle (see Chapter 8), then remove the brake pads, caliper, caliper mounting bracket and then unbolt the disc from the hub. On single rear wheel axles the disc can

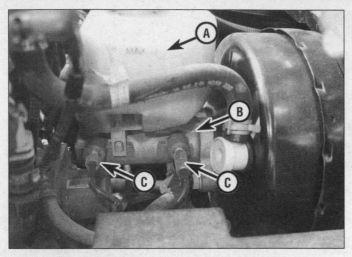

7.2a Master cylinder components

A   Reservoir
B   Master cylinder
C   Brake lines

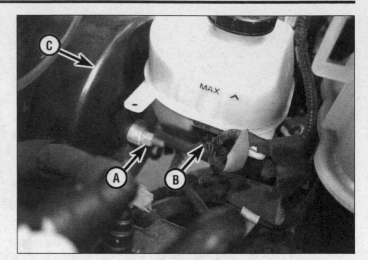

7.2b Master cylinder components

A   Mounting nut (same on the other side)
B   Electrical connector
C   Brake booster (vacuum type shown)

7.21 The best way to bleed air from the master cylinder before installing it on the vehicle is with a pair of bleeder tubes that direct brake fluid into the reservoir during bleeding

simply be slid off the wheel studs with the caliper mounting bracket removed.
Warning: *The thinner the brake disc the quicker it will heat up during stop and go traffic. Hot discs and hot pads lose their friction coefficient and take a greater amount of pedal force for the same stopping distance of cooler pads and discs. The closer the disc is to the minimum thickness the better off you are to replace them.*
10   Installation is the reverse of removal. On models so equipped, tighten the disc-to-hub bolts to the torque listed in this Chapter's Specifications.
11   Lower the vehicle and tighten the wheel lug nuts to the torque listed in the Chapter 1 Specifications.

7   Master cylinder - removal and installation

Removal

1   Place rags under the brake line fittings and prepare caps or plastic bags to cover the ends of the lines once they're disconnected.
Caution: *Brake fluid will damage paint. Cover all painted surfaces and avoid spilling fluid during this procedure.*
2   Familiarize yourself with the various components of the master cylinder (see illustrations) before proceeding.

Gas engine models

3   With the engine not running, press the brake pedal several times to release vacuum or hydraulic pressure from the brake system.
4   Remove the air cleaner assembly.
5   Remove the degas bottle.
6   Remove the electrical connections from the master cylinder.
7   Disconnect the brake lines from the master cylinder.
Note: *To prevent rounding off the flats on these nuts, a flare-nut wrench, which wraps around the nut, should be used.*
8   In models with hydro-boost systems, disconnect the two hydraulic lines from the master cylinder.
9   Remove and discard the master cylinder fasteners.

10   Installation is the reverse of removal. Bleed the brake system.

Diesel engine models

11   With the engine not running, press the brake pedal several times to release vacuum or hydraulic pressure from the brake system.
12   Remove the secondary battery from the vehicle.
13   Remove the battery tray and the degas bottle.
14   Disconnect the electrical connections from the master cylinder.
15   If equipped, disconnect the trailer brake electrical connections.
16   Remove the brake lines from the master cylinder.
Note: *To prevent rounding off the flats on these nuts, a flare-nut wrench, which wraps around the nut, should be used.*
17   In models with hydro-boost systems, disconnect the two hydraulic lines from the master cylinder.
18   Remove and discard the master cylinder fasteners.
19   Installation is the reverse of removal. Bleed the brake system.

Installation

20   Bench bleed the master cylinder before installing it. Mount the master cylinder in a vise, with the jaws of the vise clamping on the mounting flange.
21   Attach a pair of master cylinder bleeder tubes to the outlet ports of the master cylinder (see illustration).
22   Fill the reservoir with brake fluid of the recommended type (see Chapter 1).
23   Slowly push the pistons into the master cylinder (a large Phillips screwdriver can be used for this) - air will be expelled from the pressure chambers and into the reservoir. Because the tubes are submerged in fluid, air can't be drawn back into the master cylinder when you release the pistons.

24 Repeat the procedure until no more air bubbles are present.

25 Remove the bleed tubes, one at a time, and install plugs in the open ports to prevent fluid leakage and air from entering. Install the reservoir cap.

26 Install the master cylinder over the studs on the power brake booster and tighten the nuts only finger-tight at this time.

27 Thread the brake line fittings into the master cylinder. Since the master cylinder is still a bit loose, it can be moved slightly so the fittings thread in easily. Don't strip the threads as the fittings are tightened.

28 Tighten the mounting nuts securely. Tighten the brake line fittings securely.

29 Fill the master cylinder reservoir with fluid, then bleed the master cylinder and the rest of the brake system (see Section 9). To bleed the master cylinder on the vehicle, have an assistant depress the brake pedal and hold it down. Loosen the rear fitting to allow air and fluid to escape. Tighten the fitting, then allow your assistant to return the pedal to its rest position. Repeat this procedure on the other fitting until the fluid is free of air bubbles.

30 Reinstall any components that were removed for access to the master cylinder.

Warning: *If air has found its way into the HCU (Hydraulic Control Unit), the system must be bled with the use of a scan tool. If the brake pedal feels "spongy" even after bleeding the brakes, or the ABS light on the instrument panel does not go off, or if you have any doubts whatsoever about the effectiveness of the brake system, have the vehicle towed to a dealer service department or other repair shop equipped with the necessary tools for bleeding the system.*

31 Re-check the brake fluid level, then check the operation of the brake system carefully before driving the vehicle in traffic.

8 Brake hoses and lines - check and replacement

Warning: *If air has found its way into the hydraulic control unit, the system must be bled with the use of a scan tool. If the brake pedal feels "spongy" even after bleeding the brakes, or the ABS light on the instrument panel does not go off, or if you have any doubts whatsoever about the effectiveness of the brake system, have the vehicle towed to a dealer service department or other repair shop equipped with the necessary tools for bleeding the system.*

1 About every six months, with the vehicle raised and placed securely on jackstands, the flexible hoses which connect the steel brake lines with the front and rear brake assemblies should be inspected for cracks, chafing of the outer cover, leaks, blisters and other damage. These are important and vulnerable parts of the brake system and inspection should be complete. A light and mirror will be needed for a thorough check. If a hose exhibits any of the above defects, replace it with a new one.

Flexible hoses

2 Clean all dirt away from the ends of the hose. To disconnect a front brake hose from the brake lines, unscrew the brake line fittings from the hose junction block. Be careful not to bend the junction bracket or kink the lines. If necessary, soak the connections with penetrating oil.

3 To disconnect a rear brake hose from the brake line, unscrew the metal tube nut with a flare nut wrench, then remove the U-clip from the female fitting at the bracket and remove the hose from the bracket.

4 Disconnect the hose from the caliper, discarding the sealing washers on either side of the fitting.

5 Using new sealing washers, attach the new brake hose to the caliper or wheel cylinder.

6 To reattach a front brake hose to the metal lines, bolt the hose junction bracket to the frame, tighten the bolt securely, and screw in the tube nuts.

7 To reattach a rear brake hose to the metal line, insert the end of the hose through the frame bracket. Make sure the hose isn't twisted, then attach the metal line by tightening the tube nut fitting securely. Install the U-clip at the frame bracket.

Note: *The weight of the vehicle must be on the suspension, so the vehicle should not be raised while positioning the hose.*

8 Carefully check to make sure the suspension or steering components don't make contact with the hose. Have an assistant push down on the vehicle and also turn the steering wheel lock-to-lock during inspection.

9 Bleed the brake system (see Section 9).

Metal brake lines

10 When replacing brake lines, be sure to use the correct parts. Don't use copper tubing for any brake system components. Purchase steel brake lines from a dealer parts department or auto parts store.

11 Prefabricated brake line, with the tube ends already flared and fittings installed, is available at auto parts stores and dealer parts departments. These lines can be bent to the proper shapes using a tubing bender.

12 When installing the new line make sure it's well supported in the brackets and has plenty of clearance between moving or hot components.

13 After installation, check the master cylinder fluid level and add fluid as necessary. Bleed the brake system as outlined in Section 9 and test the brakes carefully before placing the vehicle into normal operation.

9 Brake hydraulic system - bleeding

Warning: *Wear eye protection when bleeding the brake system. If the fluid comes in contact with your eyes, immediately rinse them with water and seek medical attention.*

Note: *Bleeding the brake system is necessary to remove any air that's trapped in the system when it's opened during removal and installation of a hose, line, caliper, wheel cylinder or master cylinder.*

1 It will probably be necessary to bleed the system at all four brakes if air has entered the system due to low fluid level, or if the brake lines have been disconnected at the master cylinder.

2 If a brake line was disconnected only at a wheel, then only that caliper or wheel cylinder must be bled.

3 If a brake line is disconnected at a fitting located between the master cylinder and any of the brakes, that part of the system served by the disconnected line must be bled.

4 Remove any residual vacuum (or hydraulic pressure) from the brake power booster by applying the brake several times with the engine off.

5 Remove the master cylinder reservoir cap and fill the reservoir with brake fluid. Reinstall the cap.

Note: *Check the fluid level often during the bleeding operation and add fluid as necessary to prevent the fluid level from falling low enough to allow air bubbles into the master cylinder.*

6 Have an assistant on hand, as well as a supply of new brake fluid, an empty clear plastic container, a length of plastic, rubber or vinyl tubing to fit over the bleeder valve and a wrench to open and close the bleeder valve.

7 Beginning at the right rear wheel, loosen the bleeder screw slightly, then tighten it to a point where it's snug but can still be loosened quickly and easily.

8 Place one end of the tubing over the bleeder screw fitting and submerge the other end in brake fluid in the container (see illustration).

9 Have the assistant slowly depress the brake pedal and hold it in the depressed position.

9.8 When bleeding the brakes, a hose is connected to the bleed screw at the caliper and submerged in brake fluid - air will be seen as bubbles in the tube and container (all air must be expelled before moving to the next wheel)

10.12 Disconnect the electrical connector, then twist the switch to remove it from the bracket

10   While the pedal is held depressed, open the bleeder screw just enough to allow a flow of fluid to leave the valve. Watch for air bubbles to exit the submerged end of the tube. When the fluid flow slows after a couple of seconds, tighten the screw and have your assistant release the pedal.

11   Repeat Steps 9 and 10 until no more air is seen leaving the tube, then tighten the bleeder screw and proceed to the left rear wheel, the right front wheel and the left front wheel, in that order, and perform the same procedure. Be sure to check the fluid in the master cylinder reservoir frequently.

12   Never use old brake fluid. It contains moisture which can boil, rendering the brake system inoperative.

13   Refill the master cylinder with fluid at the end of the operation.

14   Check the operation of the brakes. The pedal should feel solid when depressed, with no sponginess. If necessary, repeat the entire process.

Warning: *Do not operate the vehicle if you are in doubt about the effectiveness of the brake system. It is possible for air to become trapped in the anti-lock brake system hydraulic control unit, so, if the pedal continues to feel spongy after repeated bleedings or the BRAKE or ANTI-LOCK light stays on, have the vehicle towed to a dealer service department or other qualified shop to be bled with the aid of a scan tool.*

10   Brake lights check and brake light switch - replacement

Brake lights check

1   Functioning brake lights require a good fuse, brake light switch, turn signal switch, SCCM, BCM, stop/turn bulb, bulb socket/housing, and all the connections and wiring to each component to be in working order. For more details on the various functions of the brake light switch go to the ABS - general information section of this chapter.

2   While an assistant applies the brake pedal check the stop lamps. If any of the bulbs light up the fuse is good. (All the bulbs including the CHSML are on the same fuse.) Go to step 4.

3   If ALL the brake lights are inoperative, check the stop lamp fuse (see Chapter 12).

4   Verify that the bulbs are good by removing one of them and check to see if the filament is damaged or if the bulb has a black smokey look. See Chapter 12 for bulb replacement procedures. Replace the bulb with a known good one and recheck the brake lights.

5   If the fuse is good, check for voltage at the brake light switch input wire (refer to the wiring diagrams at the end of this manual for the proper colored wire and terminal). If voltage is present continue to the next step.

6   Depress the brake pedal and check for voltage at the output wire terminal (again, refer to the wiring diagrams). If no voltage is present, replace the switch and recheck. If voltage is present continue to the next step.

7   Check for power at the taillight housings. (Follow the wiring diagram for proper wire color and location. Check for brake light voltage at the rear taillight housings with the brake pedal depressed.) If voltage is not present, repair the circuit between the BCM and the brake lights. Refer to the wiring diagrams for the proper wire terminal location. If it is determined that the lead between the BCM and the taillight housing is in good condition the BCM is most likely the problem. Take the vehicle to your local dealership or qualified independent repair facility for further diagnostics.

8   If voltage is present at the taillight housing, check for a bad ground. Using a jumper wire connected to a good ground, probe the ground wire terminal at the taillight connector. If the brake lights go on, repair the ground circuit (follow the ground wire from the taillight housing).

9   If voltage is present and the ground has been checked, and the bulb is a known good bulb, replace the bulb socket or housing. Recheck after replacing.

Note: *Keep in mind that ALL the brake light bulbs could be burned out, but the likelihood of all the bulbs being burned out at the same*

time is very slim. However, if you haven't checked the stop lamps in a while, chances are one burned out some time ago and then shortly after that the other one has burned out. This is typical when replacing the original factory bulbs that have never been changed.

Brake light switch replacement

10   Remove the driver's side knee bolster trim panel (see Chapter 11) below the steering column.

11   Unplug the electrical connector from the switch.

12   Rotate the switch 1/8 of a turn to release the switch (see illustration).

13   Installation is the reverse of removal.

11   Power brake booster - check, removal and installation

Operating check

1   Depress the brake pedal several times with the engine off to release any stored vacuum or hydraulic pressure in the brake system.

2   Depress the pedal and start the engine. If the pedal goes down slightly, operation is normal.

Airtightness check (vacuum booster only)

3   Start the engine and turn it off after one or two minutes. Depress the brake pedal several times slowly. If the pedal goes down farther the first time but gradually rises after the second or third depression, the booster is airtight.

4   Depress the brake pedal while the engine is running, then stop the engine with the pedal depressed. If there is no change in the pedal reserve travel after holding the pedal for 30 seconds, the booster is airtight.

Removal and installation

5   Disassembly of the power unit requires special tools and is not ordinarily performed by the home mechanic. If a problem develops, it's recommended that a new or factory rebuilt unit be installed.

6   In the engine compartment, remove and discard the nuts attaching the master cylinder (see Section 7), to the booster and carefully pull the master cylinder forward until it clears the mounting studs. Be careful not to bend or kink the brake lines.

Note: *There is no need in disconnecting the master cylinder hydraulic lines unless you are replacing the booster and master cylinder as an assembly.*

7   If you're working on a model with a vacuum booster, disconnect the vacuum hose where it attaches to the power brake booster (see illustration). If you're working on a model with a hydraulic booster, detach the fluid pressure and return lines from the booster.

8   If equipped, detach the power distribu-

11.7 Detach the intake manifold vacuum hose from the booster (vacuum booster units only)

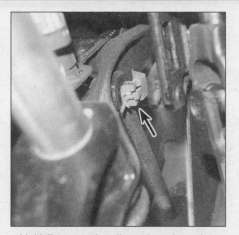

11.10 Remove the clip and retainer, then slide actuator lever off of the pedal assembly

11.12 To detach the power brake booster from the firewall, remove these four nuts

12.2a Insert a 4 mm pin into the hole to hold the parking brake lever up

12.2b A Phillips screwdriver works well in place of a 4 mm pin

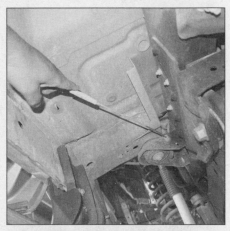

12.2c Have your assistant pull on the cable to help relieve the tension as you insert the pin into the parking brake lever. Then you can separate the cable coupling

tion auxiliary relay box and move it out of the way.

9   Remove the driver's side knee bolster trim panel (see Chapter 11).

10   Remove the retaining clip to the brake pedal pin and remove the brake pedal actuator rod from the pedal (see illustration).

11   Disconnect the brake light switch (see Section 10).

12   Remove the four nuts attaching the booster to the firewall (see illustration).

13   Carefully lift the booster unit away from the firewall and out of the engine compartment.

14   To install the booster, place it into position and tighten the retaining nuts. Connect the pushrod to the brake pedal and install the brake light switch (see Section 10).

15   Install the master cylinder. Reconnect the vacuum hose or pressure and return lines, as applicable.

16   If you're working on a model with a hydraulic booster, bleed the air from the booster unit as follows:

a) *Remove the PCM fuse from the engine compartment fuse block (fuse no. 24).*

b) *Crank the engine over for a few seconds, then check the power steering fluid level (see Chapter 1). Add fluid as necessary, then reinstall the PCM fuse.*

c) *Start the engine and allow it to idle. Turn the steering wheel from stop-to-stop two times, then turn off the engine.*

d) *Depress the brake pedal several times to discharge the accumulator (the pedal will become harder to push).*

e) *Repeat this procedure until the fluid in the power steering fluid reservoir is free of air bubbles.*

17   Carefully test the operation of the brakes before placing the vehicle in normal service.

Brake vacuum pump - removal and installation

Note: *On some models a brake vacuum pump is used to increase the vacuum to the vacuum brake booster.*

18   On the 6.7 diesel models remove the coolant fan drive assembly (see Chapter 3).

19   Remove the vacuum pump bolts.

20   Position the pump so that you can get to the vacuum line. Disconnect the vacuum line.

21   To install, use a new gasket, making sure the gasket mating surfaces are clean and that the gasket is aligned to the housing.

22   Installation is the reverse of removal.

12   Parking brake shoes - replacement

1   Release the parking brake. Loosen the rear wheel lug nuts. Raise the vehicle and support it securely on jackstands. Remove the wheels.

2   Remove the tension on the parking brake cable and disconnect both of the parking brake rear cables from the middle cable connector. (see illustrations).

Note: *To remove the tension on the parking brake cable, have an assistant pull the intermediate cable until enough slack is gained. Then from inside the vehicle, lift up on the parking brake pedal until it cannot go up any higher. Find the 4 mm hole on the side of the parking brake housing (see illustrations) and*

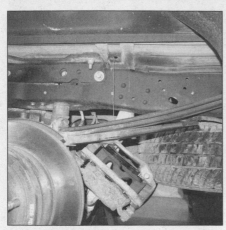

12.3a Remove the caliper and hang it by a wire out of the way

12.3b Remove the rotors from both sides

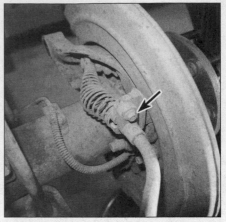

12.4a Remove the retaining bolt

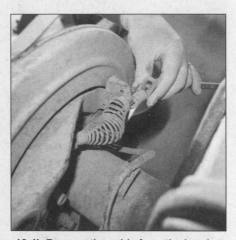

12.4b Remove the cable from the bracket

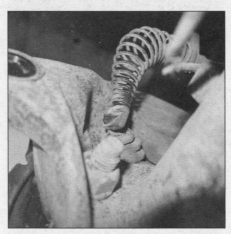

12.4c Slip the end of the cable off by rotating it 90 degrees from the hooked end of the lever

12.5 Clean the area before starting the tear down

insert a 4 mm pin (an Allen wrench works best) into the hole. Now have the assistant release the cable and the tension should be off of the cable. Remove the 4 mm pin after replacing the shoes.

3   Remove the calipers and brake rotors (see illustrations).

4   Disconnect the cables from the levers at the rear wheels by turning the cable end 90 degrees from their working position (see illustrations).

5   Clean the parking brake assembly with brake system cleaner before beginning work (see illustration).

6   Remove the hold down clips from the shoes (see illustrations).

7   Remove the upper shoe retracting spring (see illustration).

8   Remove the lower retaining spring and shoe adjuster (see illustrations).

9   Remove the shoes (see illustration).

10   Once the shoe assembly is free, remove the upper retracting spring.

All models

11   Clean the brake disc/parking brake drum and check it for score marks, deep grooves, hard spots (which will appear as small discolored areas) and cracks. If the disc/drum is worn, scored or out-of-round, it can be resurfaced by an automotive machine shop.

12   Lubricate all the areas where the shoe will make contact with the backing plate and lubricate the adjuster sleeve (see illustrations). Install the shoes by reversing the removal procedure. Turn the adjusting screw so the disc just fits over the new shoes. When the disc is installed, the shoes should not rub as the disc is turned. If you have a brake shoe adjusting gauge, adjust the diameter of the shoes 0.030-inch less than that of the drum surface of the rear brake disc.

13   Repeat this procedure for the other parking brake assembly.

14   Install the brake discs (see Section 6) and the brake calipers (see Section 5). Reconnect the parking brake cable to the actuator arm.

15   Remove the rubber plug from the brake backing plate and, using a screwdriver or brake adjusting tool, turn the adjusting screw star wheel until the parking brake shoes start to drag as the disc is turned, then back off the star wheel until the shoes don't drag.

Note: *To check the operation of the parking brake set and release the parking brake lever at least three times. Then recheck the adjustments.*

16   Install the rear wheels and lug nuts, lower the vehicle and tighten the lug nuts to the torque listed in the Chapter 1 Specifications.

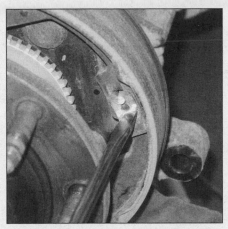

12.6a Each shoe has a center hold down clip. Press in on the clip with a flat screwdriver and slide the clip off as you push down on the clip

12.6b Do the same on the other side

12.7 Grip the spring tightly and slip the hooked end out of the shoe

12.8a Remove the lower retaining spring . . .

12.8b . . . then remove the adjuster

12.9 Remove the shoes

12.12a Lubricate the spots where the shoe will be in contact with the backing plate

12.12b Lubricate the adjuster sleeve before reinstalling

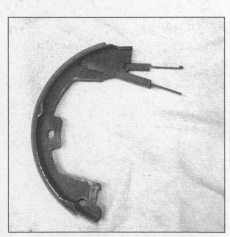

12.12c Top spring reference picture

Notes

Chapter 10
Suspension and steering systems

Contents

Specifications

General

Power steering fluid type	See Chapter 1

Torque specifications

Ft-lbs (unless otherwise indicated)

Note: *One foot-pound (ft-lb) of torque is equivalent to 12 inch-pounds (in-lbs) of torque. Torque values below approximately 15 ft-lbs are expressed in inch-pounds, because most foot-pound torque wrenches are not accurate at these smaller values.*

Front suspension (2WD models)

Lower balljoint nut	98
Upper balljoint bolt/nut	59
Coil spring lower nuts	148
Shock absorber upper nut*	46
Shock absorber lower nut*	52
Radius arm front nuts*	295
Radius arm rear nuts*	222
Stabilizer bar link nuts*	85
Stabilizer bracket bolts/nuts*	35
Tie rod end nut	85
Track bar balljoint nut	184
Track bar-to-frame bracket bolt	406
Track bar bracket to frame bolt	129

* Manufacturer recommends new fasteners. Do not reuse the original fasteners.

Torque specifications (continued)

Ft-lbs (unless otherwise indicated)

Note: *One foot-pound (ft-lb) of torque is equivalent to 12 inch-pounds (in-lbs) of torque. Torque values below approximately 15 ft-lbs are expressed in inch-pounds, because most foot-pound torque wrenches are not accurate at these smaller values.*

Front suspension (4WD models)

Lower ball joint	150
Upper ball joint	69
Hub lock screws	See Chapter 8
Wheel bearing and hub nuts	See Chapter 8
Jounce bumper nuts	26
Shock absorber upper nuts*	46
Shock absorber lower nuts*	111
Radius arm front and rear nuts*	222
Stabilizer bar bracket nuts*	35
Stabilizer bar link nuts*	59
Track bar balljoint nut*	184
Track bar bracket nut*	406
Wheel extension nuts	130

Rear suspension

Rear stabilizer bar bolts*	35
Rear stabilizer end link nuts*	76
Spring to front bracket*	340
Spring to shackle nut*	166
Shock absorber bracket nut*	30
Shock absorber lower nut*	66
Shock absorber upper nut*	52
Shackle to bracket bolt*	166
Leaf spring U bolts*	
Use an "X" pattern for the tightening sequence	
Step 1	48
Step 2	96
Step 3	148
Step 4	195

Steering

Steering wheel bolt	35
Steering column nuts to instrument panel	22
Upper steering column shaft-to-steering column bolt	21
Upper steering column shaft-to-lower steering column shaft	41
Lower steering column shaft-to-steering gear bolt	26
Tie-rod end-to-steering knuckle nut	85
Inner tie rod-to-center link	65
Drag link clamp nut	41
Outer drag link nut	85
Sector shaft arm nut	350
Sector shaft arm-to-drag link/center link nut*	129
Idler arm-to-frame bolts and center link nuts*	85
Tie-rod clamp nuts	41
Steering linkage damper-to-drag link bracket nut	66
Steering linkage damper bracket-to-frame nuts	52
Steering gear box-to-frame bolts*	148
Power steering pump-to-engine bolts	18

* Manufacturer recommends new fasteners. Do not reuse the original fasteners.

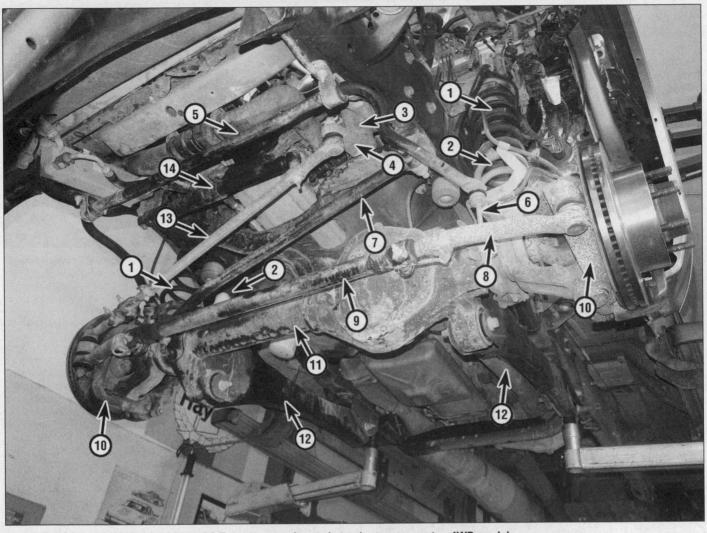

1.2 Front suspension and steering components - 4WD model

1	Coil spring	
2	Shock absorber	
3	Steering gear	
4	Pitman arm	
5	Stabilizer bar	

6	Stabilizer bar link
7	Track bar
8	Tie-rod end
9	Tie-rod
10	Steering knuckle

11	Front axle
12	Radius arm
13	Drag link
14	Steering damper

1   General Information

1   The front suspension on 2WD models covered in this manual is a twin I-beam type, which is composed of coil springs, I-beam axle arms, radius arms, upper and lower balljoints, steering knuckles, tie-rods, shock absorbers and a stabilizer bar.

2   On 4WD models the front suspension uses a Dana model 60 type axle with radius arms, a track bar, coil springs and shock absorbers. The wheel knuckles are attached using an upper and lower balljoint (see illustration).

3   The rear suspension uses shock absorbers and leaf springs (see illustration). The forward end of each spring is attached to the bracket on the side of the frame side rail. The rear of each spring is shackled to a bracket on

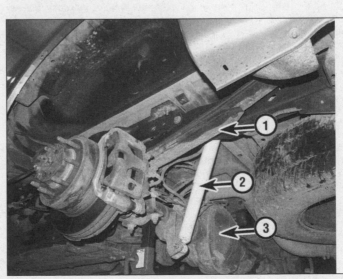

1.3 Rear suspension components

1   Leaf spring
2   Shock absorber
3   Rear axle housing

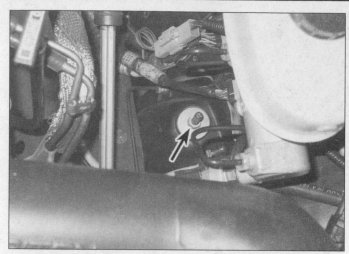

2.1 Shock absorber upper mounting nut (left-side shown) - 2WD model

2.3 Hold the shock absorber with one wrench while unscrewing the upper mounting nut with another - 4WD model

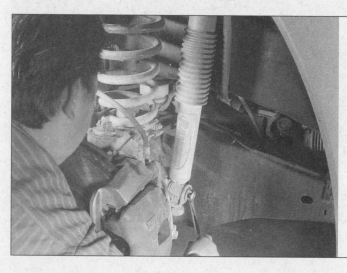

2.4 Shock absorber lower mounting nut/ bolt removal - 4WD model shown

the side of the frame rail. The rear suspension may also incorporate a rear stabilizer bar to aid in stability control.

4   The steering system consists of a steering column, intermediate shaft, steering gear, Pitman arm, drag link and tie-rod (4WD models), center link, idler arm and two tie-rods (2WD models), a steering damper, and tie-rod ends. The tie-rods are equipped with an adjusting sleeve for setting the toe-in. All models are equipped with power steering.

Note: *Frequently, when working on the suspension or steering system components, you may come across fasteners which seem impossible to loosen. These fasteners on the underside of the vehicle are continually subjected to water, road grime, mud, etc., and can become rusted or "frozen," making them extremely difficult to remove. In order to unscrew these stubborn fasteners without damaging them (or other components), be sure to use lots of penetrating oil and allow it to soak in for a while. Using a wire brush to clean exposed threads will also ease removal of the nut or bolt and prevent damage to the threads.*

Sometimes a sharp blow with a hammer and punch will break the bond between a nut and bolt threads, but care must be taken to prevent the punch from slipping off the fastener and ruining the threads. Heating the stuck fastener and surrounding area with a torch sometimes helps too, but isn't recommended because of the obvious dangers associated with fire. Long breaker bars and extension, or "cheater" pipes will increase leverage, but never use an extension pipe on a ratchet - the ratcheting mechanism could be damaged. Sometimes tightening the nut or bolt first will help to break it loose. Fasteners that require drastic measures to remove should always be replaced with new ones.

Note: *Since most procedures that are dealt with in this chapter involve jacking up the vehicle and working underneath it, a good pair of jackstands will be needed. A hydraulic floor jack is the preferred type of jack to lift the vehicle, and it can also be used to support certain components during various operations.*

Warning: *Never, under any circumstances, rely on a jack to support the vehicle while*

working under it - always support the vehicle with jackstands.

Caution: *Whenever any of the suspension or steering fasteners are loosened or removed they must be inspected and, if necessary, replaced with new ones of the same part number or of original equipment quality and design. Torque specifications must be followed for proper reassembly and component retention. Never attempt to heat or straighten suspension or steering components. Instead, replace bent or damaged parts with new ones.*

2   Shock absorbers (front) - removal and installation

Caution: *The shock absorbers are pressurized with nitrogen gas. Do not attempt to open, puncture or apply heat to the shock absorbers.*

1   If you're working on a 2WD model, open the hood and remove the shock absorber upper mounting nut, washer and bushing (see illustration).

2   Loosen the front wheel lug nuts. Raise the front of the vehicle, support it securely on jackstands, block the rear wheels and set the parking brake. Remove the front wheel.

3   Support the outer end of the axle with a floor jack. If you're working on a 4WD model, remove the shock absorber upper mounting nut (see illustration).

4   Remove the shock absorber lower mounting nut (see illustration), then remove the shock absorber.

5   Installation is the reverse of the removal steps with the following additions:

a) *Tighten the nuts and bolts to the torque listed in this Chapter's Specifications.*

b) *Install the wheel and lug nuts, lower the vehicle and tighten the lug nuts to the torque listed in the Chapter 1 Specifications.*

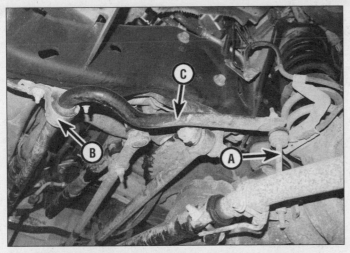

3.2 Front stabilizer bar details (4WD shown, 2WD similar)

A   Stabilizer bar link   C   Stabilizer bar
B   Stabilizer bar bracket

3.3 Front stabilizer bar bracket nuts

3.4 Remove the upper stabilizer bar link nut to remove the stabilizer bar

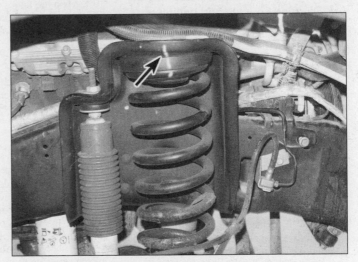

4.2 Match mark the spring and insulator to the frame

3   Stabilizer bar (front) - removal and installation

1   Raise the front of the vehicle, support it securely on jackstands, block the rear wheels and set the parking brake. Position the front wheels in the straight ahead position.

2   Remove the nuts, washers and bolts and detach the stabilizer bar links from the bar (see illustration).

3   Remove the stabilizer bar brackets and bushings (see illustration).

4   Remove the retaining nut to the stabilizer bar link (see illustration).

5   Remove the stabilizer bar.

6   Installation is the reverse of the removal steps. Tighten the bolts and nuts to the torque listed in this Chapter's Specifications.

4   Coil spring - removal and installation

Removal

1   Loosen the front wheel lug nuts. Raise the front of the vehicle, support it securely on jackstands placed under the frame rails, block the rear wheels and set the parking brake. Remove the front wheel.

2   Match mark the coil spring and upper insulator so they can be referenced back to the same spot on the frame (see illustration).

3   Place a floor jack under the front axle assembly as far out near the outer edge as possible. Detach the lower end of the shock absorber from the radius arm and discard the bolt and nut (see Section 2).

4   On 2WD models, remove the spring bracket and bolt. Discard the bolt and replace it with a new one.

5   On 4WD models, remove the stabilizer bar link from the stabilizer bar (see Section 3).

6   Lower the floor jack until the coil spring is free from the upper support.

7   On 2WD models, use an long extension to reach into the coil spring and unscrew the lower retaining nut. Discard the nut and replace with a new one.

8   Remove the spring and lower insulator/retainer.

9   Installation is the reverse of removal.

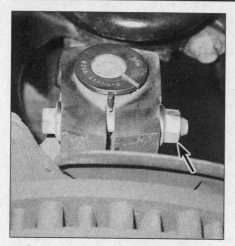

5.6 Upper balljoint pinch bolt (2WD models)

5.7 Upper balljoint nut (4WD models)

5.8 Lower balljoint nut

5   Steering knuckle and balljoints - removal and installation

Steering knuckle

Removal

Warning: *Replace all cotter pins with new ones. Do not reuse old cotter pins.*

1   Remove the front wheel and brake disc (see Chapter 9). Remove the brake dust shield. If equipped with four-wheel ABS, remove the ABS wheel speed sensor from the steering knuckle.

2   On 4WD models, remove the hub and bearing assembly and the front axleshaft (see Chapter 8).

3   On 4WD models, using a drift pin, drive the main axle seal out of the steering knuckle.

4   On 4WD models, disconnect the pulse vacuum hub hose, if applicable.

5   On all models, disconnect the tie-rod end from the steering knuckle (see Section 15).

Caution: *Don't use a fork-type separator to detach the tie-rod end. This will damage the balljoint seal.*

6   If you're working on a 2WD model, remove the nut and upper balljoint pinch bolt (see illustration).

7   If you're working on a 4WD model, remove the cotter pin and unscrew the upper balljoint nut (see illustration). Discard the bolt and nut, replace with new fasteners.

8   Loosen the lower balljoint nut a few turns, but do not remove it yet (see illustration).

9   Using a balljoint removal tool, separate the balljoints from the steering knuckle. (If needed, remove the brake disc backing plate to gain clearance for the removal tool.)

Note: *An alternative method for removing the balljoints is to strike the lower and upper edges of the axle yoke flange just where the balljoint slips through them with a large hammer. A few solid strikes should dislodge the balljoints.*

10   When the knuckle comes loose it will be caught by the lower balljoint nut. Lift up on the knuckle and finish taking the lower balljoint nut off.

11   Remove the knuckle assembly from the axle housing yoke.

12   Mark the position of the camber adjuster sleeve to the axle housing, then remove the camber adjuster sleeve. Clean the adjuster sleeve and its bore.

Installation

13   Before installation, check that the upper and lower balljoint seals were not damaged during steering knuckle removal and that they are positioned correctly. Replace if necessary.

14   Install the camber adjuster on the upper balljoint, making sure it's aligned correctly. (Install in the same position it was originally.)

15   Position the steering knuckle and balljoints in the axle arm.

16   Tighten the balljoint nut(s) to the torque listed in this Chapter's Specifications, then tighten further until the cotter pin hole lines up. Install a new cotter pin and bend it to secure

6.4 Radius arm-to-frame bracket bolt and nut

the nut.

17   If you're working on a 2WD model, install and tighten the upper balljoint clamp bolt/nut to the torque listed in this Chapter's Specifications.

18   The remainder of installation is the reverse of removal.

19   Install the wheel and lug nuts. Lower the vehicle and tighten the lug nuts to the torque listed in the Chapter 1 Specifications.

20   Have the front end alignment checked and, if necessary, adjusted.

Balljoints

21   The balljoints are a press fit in the steering knuckle, which necessitates the use of a special press tool and receiver cup to remove and install them. The tools are available at most parts stores to rent for a nominal fee. Directions on the use of the tool should be read completely before attempting to use it. If you cannot find the proper tool to perform this procedure we recommended that the steering knuckle be removed and taken to an automotive machine shop or other qualified repair facility to have the balljoints replaced.

6   Front axle I-beam and radius arm (2WD models) - removal and installation

Removal

1   Raise the front of the vehicle, support it securely on jackstands, block the rear wheels and set the parking brake. Position the front wheels in the straight ahead position.

2   Remove the coil spring (see Section 4).

3   Unscrew the radius arm-to-I-beam axle nut, then, using a hammer and punch, drive out the radius arm-to-I-beam bolt.

4   Remove the bolt and nut securing the I-beam axle arm to the frame pivot bracket (see illustration).

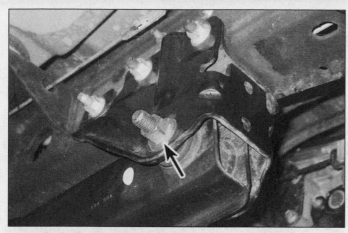

6.5 Radius arm-to-frame bracket bolt/nut

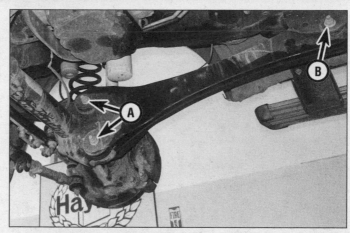

7.4 Front (A) and rear (B) radius arm fasteners (4WD models)

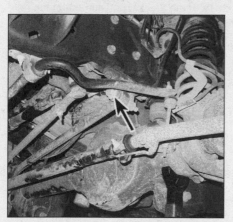

8.3 Upper track bar bolt location

9.2 Shock lower mounting bolt location

9.3 Shock upper mounting nut location

5   Remove the bolt and nut attaching the rear of the radius arm to the frame bracket, then detach the arm from the bracket (see illustration).

Installation

6   Install the I-beam axle arm and the radius arm to their brackets on the frame and install the bolts and nuts. Don't fully tighten the nuts yet.
7   Connect the radius arm to the I-beam axle and install the bolt and nut. Tighten the nut to the torque listed in this Chapter's Specifications.
8   Install the coil spring (see Section 4).
9   Install the wheel and lug nuts. Lower the vehicle and tighten the lug nuts to the torque listed in the Chapter 1 Specifications.
10   With the vehicle at curb height, tighten the I-beam axle and radius arm pivot bolts/nuts to the torque listed in this Chapter's Specifications.
Caution: *Do not tighten the I-beam or radius arm while there is no weight on them. This will damage the bushings.*
11   Have the front end alignment checked and, if necessary, adjusted.

7   Radius arm (4WD models) - removal and installation

1   Loosen the wheel lug nuts. Raise the front of the vehicle, support it securely on jackstands, block the rear wheels and set the parking brake. Remove the wheel.
2   Remove the shock absorber (see Section 2).
3   Remove the ABS speed sensor bracket bolt.
4   Remove the front and rear radius arm bolts (see illustration). Discard them and obtain new fasteners.
5   Remove the radius arm.
6   Installation is the reverse of removal. Tighten the fasteners to the torque listed in this Chapter's Specifications.
Caution: *Do not tighten the radius arm fasteners while there is no weight on them. This will damage the bushings.*

8   Track bar - removal and installation

1   Raise the front of the vehicle and place jackstands under the front axle, not the frame.

Block the rear wheels and set the parking brake.
2   Loosen the nut and disconnect the track bar ballstud from the axle bracket on the right end of the axle, then remove the nut and separate the bar from the axle bracket.
3   Remove the nut and bolt and disconnect the upper end of the track bar from the frame bracket (see illustration).
4   Installation is the reverse of removal. Tighten all fasteners to the torque values listed in this Chapter's Specifications.

9   Shock absorbers (rear) - removal and installation

1   Raise the rear of the vehicle, support it securely on jackstands and block the front wheels. Place a floor jack under the axle adjacent to the shock absorber being removed. Raise the jack just enough to take the load off the shock absorber.
2   Remove the nut and bolt securing the lower end of the shock absorber to the rear axle (see illustration).
3   Remove the nut securing the top of the shock absorber to the upper mounting bracket on the frame (see illustration).

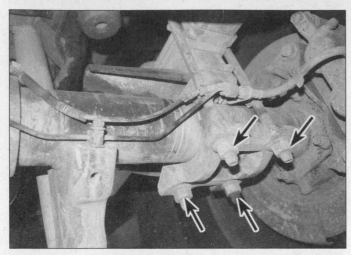

11.3 With the axle supported, remove the nuts from the U-bolts

11.4 Remove the shackle-to-frame bracket bolt and nut at the rear of the spring . . .

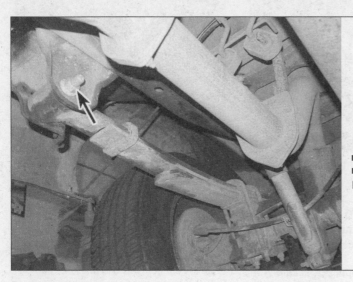

11.5 . . . then the spring hanger bolt and nut at the front of the spring, then remove the spring (it may be necessary to pry the spring out of the bracket)

4   Installation is the reverse of the removal steps. Use new fasteners. Tighten the fasteners to the torque listed in this Chapter's Specifications.

10   Stabilizer bar (rear) - removal and installation

1   Raise the rear of the vehicle, support it securely on jackstands and block the front wheels.
2   Remove the nuts from each end of the stabilizer bar links and remove the links.
3   Remove the stabilizer bar retaining brackets and bushings and detach the stabilizer bar from the axle housing.
4   Installation is the reverse of the removal steps. Use new fasteners. Tighten the bolts and nuts to the torque listed in this Chapter's Specifications.

11   Leaf spring (rear) - removal and installation

Removal

1   Loosen the rear wheel lug nuts. Raise the rear of the vehicle, support it securely on jackstands and block the front wheels. Remove the wheel.
2   Support the vehicle securely on jackstands placed under the frame rails. Support the axle with a floor jack placed under the axle tube and raise it slightly to take the weight off the spring.
3   Remove the nuts from the U-bolts, then remove the U-bolts and the spring plate from the spring (see illustration). Now lower the jack far enough to relieve tension on the spring.
Caution: *Discard the U-bolts and nuts and shackle fasteners. Replace with new fasteners.*

4   Remove the lower bolt and nut securing the shackle assembly to the frame bracket at the rear of the spring (see illustration).
5   Remove the spring hanger bolt and nut at the front of the spring (see illustration). Remove the spring.
6   Inspect the spring eye bushings for wear or distortion. If worn or damaged, have them replaced by a dealer service department or properly equipped shop.

Installation

7   Install the spring in the front bracket. Tighten the bolt and nut finger-tight.
8   Place the spring shackle in the frame bracket. Install the bolt and nut and tighten them finger-tight.
9   Raise the axle into contact with the spring and install the spring seat. Make sure the spring tie-bolt or pin is positioned in the hole in the seat, then install the U-bolts and nuts.
10   Install the wheel and lug nuts. Lower the vehicle to the ground and tighten the spring bracket bolt and nut, spring U-bolts and shackle-to-frame bracket bolt and nut to the torque values listed in this Chapter's Specifications. Tighten the lug nuts to the torque listed in the Chapter 1 Specifications.

12   Steering wheel and airbag clockspring - removal and installation

Warning: *These models are equipped with airbags. Always disable the airbag system whenever working in the vicinity of any airbag system components to avoid the possibility of accidental airbag deployment, which could cause personal injury (see Chapter 12).*

12.1a Turn the steering wheel 90 degrees to the left to expose the airbag removal slots

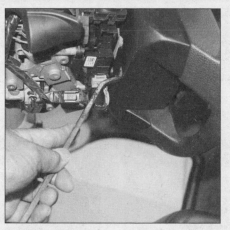

12.1b Pry the wire clips inward to release the airbag from the mounting pins. Perform this procedure with all three wire clips

12.2 Lift the airbag off the steering wheel and unplug the electrical connectors for the horn and for the airbag

12.3 Remove the steering wheel retaining bolt

12.4 Steering wheel matchmarks

12.10 Clock spring screw locations

Steering wheel

Removal

Warning: *Disconnect the cable(s) from the negative battery terminal(s) and wait at least two minutes for the airbag system back-up power supply to be depleted (see Chapter 5).*

Warning: *Carry the airbag module with the trim cover (upholstered side) facing upward, and set the airbag module in a safe location with the trim cover facing up.*

1 Remove the steering column covers (see Chapter 11). Rotate the steering wheel counterclockwise until the top of the wheel is at the 9 o'clock position. Remove the screw and the tilt lever (if equipped), then remove the steering column covers (see Chapter 11). Position a mirror behind the steering wheel to see the three wire clips, then release them one at a time, using a screwdriver or prybar to move them inward. Return the steering wheel to the straight-ahead position (see illustrations).

2 Lift the airbag off of the steering wheel far enough to unplug the electrical connectors for the airbag and horn (see illustration).

3 Remove the steering wheel bolt (see illustration). Discard the bolt and replace with new fastener.

4 Check to see if there are matchmarks on the steering wheel and the steering column shaft (see illustration). If there aren't any, mark the relationship of the steering wheel to the steering shaft. Unplug the electrical connector(s) from the steering wheel.

Caution: *Do not hammer or pry on the steering wheel to separate it from the steering shaft.*

5 Rotate the steering wheel to the straight ahead driving position if you haven't already, then remove the steering wheel by pulling straight up and off of the steering shaft.

6 Aplpy a piece of tape across the clockspring to prevent it from rotating.

Installation

7 Before installing the steering wheel, make sure the airbag clockspring is properly centered (if it isn't, turn the hub of the clockspring counterclockwise until resistance is felt, then turn the clockspring 3-1/2 turns clockwise and position the connector in the 12 o'clock position). Then, using a new steering wheel bolt, tighten the steering wheel bolt to the torque listed in this Chapter's Specifications.

Clockspring

Removal

8 Remove the steering wheel as previously outlined.

9 Stabilize the service lock on the clockspring (if needed) by installing two pieces of tape across the airbag clockspring.

Note: *When the clockspring is removed it will automatically lock in whatever position it was taken off at and will not turn at all until reinstalled.*

10 Remove the three screws securing the clockspring (see illustration).

11 Disconnect the clockspring from the SCCM (Steering Column Control Module) and remove the clockspring from the steering column.

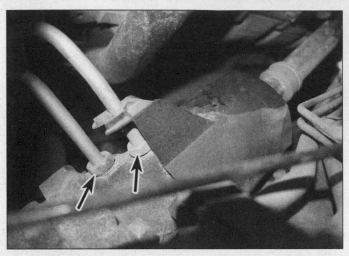

14.4 Slide the plastic protective shroud and unscrew the power steering fluid line fittings

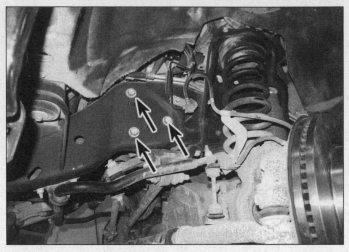

14.7 Steering gear mounting bolts

Installation

12   If the clockspring has become uncentered, center it as described in Step 7.
13   Align the electrical connector to the SCCM and the two locating pins on the back of the clockspring to the steering shaft rotation sensor.
14   With the locator pins in position press the clockspring into place seating the electrical connector.
15   Secure any wiring harnesses that were disturbed back into their retaining clips.
16   Install the three clockspring retaining screws.
17   The remainder of the installation is the reverse of the removal procedure.

13   Steering column - removal and installation

Warning: *These models are equipped with airbags. Always disable the airbag system whenever working in the vicinity of any airbag system components to avoid the possibility of accidental airbag deployment, which could cause personal injury (see Chapter 12).*

Removal

1   Park the vehicle with the wheels pointing straight ahead. Disconnect the cable(s) from the negative battery terminal(s). Wait at least two minutes before proceeding (to allow the backup power supply for the airbag system to become depleted).
2   Remove the knee bolster (see Chapter 11).
3   Remove the steering column covers (see Chapter 11).
4   If you are replacing the steering column, remove the steering wheel (see Section 12).
5   Detach the shift cable from the shift lever on the column. Also detach the shift cable from the steering column cable bracket (see Chapter 7A).

6   Remove the steering column covers (see Chapter 11).
7   Detach the electrical connectors for SCCM (Steering Column Control Module), PATS (Passive Anti-Theft System) transceiver, multi-function switch, shift interlock solenoid, and the adjustable pedal switch (if applicable).
8   Remove any harness retainers leading down the steering column.
9   Lock the steering shaft from turning.
10   Remove the shaft coupler bolt and slide the shaft off.
11   Remove the steering column to instrument panel nuts.
12   Lower the steering column and remove it from the vehicle.

Installation

13   Guide the steering column into position, connecting the steering column shaft to the intermediate shaft. Install the new instrument panel mounting nuts, but don't tighten them yet.
14   Install the new shaft coupler bolt, then tighten the nut to the torque listed in this Chapter's Specifications.
15   Tighten the column mounting nuts to the torque listed in this Chapter's Specifications.
16   The remainder of installation is the reverse of removal. On automatic transmission models, adjust the shift cable and the shift indicator cable (if necessary) by following the procedures described in Chapter 7A.

14   Steering gear - removal and installation

Warning: *These models are equipped with airbags. Always disable the airbag system before working in the vicinity of any airbag system components to avoid the possibility of accidental deployment of the airbag, which could cause personal injury (see Chapter 12).*

Removal

1   Set the front wheels to the straight-ahead position.
Warning: *Make sure that the steering shaft is not rotated with the steering gear removed or damage to the airbag clockspring assembly could occur. One method of preventing the steering shaft from rotating is to run the seat belt through the steering wheel and clip the seat belt buckles together.*
2   Loosen the left front wheel lug nuts. Raise the front of the vehicle, support it securely on jackstands, block the rear wheels and set the parking brake. Remove the wheel.
3   Remove the driver's side inner fender splash shield (see Chapter 11).
4   Place a drain pan under the steering gear. Unscrew the power steering pressure and return line fittings and cap the ends and the ports to prevent excessive fluid loss and contamination (see illustration).
Caution: *Use a flare-nut wrench, if available, to prevent rounding off the corners of the fittings.*
5   Mark the relationship of the intermediate shaft to the steering gear input shaft, then remove the pinch bolt.
6   Disconnect the drag link from the Pitman arm (see Section 15).
7   Loosen all three bolts securing the steering gear box to the frame (see illustration). Support the steering gear and remove the bolts and washers and remove the steering gear from the lower steering column shaft.
Warning: *Replace all fasteners with new fasteners. The original fasteners are coated with a dry adhesive which cannot be reused. When the fasteners are removed the adhesive is no longer effective and the bolt or nut can work loose if reinstalled.*

Installation

8   To make sure the steering gear is centered, turn the input shaft to full lock in one direction, then count and record the number

15.4a Pitman arm-to-drag link connection

15.4b Puller installed for removal

15.11a 4WD right side tie-rod shown with the drag link steering system. The tie-rod joint goes through both the drag link and center link as well as the steering knuckle. Driver's side only has one joint fitting.

of turns required to rotate it to the opposite full lock position. Turn the shaft back through one-half of the number of turns just counted to center the unit.

9   Check that the front wheels are in the straight ahead position and that the steering wheel spokes are centered. Reposition if necessary.

10   Raise the steering gear into position, connecting the intermediate shaft to the steering gear input shaft (be sure to align the marks made earlier).

11   Install the new three bolts securing the steering gear box to the frame. Tighten the bolts to the torque listed in this Chapter's Specifications.

12   Connect the drag link to the Pitman arm, tightening the new nut to the torque listed in this Chapter's Specifications.

13   Install the new bolt securing the flex coupling to the steering gear input shaft. Tighten the bolt to the torque listed in this Chapter's Specifications.

14   Install the front wheel and lug nuts.

15   Lower the vehicle and tighten the lug nuts to the torque listed in the Chapter 1 Specifications.

16   Connect the pressure and return lines to the steering gear and tighten the fittings securely.

17   The remainder of installation is the reverse of removal.

18   Fill the fluid reservoir with the specified fluid and refer to Section 17 for the power steering bleeding procedure.

---

15   Steering linkage - removal and installation

---

1   Two types of steering linkage systems are utilized on these vehicles. One system uses an idler arm and relay rod setup, the other type uses a center link and drag link setup. For description and location of the various steering linkage components see Section 1.

Drag link

Removal

2   Raise the front of the vehicle, support it securely on jackstands, block the rear wheels and set the parking brake. Position the front wheels in the straight ahead position.

3   Remove the cotter pins and loosen the nuts securing the drag link at each end.

4   Use a Pitman arm puller (or equivalent) and break loose the drag link from the Pitman arm (see illustrations). Remove the nuts and detach the ballstuds.

5   If you're replacing the drag link end, loosen the clamp bolts on the tie-rod adjusting sleeve. Unscrew the drag link end from the adjusting sleeve. Count and record the number of turns it takes to turn the drag link off the sleeve.

Installation

6   Installation is the reverse of the removal procedure. Make sure the ball studs are seated in the tapers to prevent them from rotating while tightening the nuts to the torque listed in this Chapter's Specifications. Be sure to install new cotter pins and bend the ends over completely.

Note: *If necessary, tighten the nuts a little more to align the slots of the nut with the hole in the ball stud. Never loosen the nut to achieve this alignment.*

7   Tighten the clamp bolts on the tie-rod adjusting sleeve to the torque listed in this Chapter's Specifications, if removed.

8   Have the front end alignment checked and, if necessary, adjusted.

Tie-rod/tie-rod ends

Removal

Warning: *The manufacturer recommends all new fasteners and cotter pins. Discard the old ones and replace them with new.*

9   Loosen the front wheel lug nuts on the side to be dismantled. Raise the front of the vehicle, support it securely on jackstands, block the rear wheels and set the parking

15.11b Use a puller to remove the tie-rod from the steering knuckle (2WD right side shown)

brake. Remove the front wheel.

10   Remove the cotter pin and loosen the nut on the tie-rod end stud. Discard the cotter pin.

11   Disconnect the tie-rod end from the steering knuckle with a puller (see illustrations).

12   If you're replacing a tie-rod end, loosen the clamp bolts on the tie-rod adjusting sleeve.

13   Unscrew the tie-rod end from the adjusting sleeve. Count and record the number of turns it takes to back the tie-rod end off the sleeve.

Note: *A good precaution to take is to back off the tie-rod lock nut far enough to dab some paint onto the exposed threads. Then (still counting the turns) remove the tie-rod end. Just in case you've forgotten the count or aren't sure, the paint mark will get it close enough to get the vehicle to the alignment shop.*

14   Remove the tie-rod end.

18.3 Use a press tool such as this to push the stud out of the flange

18.4 Install a spacer or washers and a lug nut on the stud, then tighten the nut to draw the stud into place

1 Axle flange	2 Spacer

Installation

15   Installation is the reverse of the removal procedure. Make sure the ballstuds are seated in the tapers to prevent them from rotating while tightening the nuts to the torque listed in this Chapter's Specifications. Be sure to install new cotter pins and bend the ends over completely.

Note: *If necessary, tighten the nuts a little more to align the slots of the nut with the hole in the ballstud. Never loosen the nut to achieve this alignment.*

16   Install the wheel and lug nuts. Lower the vehicle and tighten the lug nuts to the torque listed in the Chapter 1 Specifications.

17   Have the front end alignment checked and, if necessary, adjusted.

16   Power steering pump - removal and installation

Warning: *The manufacturer states that it is necessary to replace the power steering pump pulley with a new one after it has been removed two times. Before removing the pulley, clean the face of the pulley and apply a paint mark to it (if there are two marks, you'll have to obtain a new one).*

Removal

1   Disconnect the cable(s) from the negative battery terminal(s) (see Chapter 4A).
2   Remove the drivebelt (see Chapter 1).
3   Remove the pulley as follows:

Note: *The pump and pulley are removed together.*

Note: *On diesel models, remove the expansion tank to gain access to the power steering pump lines (see Chapter 3).*

Note: *On some models it is necessary to remove the coolant fan shroud to gain access to the power steering pump pulley (see Chapter 3).*

Note: *On some models it is necessary to remove the driver's side inner fender splash shield to gain access to the power steering fluid lines connected to the pump (see Chapter 11).*

4   Place a container or drip pan under the vehicle to catch the power steering fluid that will inevitably spill when the lines are detached.

5   Disconnect the fluid lines from the pump. Plug the lines and the pump fittings to prevent excessive fluid loss and the entry of contaminants.

6   Remove the pump mounting bolts.

Installation

7   Place the pump in position and install the mounting bolts. Tighten the bolts to the torque listed in this Chapter's Specifications.

8   Install the drivebelt (see Chapter 1).

9   Connect the pressure and return hoses to the proper fittings on the pump. New O-rings should be installed on the pressure line fitting.

10   Fill the pump reservoir with the specified fluid (see Chapter 1) and bleed the system (see Section 17).

11   The remainder of the installation is the reverse of removal.

17   Power steering system - bleeding

1   The power steering system must be bled whenever a line is disconnected. Bubbles can be seen in power steering fluid that has air in it and the fluid will often have a milky appearance. Low fluid level can cause air to mix with the fluid, resulting in a noisy pump as well as foaming of the fluid.

2   Open the hood and check the fluid level in the reservoir, adding the specified fluid necessary to bring it up to the proper level (see Chapter 1).

3   Start the engine and slowly turn the steering wheel several times from left-to-right and back again. Do not turn the wheel completely from lock-to-lock. Check the fluid level, topping it up as necessary until it remains steady and no more bubbles are visible.

18   Wheel studs - replacement

Note: *This procedure applies to both the front and rear wheel studs.*

Note: *On some models, you may need to remove the rear axle for clearance.*

1   Loosen the wheel lug nuts, raise the vehicle and support it securely on jackstands. Remove the wheel.

2   Remove the brake disc (see Chapter 9).

3   Push the stud out of the hub flange or axle flange with a press tool (see illustration).

4   Insert the new stud into the hub flange or axle flange from the back side and install some flat washers and a lug nut on the stud (see illustration).

5   Tighten the lug nut until the stud is seated in the flange.

6   Reinstall the disc and caliper (see Chapter 9).

7   Install the wheel and lug nuts. Lower the vehicle and tighten the lug nuts to the torque listed in the Chapter 1 Specifications.

19   Wheels and tires - general information

1   Most models covered by this manual are equipped with radial tires (see illustration). Use of other size or type of tires may affect the ride and handling of the vehicle. Don't mix different types of tires, such as radials and bias belted tires, on the same vehicle - handling may be seriously affected. It's recommended that tires be replaced in pairs on

the same axle, but if only one tire is being replaced, be sure it's the same size, structure and tread design as the other tire on the same axle.

2   Because tire pressure has a substantial effect on handling and wear, the pressure of all tires should be checked at least once a month or before any extended trips are taken (see Chapter 1).

3   Wheels must be replaced if they are bent, dented, leak air, have elongated bolt holes, are heavily rusted, out of vertical symmetry or if the lug nuts won't stay tight. Wheel repairs that use welding or peening are not recommended.

4   Tire and wheel balance are important to the overall handling, braking and performance of the vehicle. Unbalanced wheels can adversely affect handling and ride characteristics as well as tire life. Whenever a tire is installed on a wheel, the tire and wheel should be balanced by a shop with the proper equipment and expertise.

20   Wheel alignment - general information

Note: *Since wheel alignment requires special equipment and techniques it is beyond the scope of this manual. This section is intended only to familiarize the reader with the basic terms used and procedures followed during a typical wheel alignment.*

1   The three basic checks made when aligning a vehicle's front wheels are camber, caster and toe-in (see illustration).

2   Camber and caster are the angles at which the wheels and suspension are inclined in relation to a vertical centerline. Camber is the angle of the wheel in the lateral, or side-to-side plane, while caster is the tilt between the steering axis and the vertical plane, as viewed from the side. Camber angle affects the amount of tire tread which contacts the road and compensates for changes in suspension geometry as the vehicle travels around curves and over bumps. Caster angle affects the self-centering action of the steering, which governs straight-line stability. Camber and caster are adjusted by changing or turning an adjuster sleeve in the top of the steering knuckle.

3   Toe-in is the amount the front wheels are angled in relationship to the center line of the vehicle. For example, in a vehicle with

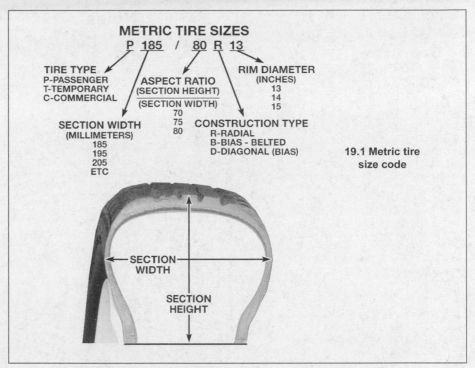

METRIC TIRE SIZES

P 185 / 80 R 13

TIRE TYPE
P-PASSENGER
T-TEMPORARY
C-COMMERCIAL

ASPECT RATIO
(SECTION HEIGHT)
(SECTION WIDTH)
70
75
80

RIM DIAMETER
(INCHES)
13
14
15

SECTION WIDTH
(MILLIMETERS)
185
195
205
ETC

CONSTRUCTION TYPE
R-RADIAL
B-BIAS - BELTED
D-DIAGONAL (BIAS)

SECTION WIDTH

SECTION HEIGHT

19.1 Metric tire size code

zero toe-in, the distance measured between the front edges of the wheels and the distance measured between the rear edges of the wheels are the same. In other words, the wheels are running parallel with the centerline of the vehicle. Toe-in is adjusted by lengthening or shortening the tie-rods. Incorrect toe-in will cause the tires to wear improperly by allowing them to "scrub" against the road surface.

4   Proper wheel alignment is essential for safe steering and even tire wear. Symptoms of alignment problems are pulling of the steering to one side or the other and uneven tire wear. If these symptoms are present, check for the following before having the alignment adjusted:

a) *Loose steering gear mounting bolts*
b) *Damaged or worn steering gear mounts*
c) *Worn or damaged wheel bearings*
d) *Bent tie-rods*
e) *Worn balljoints*
f) *Improper tire pressures*
g) *Mixing tires of different construction*

5   Front wheel alignment should be left to an alignment shop with the proper equipment and experienced personnel.

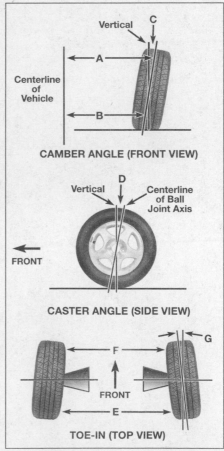

CAMBER ANGLE (FRONT VIEW)

CASTER ANGLE (SIDE VIEW)

TOE-IN (TOP VIEW)

20.1 Front wheel alignment details

A minus B = C (degrees camber)
D = degrees caster
E minus F = toe-in (measured in inches)
G = toe-in (expressed in degrees)

Notes

Chapter 11 Body

Contents

Specifications

Torque specifications
Ft-lbs (unless otherwise indicated)

Note: *One foot-pound (ft-lb) of torque is equivalent to 12 inch-pounds (in-lbs) of torque. Torque values below approximately 15 foot-pounds are expressed in inch-pounds, because most foot-pound torque wrenches are not accurate at these smaller values.*

Cab to chassis bolts	75
Bed bolts*	59
Seat belt anchor bolts	46
Seat bolts	41
Sun roof motor mounting bolts	35 in-lbs

* Manufacturer recommends that new fasteners are used. Do not reuse the original fasteners.

1 General information

Warning: *The models covered by this manual are equipped with a Supplemental Restraint System (SRS), more commonly known as airbags. Always disable the airbag system before working in the vicinity of any airbag system components to avoid the possibility of accidental deployment of the airbags, which could cause personal injury (see Chapter 12).*

Certain body components are particularly vulnerable to accident damage and can be unbolted and repaired or replaced. Among these parts are the hood, doors, tailgate, liftgate, bumpers and front fenders.

Only general body maintenance practices and body panel repair procedures within the scope of the do-it-yourselfer are included in this Chapter.

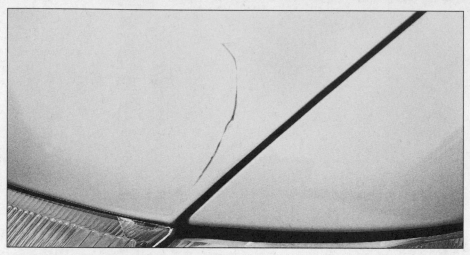

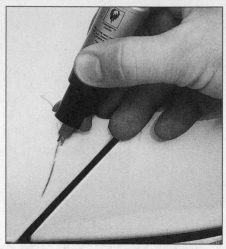

Make sure the damaged area is perfectly clean and rust free. If the touch-up kit has a wire brush, use it to clean the scratch or chip. Or use fine steel wool wrapped around the end of a pencil. Clean the scratched or chipped surface only, not the good paint surrounding it. Rinse the area with water and allow it to dry thoroughly

Thoroughly mix the paint, then apply a small amount with the touch-up kit brush or a very fine artist's brush. Brush in one direction as you fill the scratch area. Do not build up the paint higher than the surrounding paint

2   Repair minor paint scratches

No matter how hard you try to keep your vehicle looking like new, it will inevitably be scratched, chipped or dented at some point. If the metal is actually dented, seek the advice of a professional. But you can fix minor scratches and chips yourself. Buy a touch-up paint kit from a dealer parts department or an auto parts store. To ensure that you get the right color, you'll need to have the specific make, model and year of your vehicle and, ideally, the paint code, which is located on a special metal plate under the hood or in the door jamb.

3   Body repair - minor damage

Plastic body panels

The following repair procedures are for minor scratches and gouges. Repair of more serious damage should be left to a dealer service department or qualified auto body shop. Below is a list of the equipment and materials necessary to perform the following repair procedures on plastic body panels.

> *Wax, grease and silicone removing solvent*
> *Cloth-backed body tape*
> *Sanding discs*
> *Drill motor with three-inch disc holder*
> *Hand sanding block*
> *Rubber squeegees*
> *Sandpaper*
> *Non-porous mixing palette*
> *Wood paddle or putty knife*
> *Curved-tooth body file*
> *Flexible parts repair material*

Flexible panels (bumper trim)

1   Remove the damaged panel, if necessary or desirable. In most cases, repairs can be car-

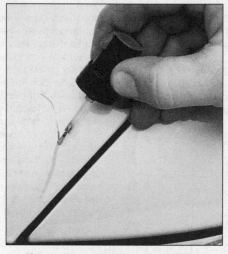

If the vehicle has a two-coat finish, apply the clear coat after the color coat has dried

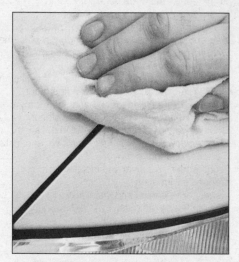

Wait a few days for the paint to dry thoroughly, then rub out the repainted area with a polishing compound to blend the new paint with the surrounding area. When you're happy with your work, wash and polish the area

ried out with the panel installed.

2   Clean the area(s) to be repaired with a wax, grease and silicone removing solvent applied with a water-dampened cloth.

3   If the damage is structural, that is, if it extends through the panel, clean the backside of the panel area to be repaired as well. Wipe dry.

4   Sand the rear surface about 1-1/2 inches beyond the break.

5   Cut two pieces of fiberglass cloth large enough to overlap the break by about 1-1/2 inches. Cut only to the required length.

6   Mix the adhesive from the repair kit according to the instructions included with the kit, and apply a layer of the mixture approximately 1/8-inch thick on the backside of the panel. Overlap the break by at least 1-1/2 inches.

7   Apply one piece of fiberglass cloth to the adhesive and cover the cloth with additional adhesive. Apply a second piece of fiberglass

cloth to the adhesive and immediately cover the cloth with additional adhesive in sufficient quantity to fill the weave.

8   Allow the repair to cure for 20 to 30 minutes at 60-degrees to 80-degrees F.

9   If necessary, trim the excess repair material at the edge.

10   Remove all of the paint film over and around the area(s) to be repaired. The repair material should not overlap the painted surface.

11   With a drill motor and a sanding disc (or a rotary file), cut a "V" along the break line approximately 1/2-inch wide. Remove all dust and loose particles from the repair area.

12   Mix and apply the repair material. Apply a light coat first over the damaged area; then continue applying material until it reaches a level

slightly higher than the surrounding finish.

13   Cure the mixture for 20 to 30 minutes at 60-degrees to 80-degrees F.

14   Roughly establish the contour of the area being repaired with a body file. If low areas or pits remain, mix and apply additional adhesive.

15   Block sand the damaged area with sandpaper to establish the actual contour of the surrounding surface.

16   If desired, the repaired area can be temporarily protected with several light coats of primer. Because of the special paints and techniques required for flexible body panels, it is recommended that the vehicle be taken to a paint shop for completion of the body repair.

Steel body panels

See photo sequence

Repair of dents

17   When repairing dents, the first job is to pull the dent out until the affected area is as close as possible to its original shape. There is no point in trying to restore the original shape completely as the metal in the damaged area will have stretched on impact and cannot be restored to its original contours. It is better to bring the level of the dent up to a point that is about 1/8-inch below the level of the surrounding metal. In cases where the dent is very shallow, it is not worth trying to pull it out at all.

18   If the backside of the dent is accessible, it can be hammered out gently from behind using a soft-face hammer. While doing this, hold a block of wood firmly against the opposite side of the metal to absorb the hammer blows and prevent the metal from being stretched.

19   If the dent is in a section of the body which has double layers, or some other factor makes it inaccessible from behind, a different technique is required. Drill several small holes through the metal inside the damaged area, particularly in the deeper sections. Screw long, self-tapping screws into the holes just enough for them to get a good grip in the metal. Now pulling on the protruding heads of the screws with locking pliers can pull out the dent.

20   The next stage of repair is the removal of paint from the damaged area and from an inch or so of the surrounding metal. This is easily done with a wire brush or sanding disk in a drill motor, although it can be done just as effectively by hand with sandpaper. To complete the preparation for filling, score the surface of the bare metal with a screwdriver or the tang of a file or drill small holes in the affected area. This will provide a good grip for the filler material. To complete the repair, see the Section on filling and painting.

Repair of rust holes or gashes

21   Remove all paint from the affected area and from an inch or so of the surrounding metal using a sanding disk or wire brush mounted in a drill motor. If these are not available, a few sheets of sandpaper will do the job just as effectively.

22   With the paint removed, you will be able to determine the severity of the corrosion and

decide whether to replace the whole panel, if possible, or repair the affected area. New body panels are not as expensive as most people think and it is often quicker to install a new panel than to repair large areas of rust.

23   Remove all trim pieces from the affected area except those which will act as a guide to the original shape of the damaged body, such as headlight shells, etc. Using metal snips or a hacksaw blade, remove all loose metal and any other metal that is badly affected by rust. Hammer the edges of the hole in to create a slight depression for the filler material.

24   Wire-brush the affected area to remove the powdery rust from the surface of the metal. If the back of the rusted area is accessible, treat it with rust inhibiting paint.

25   Before filling is done, block the hole in some way. This can be done with sheet metal riveted or screwed into place, or by stuffing the hole with wire mesh.

26   Once the hole is blocked off, the affected area can be filled and painted. See the following subsection on filling and painting.

Filling and painting

27   Many types of body fillers are available, but generally speaking, body repair kits which contain filler paste and a tube of resin hardener are best for this type of repair work. A wide, flexible plastic or nylon applicator will be necessary for imparting a smooth and contoured finish to the surface of the filler material. Mix up a small amount of filler on a clean piece of wood or cardboard (use the hardener sparingly). Follow the manufacturer's instructions on the package, otherwise the filler will set incorrectly.

28   Using the applicator, apply the filler paste to the prepared area. Draw the applicator across the surface of the filler to achieve the desired contour and to level the filler surface. As soon as a contour that approximates the original one is achieved, stop working the paste. If you continue, the paste will begin to stick to the applicator. Continue to add thin layers of paste at 20-minute intervals until the level of the filler is just above the surrounding metal.

29   Once the filler has hardened, the excess can be removed with a body file. From then on, progressively finer grades of sandpaper should be used, starting with a 180-grit paper and finishing with 600-grit wet-or-dry paper. Always wrap the sandpaper around a flat rubber or wooden block, otherwise the surface of the filler will not be completely flat. During the sanding of the filler surface, the wet-or-dry paper should be periodically rinsed in water. This will ensure that a very smooth finish is produced in the final stage.

30   At this point, the repair area should be surrounded by a ring of bare metal, which in turn should be encircled by the finely feathered edge of good paint. Rinse the repair area with clean water until all of the dust produced by the sanding operation is gone.

31   Spray the entire area with a light coat of primer. This will reveal any imperfections in the surface of the filler. Repair the imperfections with

fresh filler paste or glaze filler and once more smooth the surface with sandpaper. Repeat this spray-and-repair procedure until you are satisfied that the surface of the filler and the feathered edge of the paint are perfect. Rinse the area with clean water and allow it to dry completely.

32   The repair area is now ready for painting. Spray painting must be carried out in a warm, dry, windless and dust free atmosphere. These conditions can be created if you have access to a large indoor work area, but if you are forced to work in the open, you will have to pick the day very carefully. If you are working indoors, dousing the floor in the work area with water will help settle the dust that would otherwise be in the air. If the repair area is confined to one body panel, mask off the surrounding panels. This will help minimize the effects of a slight mismatch in paint color. Trim pieces such as chrome strips, door handles, etc., will also need to be masked off or removed. Use masking tape and several thickness of newspaper for the masking operations.

33   Before spraying, shake the paint can thoroughly, then spray a test area until the spray painting technique is mastered. Cover the repair area with a thick coat of primer. The thickness should be built up using several thin layers of primer rather than one thick one. Using 600-grit wet-or-dry sandpaper, rub down the surface of the primer until it is very smooth. While doing this, the work area should be thoroughly rinsed with water and the wet-or-dry sandpaper periodically rinsed as well. Allow the primer to dry before spraying additional coats.

34   Spray on the top coat, again building up the thickness by using several thin layers of paint. Begin spraying in the center of the repair area and then, using a circular motion, work out until the whole repair area and about two inches of the surrounding original paint is covered. Remove all masking material 10 to 15 minutes after spraying on the final coat of paint. Allow the new paint at least two weeks to harden, then use a very fine rubbing compound to blend the edges of the new paint into the existing paint. Finally, apply a coat of wax

4   Body repair - major damage

1   Major damage must be repaired by an auto body shop specifically equipped to perform body and frame repairs. These shops have the specialized equipment required to do the job properly.

2   If the damage is extensive, the frame must be checked for proper alignment or the vehicle's handling characteristics may be adversely affected and other components may wear at an accelerated rate.

3   Due to the fact that all of the major body components (hood, fenders, etc.) are separate and replaceable units, any seriously damaged components should be replaced rather than repaired. Sometimes the components can be found in a wrecking yard that specializes in used vehicle components, often at considerable savings over the cost of new parts.

These photos illustrate a method of repairing simple dents. They are intended to supplement *Body repair - minor damage* in this Chapter and should not be used as the sole instructions for body repair on these vehicles.

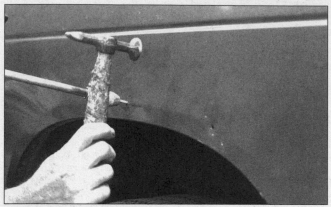

1   If you can't access the backside of the body panel to hammer out the dent, pull it out with a slide-hammer-type dent puller. Tap with a hammer near the edge of the dent to help 'pop' the metal back to its original shape, about 1/8-inch below the surface of the surrounding metal

2   Using coarse-grit sandpaper, remove the paint down to the bare metal. Clean the repair area with wax/silicone remover.

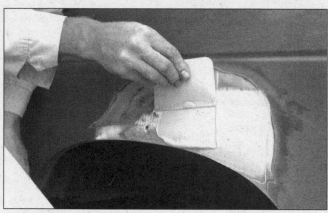

3   Following label instructions, mix up a batch of plastic filler and hardener, then quickly press it into the metal with a plastic applicator. Work the filler until it matches the original contour and is slightly above the surrounding metal

4   Let the filler harden until you can just dent it with your fingernail. File, then sand the filler down until it's smooth and even. Work down to finer grits of sandpaper - always using a board or block - ending up with 360 or 400 grit

5   When the area is smooth to the touch, clean the area and mask around it. Apply several layers of primer to the area. A professional-type spray gun is being used here, but aerosol spray primer works fine

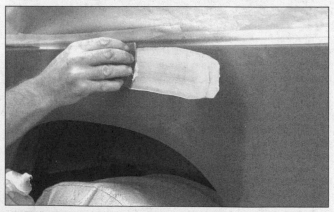

6   Fill imperfections or scratches with glazing compound. Sand with 360 or 400-grit and re-spray. Finish sand the primer with 600 grit, clean thoroughly, then apply the finish coat. Don't attempt to rub out or wax the repair area until the paint has dried completely (at least two weeks)

5 Upholstery, carpets and vinyl trim - maintenance

Upholstery and carpets

1   Every three months remove the floormats and clean the interior of the vehicle (more frequently if necessary). Use a stiff whiskbroom to brush the carpeting and loosen dirt and dust, then vacuum the upholstery and carpets thoroughly, especially along seams and crevices.

2   Dirt and stains can be removed from carpeting with basic household or automotive carpet shampoos available in spray cans. Follow the directions and vacuum again, then use a stiff brush to bring back the "nap" of the carpet.

3   Most interiors have cloth or vinyl upholstery, either of which can be cleaned and maintained with a number of material-specific cleaners or shampoos available in auto supply stores. Follow the directions on the product for usage, and always spot-test any upholstery cleaner on an inconspicuous area (bottom edge of a backseat cushion) to ensure that it doesn't cause a color shift in the material.

4   After cleaning, vinyl upholstery should be treated with a protectant.

Note: *Make sure the protectant container indicates the product can be used on seats - some products may make a seat too slippery.*

Warning: *Do not use protectant on vinyl-covered steering wheels. This can leave the steering wheel with an extremely slippery surface.*

5   Leather upholstery requires special care. It should be cleaned regularly with saddle-soap or leather cleaner. Never use alcohol, gasoline, nail polish remover or thinner to clean leather upholstery.

6   After cleaning, regularly treat leather upholstery with a leather conditioner, rubbed in with a soft cotton cloth. Never use car wax on leather upholstery.

7   In areas where the interior of the vehicle is subject to bright sunlight, cover leather seating areas of the seats with a sheet if the vehicle is to be left out for any length of time.

Vinyl trim

8   Don't clean vinyl trim with detergents, caustic soap or petroleum-based cleaners. Plain soap and water works just fine, with a soft brush to clean dirt that may be ingrained. Wash the vinyl as frequently as the rest of the vehicle.

9   After cleaning, application of a high-quality rubber and vinyl protectant will help prevent oxidation and cracks. The protectant can also be applied to weather-stripping, vacuum lines and rubber hoses, which often fail as a result of chemical degradation.

6 Fastener and trim removal

1   There is a variety of plastic fasteners used to hold trim panels, splash shields and other parts in place in addition to typical screws, nuts and bolts. Once you are familiar with them, they can usually be removed without too much difficulty.

2   The proper tools and approach can pre-

Fasteners

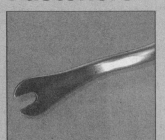

This tool is designed to remove special fasteners. A small pry tool used for removing nails will also work well in place of this tool

A Phillips head screwdriver can be used to release the center portion, but light pressure must be used because the plastic is easily damaged. Once the center is up, the fastener can easily be pried from its hole

Here is a view with the center portion fully released. Install the fastener as shown, then press the center in to set it

This fastener is used for exterior panels and shields. The center portion must be pried up to release the fastener. Install the fastener with the center up, then press the center in to set it

This type of fastener is used commonly for interior panels. Use a small blunt tool to press the small pin at the center in to release it . . .

. . . the pin will stay with the fastener in the released position

Reset the fastener for installation by moving the pin out. Install the fastener, then press the pin flush with the fastener to set it

This fastener is used for exterior and interior panels. It has no moving parts. Simply pry the fastener from its hole like the claw of a hammer removes a nail. Without a tool that can get under the top of the fastener, it can be very difficult to remove

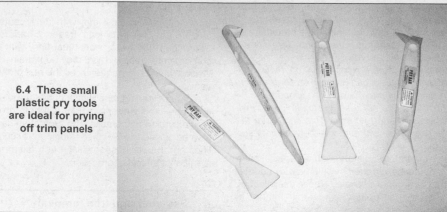

6.4 These small plastic pry tools are ideal for prying off trim panels

vent added time and expense to a project by minimizing the number of broken fasteners and/or parts.

3   The illustration on page 11-5 shows various types of fasteners that are typically used on most vehicles and how to remove and install them (see illustration). Replacement fasteners are commonly found at most auto parts stores, if necessary.

4   Trim panels are typically made of plastic and their flexibility can help during removal. The key to their removal is to use a tool to pry the panel near its retainers to release it without damaging surrounding areas or breaking-off any retainers. The retainers will usually snap out of their designated slot or hole after force is applied to them. Stiff plastic tools designed for prying on trim panels are available at most auto parts stores (see illustration). Tools that are tapered and wrapped in protective tape, such as a screwdriver or small pry tool, are also very effective when used with care.

Caution: *Age and weather as well as direct sunlight on just about any plastic or composite type trim panel can turn what used to be a pliable plastic panel into a brittle, crumbling and often cracking piece of faded trim. When you are trying to remove a trim section on an older vehicle with sun or weather damage to the trim, take extra precautions not to damage the trim panel any more than you have too. Make sure your trim removing tool is in the correct area to pry and listen closely for any unusual sounds that might indicate you're about to crack the plastic. Readjust your trim tool to a different area and try again. Sometimes, even with the best efforts, an old dried out plastic trim panel can't be prevented from cracking or splitting.*

7   Hinges and locks - maintenance

1   Once every 3000 miles, or every three months, the hinges and latch assemblies on the doors, hood and trunk should be given a few drops of light oil or lock lubricant. The door latch strikers should also be lubricated with a thin coat of grease or white lithium lubricant to reduce wear and ensure free movement. Lubricate the door and trunk locks with spray-on graphite lubricant.

8   Windshield and fixed glass - replacement

1   Replacement of the windshield and fixed glass requires the use of special fast-setting adhesive/caulk materials and some specialized tools and techniques. These operations should be left to a shop specializing in glass replacement.

9   Headlight lens refurbishing

1   The plastic lens on most headlight assemblies are susceptible to U.V. damage from the sun as well as being out in the weather. There are several companies that make restoration kits to restore the plastic lens back to their original condition. There are also a few home remedies that have been proven somewhat effective too, such as toothpaste buffed onto the lens as well as some bug sprays that dissolve the opaque and clouded surface of the headlight lens (see illustration).

2   Most restoration kits are a three-step process that requires a variable speed hand drill to perform the various steps. When starting a headlight restoration project, be sure to be in a well ventilated area and where you don't have anything you don't want the buffing chemicals and compounds to be sprayed on.

3   Mask off the area around the headlight assembly to avoid any painted surfaces from being damaged with the drill or the compounds in the restoration kit. Have plenty of clean rags handy to wipe up any spills and to perform the final buffing (see illustration).

4   Follow all the directions in the restoration kit as described in the instructions. We also recommend adding a top coat of wax or a suitable U.V. protectant as a final finish coat (see illustration).

9.1 Cloudy headlight lenses are ugly and reduce the amount of light projected

9.3 Mask off the surrounding area and apply the compound with the buffer

9.4 The result should be a lens that is almost as clear as new

10 Hood support struts - removal and installation

Note: *The hood is heavy and somewhat awkward to hold - at least two people should perform this procedure.*

1   Open the hood and support it securely.
2   Using a small screwdriver, detach the retaining clips at both ends of the support strut. Then pry or pull sharply to detach it from the vehicle (see illustrations).
3   Installation is the reverse of removal.

11 Hood - removal, installation and adjustment

Note: *The hood is heavy and somewhat awkward to remove and install - at least two people should perform this procedure.*

Removal and installation

1   Use blankets or pads to cover the cowl area of the body and fenders. This will protect the body and paint as the hood is lifted off.
2   Make marks or scribe a line around the hood hinge to ensure proper alignment during installation (see illustration).
3   Disconnect any cables or wires that will interfere with removal.
4   Have an assistant support one side of the hood while you support the other. Simultaneously remove the hinge-to-hood bolts (see illustration).
5   Lift off the hood.
6   Installation is the reverse of removal.

Adjustment

7   Fore-and-aft and side-to-side adjustment of the hood is done by moving the hinge plate slot after loosening the bolts or nuts.
8   Scribe a line around the entire hinge plate so you can determine the amount of movement (see illustration 11.2).
9   Loosen the bolts or nuts and move the hood into correct alignment. Move it only a little at a time. Tighten the hinge bolts and carefully lower the hood to check the position.
10   If necessary after installation, the entire hood latch assembly can be adjusted up-and-down as well as from side-to-side on the radiator support so the hood closes securely and flush with the fenders. To make the adjustment, scribe a line or mark around the hood latch mounting bolts to provide a reference point, then loosen them and reposition the latch assembly, as necessary (see illustration). Following adjustment, retighten the mounting bolts.
11   Finally, adjust the hood bumpers on the radiator support so the hood, when closed, is flush with the fenders (see illustration).
12   The hood latch assembly, as well as the hinges, should be periodically lubricated with white, lithium-base grease to prevent binding and wear.

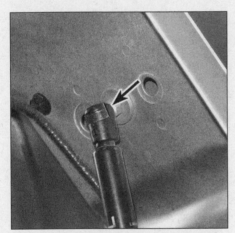

10.2a Use a small screwdriver to pry the clip out of its locking groove, then detach the end of the strut from the locating stud. Have an assistant support the hood before attempting to remove the support strut completely.

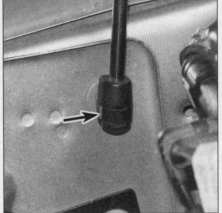

10.2b Use the small screwdriver again to pry the bottom retaining clip off, then remove the strut from the vehicle

11.2 Before removing the hood, draw a mark around the hinge plate

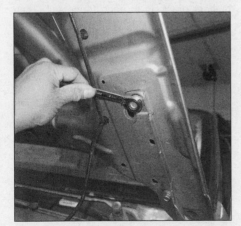

11.4 Remove the hinge-to-hood retaining bolts and lift off the hood with the help of an assistant

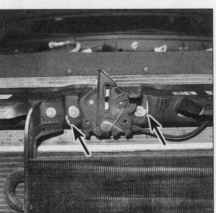

11.10 Scribe a line around the latch to use as a reference point. To adjust the hood latch, loosen the retaining bolts, move the latch and retighten bolts, then close the hood to check the fit

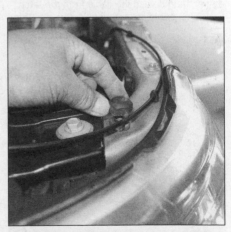

11.11 Adjust the hood closing height by turning the hood bumpers in or out

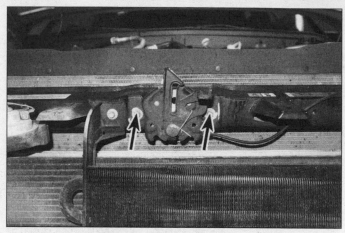

12.2 Remove the latch mounting fasteners

13.1 Radiator grille bracket bolts and radiator grille push pin locations

A   *Radiator grille bracket bolts (4)*
B   *Radiator grille push pins (5)*
C   *Grille to radiator bracket screw (1)*

12   Hood latch and release cable - removal and installation

Latch

1   Scribe a line around the latch to aid alignment when reinstalling the latch assembly.
2   Remove the latch mounting fasteners securing the latch to the radiator support and remove the latch (see illustration).
3   Disconnect the hood release cable by disengaging the cable from the back of the latch assembly.
4   Installation is the reverse of the removal procedure.
Note: *Adjust the latch so the hood engages securely when closed and the hood bumpers are slightly compressed.*

Cable

5   Remove the hood latch as described earlier, then detach the cable from the latch.
6   The cable is actually in two sections

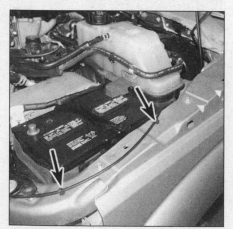

15.1 Hood release cable retainers

joined at a junction box on the driver's side of the core support.
7   Disconnect the cable sections from the junction box.
8   For the shorter cable that runs from the latch to the junction box just remove the tie down clips to remove it.
9   For the longer cable section that runs from the junction box to the inside latch handle, remove the tie down fasteners and follow the cable up to the battery junction box (underhood fuse box).
10   Remove the screws securing the battery junction box to the fender and then tip it out of the way.
11   Continue disconnecting the tie down clips to the hood cable up to the firewall.
12   Remove the driver's side cowling trim (see Section 36.
13   From inside the vehicle, remove the three nuts securing the emergency brake lever to the body.
14   Disconnect the cable from the inside handle.
15   From the cowling area pull the cable through the firewall to remove.
16   The remainder of installation is the reverse of removal.

13   Radiator grille - removal and installation

1   The radiator grille can be removed with the upper radiator grille bracket and the front hood seal as an assembly or without (see illustration).
Note: *If you choose to remove only the grille and not the grille bracket assembly, match mark the grille to the bracket for later installation to aid in lining it up.*
2   Detach the bolts for the grille bracket to remove the assembly or just remove the push pins (5) for only the grille removal.
3   If equipped, disconnect the push pins

securing the air deflectors to the grille assembly.
4   Tilt the top of the radiator grille away from the core support, then push down on the lower clips to release the grille fasteners.
5   Installation is the reverse of removal.

14   Bumpers - removal and installation

Caution: *The bumpers are heavy and awkward. Have an assistant help you with removal or installation.*
1   Apply the parking brake, raise the vehicle and support it securely on jackstands.

Front bumper

2   Working from under the vehicle, disconnect the fog light electrical connections, if equipped.
3   Working from the backside of the bumper, remove the retaining bolts securing the bumper brackets to each frame rail. Then remove the bumper from the vehicle.
4   Installation is the reverse of removal.

Rear bumper

5   Unplug any electrical connectors which would interfere with bumper removal.
6   Working from the backside of the bumper, remove the retaining bolts securing the bumper brackets to each frame rail. Then remove the bumper from the vehicle.
7   Installation is the reverse of removal.

15   Front fender - removal and installation

1   For the driver's side fender, disconnect the hood release cable retainers (see illustration). On the passenger's side fender, remove the passenger's side battery tray. Discon-

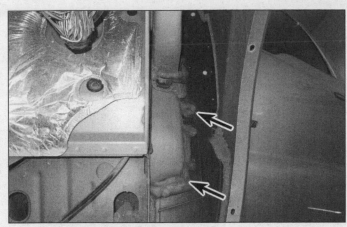

15.5 Remove as much of the foam needed to gain access to the baffle and bolts

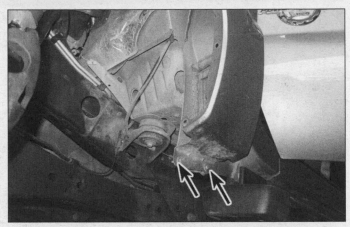

15.8 Remove the fender to rocker panel fasteners

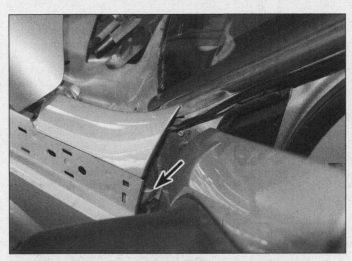

15.10 Fender to door pillar (A pillar) fastener

17.6 Fastener locations

nect the cowling clip from the top edge of the fender.

2   Remove the headlight and turn signal housings (see Chapter 12).

3   Remove the radiator grille (see Section 13).

4   Remove the inner fender liner (see Section 17).

5   From inside the wheel opening, carve out the expanded foam covering the rear inner fender bolts and the expanded foam securing the inside baffle (see illustration). Remove the bolts and baffle.

6   If equipped, remove the running board.

7   Remove the front bumper (see Section 14).

8   Remove the fender-to-rocker panel bolts (see illustration) and splash shield (if applicable).

9   Remove the lower front bolts securing the fender to the core support area.

10   Open the door and remove the fender-to-door pillar bolts (see illustration).

11   Remove the remaining fender mounting bolts across the top edge.

12   Detach the fender.

Note: *It's a good idea to have an assistant support the fender while it's being moved*

away from the vehicle to prevent damage to the surrounding body panels.

13   Installation is the reverse of removal. Respray the appropriate areas inside the fender well area with expanding foam to match what was there before.

16   Rear extension fender - removal and installation

1   Remove the splash guard shield. There are 14 push pins that need to be removed and two screws that secure the splash guard to the fender.

2   Disconnect the front and rear side marker electrical connections.

3   Remove the bolts on the bottom edge of the fender, two on the front edge and two on the rear edge.

4   From inside the fender well, remove the six nuts securing the fender to the truck bed, three in the front area, three in the rear area.

5   Remove the six "T" bolts, three in front and three in the rear section.

6   Remove the six bolts around the circum-

ference of the inner side of the fender well.

7   Lift the fender off of the center locator push pin.

8   Installation is the reverse of removal.

17   Inner fender liner - installation and removal

1   Loosen the appropriate wheel lug nuts, then raise and support the vehicle. Remove the wheel.

2   Release the wire retainers from the inner fender.

3   Remove the lower front bolts.

4   Disconnect the retaining clips from the lower front and rear wheel lip edges.

5   Remove the three bolts securing the inner fender to the fender.

6   On the driver's side, lower the liner slightly to disconnect the retaining clips that are attached on the engine side of the liner (see illustration).

7   Remove the liner by collapsing it inward as you draw it out of the wheel well.

8   Installation is the reverse of removal.

18.2 Pry the sail trim off with a flat trim tool

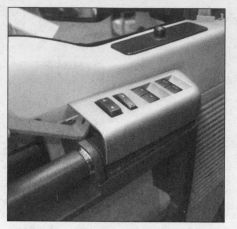

18.5a Pry up on the retaining clip at the edge and lift the armrest switch control plate out . . .

18.5b . . . then turn the switch over and disconnect the electrical connector

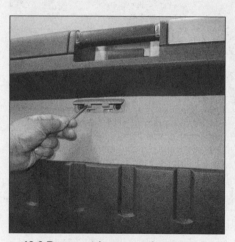

18.6 Remove trim cover, then remove the fasteners

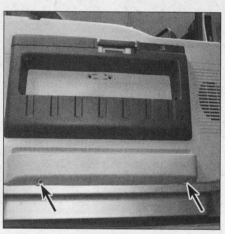

18.7 Lower screw locations

18.8 Lift up to clear the plastic hooks then pull the door panel away from the door

18.9 Carefully peel the watershield from the door

18 Door trim panels - removal and installation

Front door trim panel removal

1   Disconnect the cable(s) from the negative battery terminal(s) (see Chapter 5).
2   Remove the sail trim, if equipped disconnect the electrical connector (see illustration).
3   If equipped, lift up the outside mirror switch and disconnect it.
4   On manual window equipped models, pull the window crank cover aside, remove the screw and detach the handle.
5   On power window equipped models, pry out the rear edge of the armrest switch control plate and disconnect the electrical connections (see illustrations).
6   Remove the center bolt trim cover concealing the two bolts just below the grab handle (see illustration). Remove the fasteners.

7   Remove the two lower edge screws (see illustration).
8   Once all of the bolts are removed, remove the trim panel by gently pulling it up and out (see illustration). Detach the courtesy light bulb housing from the panel (if equipped).
9   For access to the inner door components, remove the speaker and carefully separate the water shield from the door (see illustration).
10   Installation is the reverse of removal.

Rear door trim panel removal (Super cab models)

11   Remove the shoulder belt upper retaining cover by prying it upward, then remove the retaining bolt.
12   Remove the two screws securing the window latch cover to the door.

19.3 Pull down on the clip handle and then pull the harness free from the A pillar

19.4 Door check rod bolts

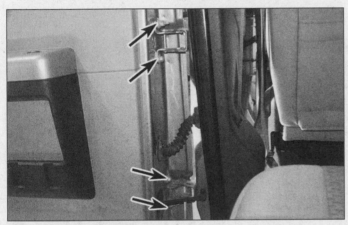

19.5 Take care to balance and control the door as you remove the hinge bolts

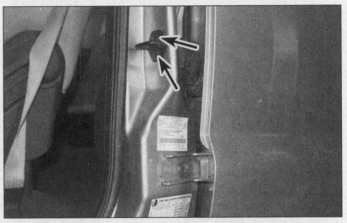

19.10 Adjust the door lock striker by loosening the mounting screws and gently tapping the striker in the desired direction

13   Remove the upper trim panel from the door by pulling it away from the door.

14   Lift up on the seat belt guide trim panel mounted in the lower door panel. Slip the seat belt out of the guide.

15   Remove the lower seat belt retaining bolt concealed behind the seat belt guide trim.

16   Remove the screws across the top edge of the lower door panel assembly. (On some models it is riveted. The rivets will need to be drilled out.)

17   Lift straight up on the door panel to unfasten the lower hooks.

18   Guide the seat belt through the door panel as you remove the panel from the vehicle.

19   Installation is the reverse of removal.

Rear door trim panel removal (Crew cab models)

20   The crew cab rear door panels are removed in a similar manner as the front door trim panels. Follow the procedure for the front door panels for removal and installation instructions.

19   Door - removal, installation and adjustment

Note: *The door is heavy and somewhat awkward to remove and install - at least two people should perform this procedure.*

Removal and installation

1   Lower the window completely in the door.

2   Open the door all the way and support it on jacks covered with rags to prevent damaging the paint.

3   Disconnect the electrical connection in the door jam at the A pillar (see illustration).

4   Disconnect the two bolts securing the door check rod (see illustration).

5   Mark around the door hinges with a marking pen or scribe to facilitate alignment during reassembly. With an assistant supporting the door, remove the hinge-to-door bolts and remove the door (see illustration).

6   Installation is the reverse of the removal.

Adjustment

7   Having proper door-to-body alignment is a critical part of a well functioning door assembly. First check the door hinge pins for excessive play. Fully open the door and lift up and down on the door without lifting the body. If a door has 1/16-inch or more excessive play, the hinges should be replaced.

8   Door-to-body alignment adjustments are made by loosening the hinge-to-body or hinge-to-door bolts and moving the door. Proper body alignment is achieved when the top of the door is aligned with the top of the front fender and rear door or quarter panel and the bottom of the door is aligned with the lower rocker panel. If these goals can't be reached by adjusting the hinge-to-body or hinge-to-door bolts, body alignment shims may have to be purchased and inserted behind the hinges to achieve correct alignment.

9   To adjust the door closed position, first check that the door latch is contacting the center of the latch striker. If not, remove the striker and add or subtract shims to achieve correct alignment.

10   Finally, adjust the latch striker as necessary to provide positive engagement with the latch mechanism (see illustration) and the door panel is flush with the rear door or quarter panel.

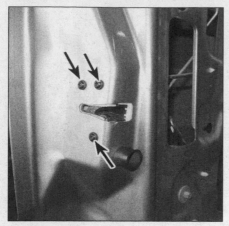

20.3 Remove the latch retaining screws from the end of the door

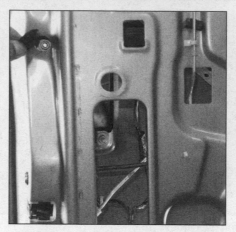

20.7 Remove the rubber plugs for access to the outside handle retaining nuts

20.8 Detach the lock actuator rod and the door handle actuator rods, then remove the outside handle from the door

20.9 Remove the wire clip and withdraw the cylinder from the handle assembly

20.12 Remove the two inner door handle retaining nuts

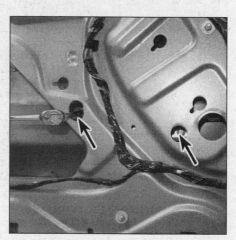

21.3 Remove the nuts securing the glass

20 Door latch, lock cylinder and handles - removal and installation

Door latch

1   Remove the outside door handle (see Section 20).
2   Working from inside the door, disconnect the actuator rod and cable from the latch assembly.
3   Remove the three screws securing the latch to the door edge (see illustration).
4   Working through the large access hole, position the latch as necessary to disconnect the electrical connector.
5   Installation is the reverse of removal.

Outside handle and door lock cylinder

6   To remove the outside handle and door lock cylinder assembly, raise the window and remove the door trim panel and watershield as described in Section 18.
7   Remove the rubber plugs in the end of the door and remove the two outside handle retaining nuts (see illustration).
8   Pull the handle and lock cylinder assembly from the door and disengage the plastic clips that secure the lock cylinder to latch rod and door handle to latch rod (see illustration).
9   To remove the lock cylinder from the handle assembly, detach the clip and withdraw the cylinder (see illustration).
10   Installation is the reverse of removal.

Inside handle and cable

11   Remove the door trim panel and watershield as described in Section 18.
12   Remove the nuts securing the inside handle (see illustration). Pull rearward on the handle to disengage it from the inner door panel.
13   Detach the latch rod from the backside of the handle and remove it from the vehicle.
14   Installation is the reverse of removal.

21 Door window glass - removal and installation

1   Remove the door trim panel and the watershield (see Section 18).
2   Leave the window switch connected (or reinstall the window crank arm temporarily).
3   Lower the window glass for access to the glass retaining nuts and remove the nuts (see illustration).
4   Now detach the window switch (or the window crank arm) completely and set them aside for reinstalling.
5   Remove the inner and outer window seal strips across the top of the door. Pull straight up to remove.
6   If needed, slide the glass channel forward enough to detach it from the glass.
7   Remove the glass by tilting it forward, then lifting it out of the door.
8   Installation is the reverse of removal.

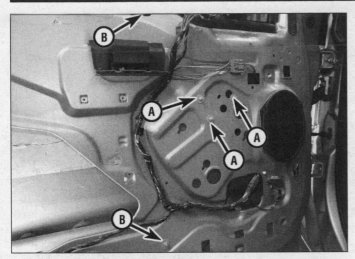

22.4 Remove the bolts securing the window regulator to the door frame

A   Window regulator bolts   B   Track guide bolts

23.1 Carefully pry off the sail trim panel

23.3 Remove the mirror mounting fasteners

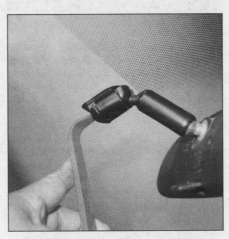

24.2a Slide the tool in far enough to engage the retaining clip. If your trim tool doesn't fit into the groove, file the tool edges narrower until it will fit

24.2b As you pry the retaining clip upward, apply pressure to the mirror to slide it off of the base

22 Door window glass regulator - removal and installation

Warning: *The regulator arms are under extreme pressure and can cause serious injury if the motor is removed without locking the sector gear. This can be done by inserting a bolt and nut through the holes in the backing plate and sector gear to lock them together.*

1   Remove the door trim panel and the plastic watershield (see Section 18).
2   Detach the window glass from the regulator (see Section 21). Push it up all the way and tape it to the top of the door frame or remove the glass completely.
3   On power-operated windows, disconnect the electrical connector from the window regulator motor.
4   Remove the regulator retaining bolts and track guide bolts (see illustration).
5   Fold the track guide to the regulator

assembly making it small enough to fit through the service hole in the door frame to remove it.
6   Installation is the reverse of removal.

23 Sideview mirrors - removal and installation

1   Remove the mirror trim cover (sail trim panel) above the front of the door panel (see illustration).
2   If you're working on a model with electric mirrors, remove the door trim panel and the watershield (see Section 18). Follow the harness and disconnect the electrical connector from the mirror.
3   Remove the dust shield and rubber plugs for access, then remove the mirror retaining nuts and detach the mirror from the vehicle (see illustration).
4   Installation is the reverse of removal.

24 Interior mirror - removal and installation

Caution: *The windshield should be at room temperature before proceeding with the repair.*
1   A special removal tool is recommended by the manufacturer: Ford tool number 501-190, but it can be done with a narrow flat bladed trim tool.

Interior mirror removal without the recommended tool

2   Slide the narrow bladed trim tool up into the slot at the bottom of the mirror mounting base (see illustrations), then apply light pressure to disengage the retaining clip and slide the mirror off of the base.

25.3 Pry up the clip and detach the tailgate support cable from the striker

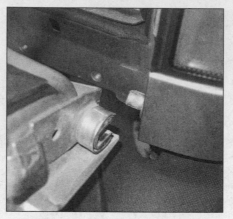

25.4 The right side of the tailgate has a slotted hinge opening for removal

25.9 Loosen the striker and adjust as necessary to provide positive latch engagement. Arrows indicate the directions that you can adjust the striker

26.1a If equipped with a liner, remove the liner to gain access to the tailgate cover

26.1b Remove the cover screws to access the inside of the tailgate

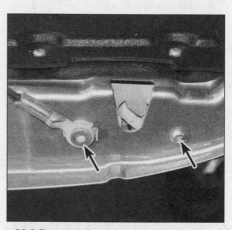

26.2 Remove the tailgate latch retaining screws (arrows)

Interior mirror removal with the recommended tool

3   Tilt the mirror as far up as possible.

4   Disconnect any electrical connections to the mirror (if applicable).

5   Install the removal tool into the bottom section of the mirror base and lightly clamp the tool in place. (This releases the internal spring clamp.)

6   With one hand holding onto the tool, use your other hand to gently bump the bottom of the tool upward causing the mirror to disengage from the spring clamp.

7   Remove the tool and slide the mirror off of the windshield bracket.

8   Installation only requires sliding the mirror back on to the bracket and listening for the "click" as the spring clamp locks the mirror into place.

9   Reconnect the electrical connection.

Note: *Vehicles with the compass feature may require re-calibrating after the mirror is disconnected. The procedure is listed in the owner's manual.*

25   Tailgate - removal, installation and adjustment

Note: *The tailgate is heavy and somewhat awkward to remove and install - at least two people should perform this procedure.*

Removal and installation

1   Open the tailgate.

2   Cover the lower bumper area around the opening with pads or cloths to protect the painted surfaces when the tailgate is removed.

3   While an assistant supports the tailgate, detach the tailgate support cables (see illustration).

4   With the help of your assistant, raise the tailgate about 41 degrees from vertical, then pull the right side off its hinge (see illustration).

5   While holding the tailgate right side hinge pin close to the hinge enclosure, raise the tailgate to about 10 degrees from vertical and pull the tailgate to the right hand side, releasing the left hinge pin from the hinge.

6   Now remove the tailgate from the vehicle.

7   Installation is the reverse of removal.

Adjustment

8   To adjust the tailgate closed position, first check that the latch is contacting the center of the latch striker. If not, remove the striker and add or subtract shims to achieve correct alignment.

9   Finally, adjust the latch striker assembly as necessary (up and down or sideways) to provide positive engagement with the latch mechanism and the outside of the tailgate is flush with the rear of the bed (see illustration).

26   Tailgate latch, handle and lock cylinder - removal and installation

1   Lower the tailgate and remove the tailgate access cover (see illustrations).

Latch

2   Remove the latch mounting screws (see illustration). It may be necessary to use an impact-driver to loosen them.

3   Disconnect the control rods from the

26.3 Disconnect the control rods by rotating the plastic clip off of the rod, then lift the rod from the tailgate latch handle

26.5 Disengage the handle-to-latch rods (A) and remove the handle retaining nuts (B)

27.4 Follow the wire harness to the main connector, then disconnect it.

27.7a You'll find some of the bolts are easy to reach while others may be covered by part of a trailer hitch bracket. If so, remove the trailer hitch bracket to gain access to the bolts

27.7b Follow along the frame to find the rest of the bolts (not all the bolts are shown)

27.7c Bolts are along both sides of the frame and are straight across from each other

latch mechanism and remove the latch by pulling on the control rods to remove it from the door (see illustration).

4   Installation is the reverse of removal.

Handle and lock cylinder

5   Disconnect the control rods (see illustration).

6   Detach the retaining nuts and remove the handle and lock cylinder assembly from the tailgate.

7   Installation is the reverse of removal.

27   Truck bed - removal and installation

1   Place the truck on a flat surface with the parking brake applied.

2   Disconnect the cable from the negative battery terminal (see Chapter 5).

3   Remove the filler caps and screws secur-

ing the fuel tank filler neck to the truck bed. On 8-foot beds there is an additional filler neck bracket for both the fuel and reductant tank filler neck that will need to be disconnected from underneath. On diesel models, disconnect the reductant tank filler neck as well.

4   Disconnect the rear section electrical connector running from the frame to the rear lighting (see illustration). On 5th-wheel equipped vehicles, disconnect the electrical connector also.

5   Disconnect any non-factory wiring such as an additional trailer wiring, tag light wiring, or accessory leads.

6   Disconnect the braided ground strap located near the fuel tank filler neck from the truck bed (if applicable).

7   Remove the bolts that secure the truck bed to the frame. Six bolts, three per side (see illustrations).

Note: *On these trucks the bed bolts are different lengths. Be sure to relocate the proper length bolts back into their respective locations.*

Warning: *The manufacturer recommends replacing the bed bolts and the attachment clips. They are a one-time use bolt. Torque the replacement bolts to 59 ft. lbs. / 80 Nm.*

Note: *The truck bed bolts can be fairly stubborn to remove. Dirt and debris as well as rust can make them even harder to remove. An impact gun with an impact socket works the best rather than trying to remove them by hand.*

Note: *If the bolts are stuck use some rust penetrating spray on them. Leave the penetrating spray soak in for a few minutes before trying them again. If a penetrating spray isn't available a spray of plain water actually will loosen them up as well.*

8   With the aid of a few helpers (one on each corner) lift the bed off of the frame.

Note: *If you have access to a car hoist you can strap the truck bed to the hoist arms and use the hoist to lift the truck bed up. Then, push the truck out from under it.*

9   Installation is the reverse of removal.

28.12 Disconnect the electrical connectors

28.19 Disconnect the ground strap

28.31a The cab bolts all have a large washer and rubber grommet

28 Cab - removal, installation and mount replacement

Note: *Removal of the cab is recommended for certain engine repair procedures.*
Warning: *A two-post lift or suitable substitute is required for this procedure. Do not attempt this procedure without taking the proper safety precautions.*
Caution: *Find the appropriate lifting spots as far apart as possible on the cab to support it when it's disconnected.*
Warning: *Make sure the weight is carried equally on all four lift arms.*
Note: *For more details on engine component removal see Chapters 2A, 2B .*
Warning: *The air conditioning system is under high pressure. Do not loosen any hose fittings or remove any components until after the system has been discharged. Air conditioning refrigerant should be properly discharged into an EPA-approved recovery/recycling unit at a dealer service department or an automotive air conditioning repair facility. Always wear eye protection when disconnecting air conditioning system fittings.*

1   Have the air conditioning system discharged by a licensed automotive air conditioning technician.
2   Disconnect the negative and positive battery terminals and remove the battery(s) from the vehicle (see Chapter 5).
3   Disconnect the positive battery charging post nut and remove the cable that leads to the starter motor. Tie the starter cable out of the way so it won't get caught on anything as you lift the cab (see Chapter 5).
4   On 4WD drive models with manual transfer case, remove the dust boot and disconnect the transfer case shift arm (see 7B).
5   Drain the coolant, A/C system and power steering of all their fluids. Cap off open lines to prevent debris from entering any of the hoses.

6   Remove the driver's side battery tray (see Chapter 5).
7   Disconnect the upper radiator hose from the radiator and remove the hose from the degas bottle. Remove the degas bottle from the vehicle (see Chapter 3).
8   Disconnect the CAC (Charge Air Cooler) inlet and outlet tubes, if applicable, and remove them from the vehicle (see Chapter 4B).
9   Disconnect the CACT (Charge Air Coolant Temp) sensor and retaining clip (see Chapter 6B).
10   Disconnect the CVH (Constant Vacuum Hublock) solenoid and regulator hose, if applicable (see Chapter 8).
11   Disconnect the vacuum hose to the brake booster, if applicable (see Chapter 9).
12   Disconnect the three large electrical connections near the brake master cylinder and battery junction box (see illustration).
13   Disconnect the lines from the power steering reservoir (see Chapter 10).
14   Disconnect the two power steering lines from the power steering pump to the Hydroboost unit at the brake booster, if applicable.
15   Disconnect the brake level sensor electrical connector and retaining clip from the master cylinder reservoir.
16   Disconnect the brake master cylinder from the brake booster (see Chapter 9). (The master cylinder will stay with the frame section.)
17   Disconnect the "T" fittings for the fuel cooler lines (see Chapter 4B).
18   Disconnect the lower steering shaft from the upper shaft (see Chapter 10). Do not allow the steering wheel to turn once the bolts have been removed.
19   Remove the ground strap from the center of the cowling area to the engine and the lower ground strap below the center of the radiator support (see illustration).
20   Disconnect the MAF (Mass Air Flow) sensor and remove the air filter box assembly

(see Chapter 4B).
21   Disconnect the PCM (Powertrain Control Module) and the five large electrical connectors in the same area (see Chapter 6A or Chapter 6B).
22   Disconnect the A/C lines from the compressor and disconnect the line from the compressor to the evaporator core (see Chapter 3).
23   Disconnect the heater hoses from the firewall to the engine (see Chapter 3).
24   Disconnect the EGR cooler lines and EGR vent hose (see Chapter 6B).
25   Disconnect the upper and lower radiator hoses from the water pump (see Chapter 3).
26   Disconnect the transmission and power steering cooler lines from the radiator (see Chapter 3).
27   Disconnect the lower radiator hose support bracket.
28   Disconnect the emergency brake cable at the cable union and unclip the guide tube from the frame support bracket. Then thread the cable out of the frame.
29   Disconnect the shift linkage arm from the transmission and unclip the linkage rod from the bracket (see Chapter 7A). Then secure the cable to the cab so that it won't get caught on anything as you raise the cab.
30   Disconnect the running board electrical connections, if applicable.
31   Remove the eight bolts securing the cab to the frame (four per side) (see illustrations).
32   Position the two-post lift arms as far apart as possible to aid in the balancing of the cab.
33   Before lifting, check things like battery cables, electrical connectors, or hoses that you may have missed. Disconnect those now, before lifting the cab.
34   Using the appropriate hold down devices or equivalent to lock the cab to the lift, secure the cab to the lift arms.
35   Slowly raise the cab an inch or two. Stop and check for any connections you may have

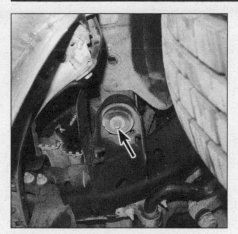

28.31b Cab bolts run along the frame on both sides

28.31c Wear eye protection when removing the cab bolts because a lot of debris will be between the washers and rubber grommets

29.1 Remove the lower console bolts (seats removed for clarity)

29.2 Electrical connector on passenger's side needs to be disconnected

29.3a Pry the center trim panel up to remove it

29.3b Remove the cup holders

missed. Also, bounce the cab up and down and check that the balance is good and that it is not binding on anything.

36   Once you are sure it is balanced and nothing else is connected, continue to raise the cab off of the frame.

37   Mark the location of the tires to the floor with paint or chalk in order to roll the cab to the exact same spot for installation.

38   Now roll the chassis out from under the cab.

Caution: *It's advised that you lower the cab to a safe level and put up a barrier to keep anyone from disturbing the cab.*

39   Installation is the reverse of removal. Tighten the cab bolts to 75 ft. lbs.

Caution: *Take your time, work safely, and have plenty of assistance on hand. Safety first.*

Warning: *The manufacturer recommends replacing the master cylinder retaining nuts and the steering shaft bolt with new bolts. Replace all "O" rings with new "O" rings.*

29   Center console - removal and installation

Warning: *These models are equipped with airbags. Always disable the airbag system before working in the vicinity of any airbag system component to avoid the possibility of accidental deployment of the airbag(s), which could cause personal injury (see Chapter 12).*

Warning: *Do not use a memory saving device to preserve the ECM's memory when working on or near airbag system components.*

Floor console

1   Remove the four lower console bolts (one on each outside corner on the seat sides of the console) (see illustration).

2   Disconnect the electrical connector on the outside of the console on the passenger's side, if applicable (see illustration).

3   Open the console lid and remove the storage bins and cup holders (see illustrations) for access to the retaining screws.

29.3c Upper retaining bolts for the console need to be removed

30.1 Pry the panel off far enough to get your finger tips behind and then pull it completely off

30.16 Use a trim tool to help lift the edge of the sill plate if needed

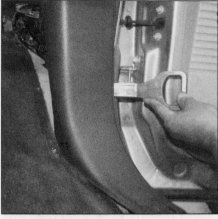

30.22 Pry the kick panel free with a flat trim tool

4   Lift the console upward to gain access to any of the electrical connectors that are still attached to the console.

5   Disconnect any electrical connections and remove the console from the vehicle.

6   Installation is the reverse of removal.

Overhead console

7   Pull the rear edge of the console down by prying on the outer edges, then pull rearward.

8   Drop the console down to disconnect the electrical connections.

9   Remove the overhead console.

10   Installation is the reverse of removal.

Rear storage console

11   Raise the rear seat bottom to its upright position.

12   Remove the partition from the inside of the storage console.

13   Remove the six bolts securing the console to the cab.

14   Pull the console out far enough to disconnect the electrical connector to the power outlet.

15   Remove the console.

16   Installation is the reverse of removal.

30   Interior trim - removal and installation

Knee bolster trim panel

1   Grasp the outer edges of the lower section of the trim panel and pull outward (see illustration).

2   Angle the trim panel out far enough to get your fingers close to the upper corners and pull outward on those corners.

3   Remove the trim panel.

4   Installation is the reverse of removal.

A pillar trim

5   Remove the trim cover screw caps that

are concealing the screws.

6   Remove the screws from the A pillar trim. Passenger side remove the screws from the grab handle (if applicable).

7   Pull the A pillar trim free from the body.

8   Installation is the reverse of removal.

B pillar trim

9   Remove the door sill plate trim.

10   Lift the trim cap upward to expose the seat belt retaining bolt. Remove the seat belt retaining bolt and the lower retaining bolt mounted on the seat.

11   Remove the lower B pillar first. Grasp the edges of the trim panel and pull outward.

12   Guide the seat belt out of the slot in the lower B pillar.

13   Now grasp the upper B pillar by the edges and pull outward to remove it.

14   Installation is the reverse of removal.

C pillar

15   The C pillar removal and install procedures are similar to the B pillar procedures.

Door sill plates

16   Grasp the outer edges of the sill plate and pull straight up (see illustration).

17   Installation is the reverse of removal.

Sun visors

18   Remove the two screws securing the sun visor to the roof line.

19   Pull down the sun visor to gain access to the electrical connection and disconnect it.

20   Remove the single screw securing the sun visor hook. Then, pull down to release the sun visor hook.

21   Installation is the reverse of removal.

Kick panel trim

22   Pull the door seal away from the kick panel area. Then, use a flat trim removal tool to pry the kick panel trim off of the retaining clips (see illustration).

23   After removing, check to be sure the clip itself came off with the kick panel. Otherwise, use the trim tool to remove the clip and reinstall it onto the kick panel.

24   Installation is the reverse of removal.

31   Headliner - removal and installation

1   Remove the overhead console, if applicable (see Section 29).

2   Remove the left and right sun visors and sun visor hooks (see Section 30).

3   Remove the A, B, and C pillars, if applicable (see Section 30).

4   Pull down all the door weatherstripping far enough to expose the edge of the headliner.

5   Remove the hanger hooks, if applicable.

6   Disconnect the headliner electrical connection which is located behind the passenger's side B pillar. For Crew cab and Super cab models the electrical connection is behind the passenger side C pillar.

7   Remove the overhead grab handles, if applicable.

8   Recline the front seat backs to their lowest position.

9   With the aid of an assistant carefully lower the headliner and work it out the passenger front door. Be careful not to crease or damage the headliner.

10   Installation is the reverse of removal.

32   Sunroof - removal and installation

Caution: *When any repairs are carried out on the sun roof system the sun roof motor will need to be initialized. Follow the procedures listed in the Sunroof motor initialization in this section.*

33.3 Trim fastener locations

33.6 Press in on the stops on either side of the glove box

Sunroof motor initialization procedure

1   For a replacement sunroof motor installation, start on step 2. For reinstalling the original motor, start on step 3.

2   Move the sunroof to its fully open position and hold the switch in the On position for an additional 10 seconds. To confirm, the sunroof will jostle rearward and then back to the full open position when the initialization has been accepted. Quickly release the switch then press and hold the switch on. The sunroof should move to the Vent position and then to the completely Closed position. These initialization steps will start within 3 seconds of the switch being depressed. When the initialization steps have been completed go to step 4.

3   Press and hold the switch until the sunroof is fully opened. Now release the switch. Press and hold the switch for an additional 10 seconds. The sunroof should jostle rearward and then back to the full open position indicating the initialization process has been accepted. Quickly release the switch then press and hold the switch on. The sunroof should move to the Vent position and then to the completely Closed position. These initialization steps will start within 3 seconds of the switch being depressed.

4   Release the switch and check that all the functions are working normally.

Sunroof glass panel removal

Note: *Sunroof needs to be in the closed position in order to properly remove the glass panel.*

5   Open sun shade and locate the four bolts securing the glass panel to the sun roof frame.

6   Remove the four bolts and carefully lift the glass panel out of the opening.

7   Installation is the reverse of removal. Adjust the height of each corner by loosening the bolts and pressing up or down on the

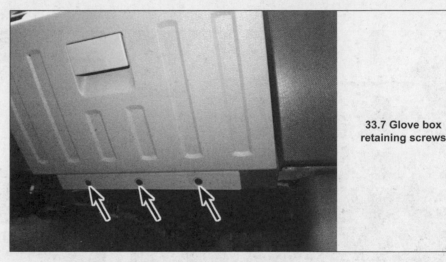

33.7 Glove box retaining screws

glass panel to align it with the roof opening.

Sunroof motor replacement

8   Remove the C pillar trim panel (see Section 30).

9   Remove the clothes hanger hooks (if applicable).

10   Lower the back portion of the headliner.

11   Disconnect the sunroof motor electrical connection and remove the three bolts securing the motor to the assembly.

12   Installation is the reverse of removal. Perform the sunroof motor initialization procedure.

33   Dashboard trim panels - removal and installation

Warning: *These models are equipped with airbags. Always disable the airbag system before working in the vicinity of any airbag system component to avoid the possibility of accidental deployment of the airbag(s), which could cause personal injury (see Chapter 12).*

Warning: *Do not use a memory saving device to preserve the ECM's memory when working on or near airbag system components.*

Instrument cluster trim panel

1   Remove the driver's side knee bolster trim panel (see Section 30).

2   Remove the two access trim caps concealing the lower center trim screws. Disconnect the electrical connectors if applicable.

3   Remove the retaining screws (see illustration).

4   With the screws removed, pull rearward to disengage the upper retaining clips.

5   Installation is the reverse of removal.

Glove box

6   Open the glove box door. Press inward on the door stops to release the upper half of the glove box (see illustration).

7   Remove the door hinge screws (see illustration) and detach the glove box.

8   Installation is the reverse of removal.

33.9a Flat trim tools are a must have tool for removing interior trim panels

33.9b Both sides come off the same way

33.13 Remove the fasteners

33.14 Pry around the edges to release the clips

Dashboard end trim panel

9   A flat trim tool works best to pry the end trim panels free (see illustrations).

10   Carefully apply pressure and work your way around the trim panel to free all the clips.

11   Installation is the reverse of removal.

Center trim panel

12   Remove the headlight switch and the auxiliry power outlet (see Chapter 12).

13   Remove the fasteners from the bottom of the center trim panel (see illustration).

14   Carefully pry around the edges of the center trim panel to release the clips (see illustration).

15   Remove the trim panel.

16   Installation is the reverse of removal.

34   Instrument panel - removal and installation

Warning: *Models covered by this manual are equipped with a Supplemental Restraint System (SRS), more commonly known as airbags. Always disable the airbag system before working in the vicinity of any airbag system component to avoid the possibility of accidental deployment of the airbag, which could cause personal injury (see Chapter 12).*

Warning: *The air conditioning system is under high pressure. Do not loosen any hose fittings or remove any components until after the system has been discharged. Air conditioning refrigerant should be properly discharged into an EPA-approved recovery/recycling unit at a dealer service department or an automotive air conditioning repair facility. Always wear eye*

protection when disconnecting air conditioning system fittings.

Note: *This is a difficult procedure for the home mechanic. There are many hidden fasteners, difficult angles to work in and many electrical connectors to tag and disconnect/connect. We recommend that this procedure be done only by an experienced do-it-yourselfer.*

Note: *During removal of the instrument panel, make careful notes of how each piece comes off, where it fits in relation to other pieces and what holds it in place. If you note how each part is installed before removing it, getting the instrument panel back together again will be much easier.*

Note: *It is not necessary, but it is suggested to remove both front seats to allow additional working space and lessen the chance of damage to the seats during this procedure.*

1   Have the air conditioning system discharged by a licensed automotive air condi-

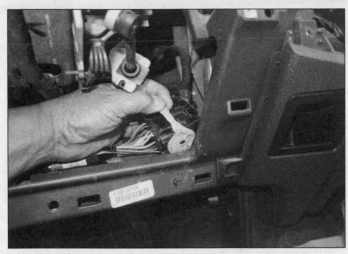

34.10 Disconnect the electrical connectors

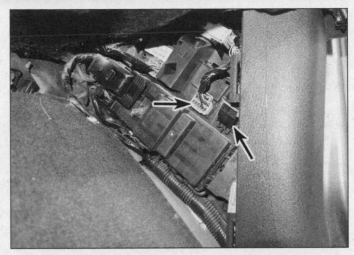

34.15 Disconnect the electrical connectors

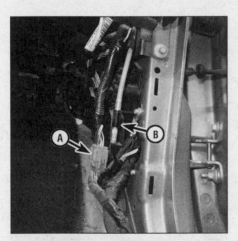

34.16 Disconnect the bulkhead electrical connector (A) and the radio antenna cable (B)

34.17 Disconnect the electrical connectors

34.18a Remove the fastener securing the cover . . .

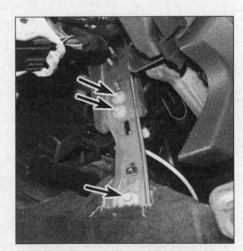

34.18b . . . then remove the fasteners securing the center braces (left side shown, right side similar)

tioning technician.

2   Disconnect the negative battery terminal (see Chapter 5).

3   Drain the engine coolant (see Chapter 3).

4   Remove the center console (see Section 29).

5   Have the front wheels straight forward and lock the steering wheel into that position.

6   Remove the driver's side knee bolster trim panel (see Section 30).

7   Disconnect the steering column upper shaft bolt (see Chapter 10). (Discard bolt. Manufacturer requires the bolt to be replaced whenever it is removed.)

8   Slide the upper steering column upper shaft off of the steering column.

9   Remove the gear selector cable from the steering column and from the bracket (see Chapter 10).

10   Disconnect the electrical connectors in the knee bolster area (see illustration). If equipped with the satellite radio system dis-

connect the satellite antenna cable located in the same area.

11   Remove the emergency brake lever and cable (see Chapter 9).

12   Remove the passenger side knee bolster trim panel (see Section 33).

13   Remove the glove box (see Section 33).

14   Remove the right side kick panel (see Section 33).

15   Disconnect the two electrical connectors on the top of the smart junction box that feed to the dash area (see illustration).

16   Disconnect the single main bulkhead electrical connector near the smart junction box and the radio antenna cable coupling. Disconnect the antenna cable coupling as well (see illustration).

17   From inside the glove box opening disconnect the two electrical connections attached to the right side (see illustration).

18   Remove the center brace covers, then remove the braces from each side of the center of the instrument panel (see illustrations).

34.19a Carefully pry off the trim cover(s) from the top of the instrument panel . . .

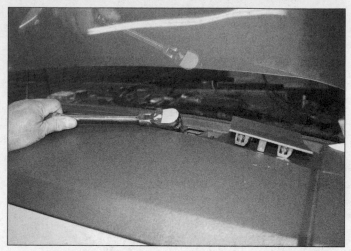

34.19b . . . then remove the fasteners from the opening(s)

34.20 Carefully pry off the door trim panel

34.21a Remove this plug . . .

34.21b . . . then remove the side mounting bolt

34.23a Remove the mounting fasteners from the left side . . .

34.23b . . . and the right side of the panel

19   Remove the three upper mounting bolts next to the windshield (see illustrations).

20   Remove the driver's side door trim panel in the door jam area by prying the four plastic retaining pins out (see illustration).

21   Remove the plastic plug, then remove the side bolt from the instrument panel (see illustrations).

22   Remove the right side dash end trim panels by prying them out starting at the edge of the trim panel.

23   Remove the dash to frame bolts on the driver's side and from the passenger's side (see illustrations).

24   Pull the instrument panel away from the firewall and detach any electrical connectors interfering with removal.

25   With the assistance of a helper, lift the dash assembly up off of the hooks and carefully remove it from the vehicle.

26   Installation is the reverse of removal.

27   If you are removing the dash to gain

36.3 Pull off the weatherstrip

36.4a Pry up on the edge to free the cowling trim from the cowling

36.4b Lift the cowling trim off

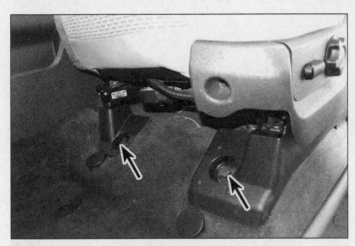

37.2a Front bolt locations

37.2b Rear bolt locations

access to the heater core or evaporator core, a mechanic's trick is to leave the driver's side dash bolts in place by a thread or two and then using the driver's side as a hinge, rotate the passenger's side away from the windshield area.

28   Installation is the reverse of removal.

35   Steering column cover - removal and installation

1   Remove the knee bolster (see Section 30).

2   Set the steering wheel in the full down tilt position and fully extended rearward for the telescopic steering column models.

3   For automatic models, depress the brake pedal and turn the key to On. Place shifter lever into Drive.

4   Push in on the rubber boot for the shift lever on the automatic models to gain a finger hold onto the edge of the upper steering column cover. All others, grab the rear section of the upper cover near the split line.

5   Pull the upper steering column cover up

and forward at the same time to disengage the clips.

6   Rotate the upper steering column cover upward to clear the hazard switch then rotate it off of the rear hooks.

7   For automatic models, return the shift lever to Park and remove the key.

8   Remove the three screws from the lower steering column cover and remove the cover.

9   Installation is the reverse of removal.

36   Cowl cover - removal and installation

1   Remove the windshield wiper arms (see Chapter 12).

2   Remove the antenna (see Chapter 12).

3   Remove the weatherstrip (see illustration).

4   Tip the front edge of the cowling sections up and pull the cowling sections forward to remove (see illustrations).

5   Installation is the reverse of removal.

37   Seats - removal and installation

Warning: *Some models are equipped with seat belt pre-tensioners, which are pyrotechnic (explosive) devices that tighten the seat belts during an impact of sufficient force. Always disable the airbag system before working in the vicinity of any restraint system component to avoid the possibility of accidental deployment of the airbag(s) and seat belt pre-tensioners, which could cause personal injury (see Chapter 12).*
Warning: *Do not use a memory saving device to preserve the ECM's memory when working on or near restraint system components.*

Front seat

1   Position the seat all the way forward and all the way to the rear to access the front seat retaining bolts.

2   Detach any bolt trim covers and remove the retaining bolts (see illustrations).

3   Tilt the seat upward to access the underneath, then disconnect any electrical connectors and lift the seat from the vehicle.

4   Installation is the reverse of removal.

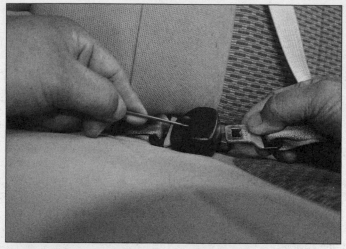

37.5 Insert a suitable tool into the latch to release the belt

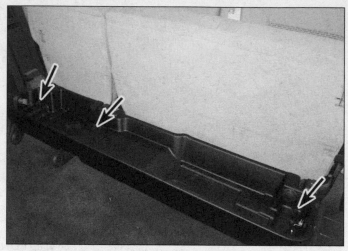

37.6 Remove the fasteners securing the storage compartment

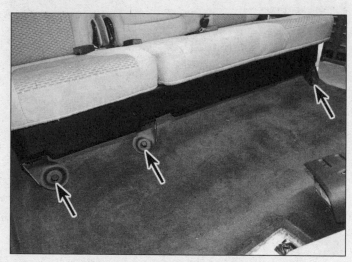

37.7a Remove the fasteners at the front . . .

37.7b . . . and under the seat

Rear seat

5   Using a suitable tool, insert it into the latch to release the belt (see illustration).

6   Remove the storage compartment (see illustration).

7   Remove the seat-to-floor mounting bolts, then detach the seat and lift it out of the vehicle (see illustrations).

8   Installation is the reverse of removal.

Chapter 12
Chassis electrical system

Contents

Specifications

Torque specifications

Ft-lbs (unless otherwise indicated)

Note: *One foot-pound (ft-lb) of torque is equivalent to 12 inch-pounds (in-lbs) of torque. Torque values below approximately 15 foot-pounds are expressed in inch-pounds, because most foot-pound torque wrenches are not accurate at these smaller values.*

Passenger airbag mounting nuts* .... 80 in-lbs

* Manufacturer recommends replacing the original fasteners with new fasteners.

1   General information

1   The electrical system is a 12-volt, negative ground type. Power for the lights and all electrical accessories is supplied by a lead/acid-type battery, which is charged by the alternator.

2   This chapter covers repair and service procedures for the various electrical components not associated with the engine. Information on the battery, alternator and starter motor can be found in Chapter 5.

3   It should be noted that when portions of the electrical system are serviced, the negative battery cable should be disconnected from the battery to prevent electrical shorts and/or fires.

2   Electrical troubleshooting - general information

1   A typical electrical circuit consists of an electrical component, any switches, relays, motors, fuses, fusible links or circuit breakers related to that component and the wiring and connectors that link the component to both the battery and the chassis. To help you pinpoint an electrical circuit problem, wiring diagrams are included at the end of this chapter.

2   Before tackling any troublesome electrical circuit, first study the appropriate wiring diagrams to get a complete understanding of what makes up that individual circuit. Noting whether other components related to the circuit are operating correctly, for instance, can often narrow down the location of potential trouble spots. If several components or circuits fail at one time, chances are the problem is in a fuse or ground connection, because several circuits are often routed through the same fuse and ground connections.

3   Electrical problems usually stem from simple causes, such as loose or corroded connections, a blown fuse, a melted fusible link or a failed relay. Visually inspect the condition of all fuses, wires and connections in a problem circuit before troubleshooting the circuit.

4   If test equipment and instruments are going to be utilized, use the diagrams to plan ahead of time where you will make the necessary connections in order to accurately pinpoint the trouble spot.

5   For electrical troubleshooting, you'll need a circuit tester, voltmeter or a 12-volt bulb with

a set of test leads, a continuity tester and a jumper wire, preferably with a circuit breaker incorporated, which can be used to bypass electrical components (see illustrations). Before attempting to locate a problem with test instruments, use the wiring diagram(s) to decide where to make the connections.

Voltage checks

6   Voltage checks should be performed if a circuit is not functioning properly. Connect one lead of a circuit tester to either the negative battery terminal or a known good ground. Connect the other lead to a connector in the circuit being tested, preferably nearest to the battery or fuse (see illustration). If the tester bulb lights, voltage is present, which means that the part of the circuit between the connector and the battery is problem free. Continue checking the rest of the circuit in the same fashion. When you reach a point at which no voltage is present, the problem lies between that point and the last test point with voltage. The problem can usually be traced to a loose connection.
Note: *Keep in mind that some circuits receive voltage only when the ignition key is in the ACC or RUN position.*

Finding a short

7   One method of finding shorts in a live circuit is to remove the fuse and connect a test light in place of the fuse terminals (fabricate two jumper wires with small spade terminals, plug the jumper wires into the fuse box and connect the test light). There should be no voltage present in the circuit. Move the suspected wiring harness from side-to-side while watching the test light. If the bulb lights, there is a short to ground somewhere in that area, probably where the insulation has rubbed through.

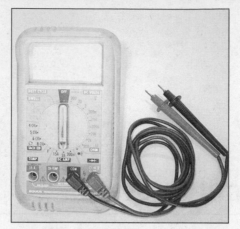

2.5a The most useful tool for electrical troubleshooting is a digital multimeter that can check volts, amps, and test continuity

Ground check

8   Perform a ground test to check whether a component is properly grounded. Disconnect the battery and connect one lead of a continuity tester or multimeter (set to the ohm scale), to a known good ground. Connect the other lead to the wire or ground connection being tested. If the resistance is low (less than 5 ohms), the ground is good. If the bulb on a self-powered test light does not light, the ground is not good.

Continuity check

9   A continuity check determines whether there are any breaks in a circuit (whether it can no longer carry current from the voltage source to ground). With the circuit off (no power in the circuit), a self-powered continuity tester or multimeter can be used to check the circuit. Connect the test leads to both ends of the circuit (or to the power end and

2.5b A test light is a very handy tool for checking voltage

a good ground), and if the test light comes on the circuit is passing current properly (see illustration). If the resistance is low (less than 5 ohms), there is continuity; if the reading is 10,000 ohms or higher, there is a break somewhere in the circuit. The same procedure can be used to test a switch, by connecting the continuity tester to the switch terminals. With the switch turned to ON, the test light should come on (or low resistance should be indicated on a meter).

Finding an open circuit

10   When diagnosing for possible open circuits, it is often difficult to locate them by sight because the connectors hide oxidation or terminal misalignment. Merely wiggling a connector on a sensor or in the wiring harness may correct the open circuit condition. Remember this when an open circuit is indicated when troubleshooting a circuit. Intermit-

2.6 In use, a basic test light's lead is clipped to a known good ground, then the pointed probe can test connectors, wires or electrical sockets - if the bulb lights, the part being tested has battery voltage

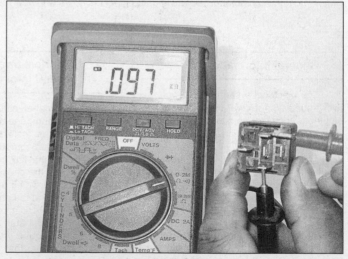

2.9 With a multimeter set to the ohms scale, resistance can be checked across two terminals - when checking for continuity, a low reading indicates continuity, a high reading indicates lack of continuity

tent problems may also be caused by oxidized or loose connections.

11 Electrical troubleshooting is simple if you keep in mind that all electrical circuits are basically electricity running from the battery, through the wires, switches, relays, fuses and fusible links to each electrical component (light bulb, motor, etc.) and to ground, from which it is passed back to the battery. Any electrical problem is an interruption in the flow of electricity to and from the battery.

Connectors

12 Most electrical connections on these vehicles are made with multi-wire plastic connectors. The mating halves of many connectors are secured with locking clips molded into the plastic connector shells. The mating halves of large connectors, such as some of those under the instrument panel, are held together by a bolt through the center of the connector.

13 To separate a connector with locking clips, use a small screwdriver to pry the clips apart carefully, then separate the connector halves. Pull only on the shell; never pull on the wiring harness as you may damage the individual wires and terminals inside the connectors. Look at the connector closely before trying to separate the halves. Often the locking clips are engaged in a way that is not immediately clear. Additionally, many connectors have more than one set of clips.

14 Each pair of connector terminals has a male half and a female half. When you look at the end view of a connector in a diagram, be sure to understand whether the view shows the harness side or the component side of the connector. Connector halves are mirror images of each other, and a terminal that is shown on the right side end-view of one half will be on the left side end-view of the other half.

Electrical connectors

Most electrical connectors have a single release tab that you depress to release the connector

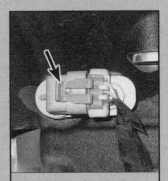

Some electrical connectors have a retaining tab which must be pried up to free the connector

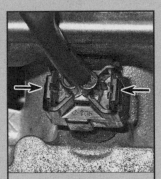

Some connectors have two release tabs that you must squeeze to release the connector

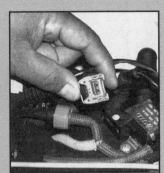

Some connectors use wire retainers that you squeeze to release the connector

Critical connectors often employ a sliding lock (1) that you must pull out before you can depress the release tab (2)

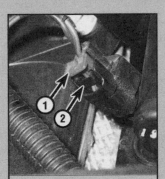

Here's another sliding-lock style connector, with the lock (1) and the release tab (2) on the side of the connector

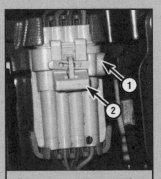

On some connectors the lock (1) must be pulled out to the side and removed before you can lift the release tab (2)

Some critical connectors, like the multi-pin connectors at the Powertrain Control Module employ pivoting locks that must be flipped open

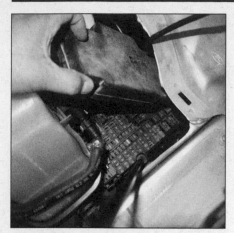

3.1a The fuse/relay box is mounted to the left inner fenderwell in the engine compartment - it contains miniaturized fuses, cartridge-type fusible links, relays and circuit breakers

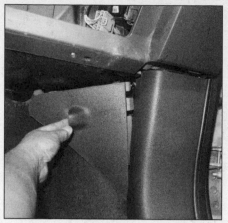

3.1b Remove the access panel from the right kick panel . . .

3.1c . . . then remove the fuse box cover to gain access to the fuses in the BCM

3.3 When a fuse blows, the element between the terminals melts - the fuse on the left is blown, the fuse on the right is good

3   Fuses and fusible links - general information

Fuses

1   The electrical circuits of the vehicle are protected by a combination of fuses, circuit breakers and fusible links. Fuse blocks are located under the instrument panel and in the engine compartment (see illustrations).
2   Each of the fuses is designed to protect a specific circuit, and the various circuits are identified on the fuse panel cover.
3   Miniaturized fuses are employed in the fuse blocks. These compact fuses, with blade terminal design, allow fingertip removal and replacement. If an electrical component fails, always check the fuse first. The best way to check a fuse is with a test light. Check for power at the exposed terminal tips of each fuse. If power is present on one side of the fuse but not the other, the fuse is blown. A

blown fuse can also be confirmed by visually inspecting it (see illustration).
4   Be sure to replace blown fuses with the correct type. Fuses of different ratings are physically interchangeable, but only fuses of the proper rating should be used. Replacing a fuse with one of a higher or lower value than specified is not recommended. Each electrical circuit needs a specific amount of protection. The amperage value of each fuse is molded into the fuse body.
5   If the replacement fuse immediately fails, don't replace it again until the cause of the problem is isolated and corrected. In most cases, this will be a short circuit in the wiring caused by a broken or deteriorated wire.

Fusible links

6   Some circuits are protected by fusible links. The links are used in circuits which are not ordinarily fused, or which carry high current.
7   Cartridge type fusible links are located in the engine compartment fusible link box and are similar to a large fuse. After disconnecting the negative battery cable(s), simply unplug and replace a fusible link with one of the same amperage.

4   Circuit breakers - general information

1   Circuit breakers protect certain circuits, such as the power windows or heated seats. Depending on the vehicle's accessories, there may be one or two circuit breakers, located in the fuse/relay box in the engine compartment (see illustration 3.1a).
2   Because the circuit breakers reset automatically, an electrical overload in a circuit-breaker-protected system will cause the circuit to fail momentarily, then come back on. If the circuit does not come back on, check it immediately.
3   For a basic check, pull the circuit breaker up out of its socket on the fuse panel, but just far enough to probe with a voltmeter. The

breaker should still contact the sockets.
4   With the voltmeter negative lead on a good chassis ground, touch each end prong of the circuit breaker with the positive meter probe. There should be battery voltage at each end. If there is battery voltage only at one end, the circuit breaker must be replaced.
5   Some circuit breakers must be reset manually.

5   Relays - general information and testing

General information

1   Several electrical accessories in the vehicle, such as the fuel injection system, horns, starter, and fog lamps use relays to transmit the electrical signal to the component. Relays use a low-current circuit (the control circuit) to open and close a high-current circuit (the power circuit). If the relay is defective, that component will not operate properly. Most relays are mounted in the engine compartment fuse/relay box, with some specialized relays located above the interior fuse box in the dash (see illustrations 3.1a and 3.1c). If a faulty relay is suspected, it can be removed and tested using the procedure below or by a dealer service department or a repair shop. Defective relays must be replaced as a unit.

Testing

2   Most of the relays used in these vehicles are of a type often called "ISO" relays, which refers to the International Standards Organization. The terminals of ISO relays are numbered to indicate their usual circuit connections and functions. There are two basic layouts of terminals on the relays used in these vehicles (see illustrations).
3   Refer to the wiring diagram for the circuit to determine the proper connections for the relay you're testing. If you can't determine the correct connection from the wiring diagrams, however, you may be able to determine the test connections from the information that follows.

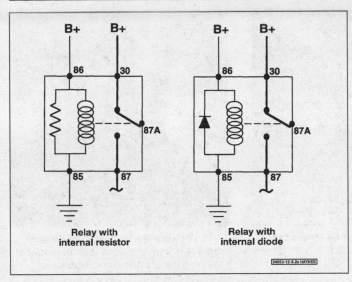

5.2a Typical ISO relay designs, terminal numbering and circuit connections

5.2b Most relays are marked on the outside to easily identify the control circuits and the power circuits - four terminal type shown

4   Two of the terminals are the relay control circuit and connect to the relay coil. The other relay terminals are the power circuit. When the relay is energized, the coil creates a magnetic field that closes the larger contacts of the power circuit to provide power to the circuit loads.

5   Terminals 85 and 86 are normally the control circuit. If the relay contains a diode, terminal 86 must be connected to battery positive (B+) voltage and terminal 85 to ground. If the relay contains a resistor, terminals 85 and 86 can be connected in either direction with respect to B+ and ground.

6   Terminal 30 is normally connected to the battery voltage (B+) source for the circuit loads. Terminal 87 is connected to the ground side of the circuit, either directly or through a load. If the relay has several alternate terminals for load or ground connections, they usually are numbered 87A, 87B, 87C, and so on.

7   Use an ohmmeter to check continuity through the relay control coil.

a) Connect the meter according to the polarity shown in the illustration for one check; then reverse the ohmmeter leads and check continuity in the other direction.

b) If the relay contains a resistor, resistance will be indicated on the meter, and should be the same value with the ohmmeter in either direction.

c) If the relay contains a diode, resistance should be higher with the ohmmeter in the forward polarity direction than with the meter leads reversed.

d) If the ohmmeter shows infinite resistance in both directions, replace the relay.

8   Remove the relay from the vehicle and use the ohmmeter to check for continuity between the relay power circuit terminals. There should be no continuity between terminal 30 and 87 with the relay de-energized.

9   Connect a fused jumper wire to terminal

86 and the positive battery terminal. Connect another jumper wire between terminal 85 and ground. When the connections are made, the relay should click.

10   With the jumper wires connected, check for continuity between the power circuit terminals. Now, there should be continuity between terminals 30 and 87.

11   If the relay fails any of the above tests, replace it.

---

6   Key fob - programming, battery replacement, key fob strength testing. Driver's door keypad - removal and installation

---

Key fob - programming

Note: *This procedure is for the RKE (Remote Key Entry) type key fobs and not for the IKT (Integrated Key Transmitter) type. The IKT is automatically programmed when the PATS (Passive Anti-Theft System) is programmed.*

Note: *IKT transmitters can only be programmed with a scan tool. Two working keys are required.*

Note: *When programming the RKE transmitter ALL remotes must be present and in working order.*

1   Programming can be done with a scanner or with the following procedure.

Caution: *DO NOT touch the brake pedal when programming the RKE or this will end the program procedure.*

2   Using the door lock switch on either door, cycle the locks to the Unlock position.

3   Rapidly cycle the ignition key from Off to On 8 times within 10 seconds ending with the 8th time in the On position. If successful, the door locks will Lock and Unlock once.

4   Within 20 seconds press any button on

the RKE transmitter that is being programmed to the vehicle.

5   The doors should Lock and then Unlock to confirm the programming was successful.

6   Repeat the same procedure of pressing any button on the next remote to be programmed up to a maximum of four remotes. Each time a successful programming of the next remote is accepted the door locks will Lock and Unlock.

7   To end the programming session turn the key to Off. The session will automatically end when a maximum of four RKE fobs are programmed or 20 seconds have passed without any RF input from the remote(s).

Key fob - battery replacement

8   Use a thin coin or small flat bladed screwdriver to separate the two halves of the key fob (see illustration).

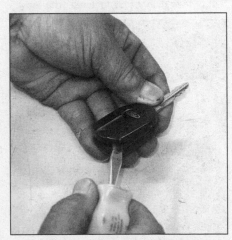

6.8 Gently pry the two halves of the key apart

6.9 Observe the battery polarity (+/- sign), then remove the battery

6.17 Pry the retainer clip forward to remove it

6.18 Grasp the keypad module and remove it from the door, then disconnect the electrical connector

9   Check that the polarity is correct, then remove the old battery (see illustration).
10   Use the correct battery (Battery number CR2032).
11   Replace the battery and check operation of the remote. If needed, reprogram the remote.

Key fob - strength testing

12   The key fob remote has an operating distance of 30 feet. Each time a button on the fob is pushed a signal is received by the TPMS (Tire Pressure Monitor System). This signal is then passed onto the BCM (Body Control Module) where it is then processed to unlock or lock the doors.
13   The fob battery can be checked with an volt meter after it has been removed from the key fob. It should have 2.5 volts.
14   If for some reason your key fob fails to operate even after you've tried a replacement battery, take your key fob to a repair shop that has a tire monitor reset tool. Most of these tools have an RF strength indicator test as part of their diagnostics features. The techni-

8.3 Multi-function switch mounting fasteners (steering wheel removed for clarity)

cian can check whether or not your key fob is in working order or not. If there is no RF signal, the key fob is more than likely faulty. If the signal is weak, chances are the battery needs to be replaced.

Driver's door key pad - removal and installation

15   Remove the driver's door panel (see Chapter 11).
16   Raise the window to the fully closed position.
17   Pull the retainer clip toward the front of the door to remove it (see illustration).
18   Slightly push the keypad module out of the door far enough to get your fingers behind it from the outside (see illustration).
19   Leave the connector to hang outside of the door so that it doesn't fall back inside.
20   Installation is the reverse of removal.

---

7   Turn signal and hazard flasher - check and replacement

Warning: *The models covered by this manual are equipped with a Supplemental Restraint System (SRS), more commonly known as airbags. Always disconnect the negative battery cable(s) and wait two minutes before working in the vicinity of any airbag system component to avoid the possibility of accidental deployment of the airbag, which could cause personal injury (see Section 26).*

1   The turn signal and hazard flashers is an integral part of the BCM (Body Control Module) and is not serviced separately. If the BCM has been determined as the fault, the unit will need to be replaced and programmed to the vehicle. See a qualified independent repair facility or dealer repair department for replacement and programming.
2   When the flasher unit is functioning properly, an audible click can be heard during its operation. If the turn signal indicator (on the

instrument panel) on one side of the vehicle flashes much more rapidly than normal, a faulty turn signal bulb is indicated.
3   If both turn signals fail to blink, the problem may be due to a blown fuse, a faulty flasher unit, a broken switch or a loose or open connection. If a quick check of the fuse box indicates that the turn signal fuse has blown, check the wiring for a short before installing a new fuse.

---

8   Steering column switches - replacement

Warning: *The models covered by this manual are equipped with a Supplemental Restraint System (SRS), more commonly known as airbags. Always disconnect the negative battery cable(s) and wait two minutes before working in the vicinity of any airbag system component to avoid the possibility of accidental deployment of the airbag, which could cause personal injury (see Section 26).*

Multi-function switch

1   Disconnect the battery cable(s) from the negative battery terminal(s).
2   Remove the steering column covers (see Chapter 11).
Note: *On models with adjustable pedals, remove the adjustable pedal switch.*
3   Turn the steering wheel from straight to about the 9 o'clock position (a quarter turn to the left). The torx screws for the multi-function switch can than be removed. Pull the multi-function switch off of the steering column (see illustration).
4   Installation is the reverse of removal.

Hazard switch

5   The hazard switch is directly mounted onto the steering column and is separate from the multi-function switch.
6   Remove the steering column covers (see Chapter 11).

8.7 Release the clips to remove the hazard swtich

8.10 Adjustable pedal switch

8.14 A - SCCM (Steering Column Control Module)

7   Release the hooked clips that secure the switch to the steering column. Pull the switch upward to remove it (see illustration).

8   Installation is the reverse of removal.

Adjustable pedal switch

9   Remove the steering column covers (see Chapter 11).

10   Carefully lift the hooked clips that secure the adjustable pedal switch to the steering column. Remove it by pulling it away from the steering column (see illustration).

11   Installation is the reverse of removal.

Cruise control switch

12   The cruise control switch is mounted on the steering wheel. To remove the switch use a flat bladed trim tool and pry the switch out of the steering wheel starting on the outside edge. Once the switch is free disconnect the electrical connectors but don't let the wires fall back into the wheel.

13   Installation is the reverse of removal.

Steering column control module - SCCM

14   The SCCM (Steering Column Control Module) collects the input signals from all of the various steering column switches and routes the data signals to the appropriate next module(s). The adjustable pedals also run through the SCCM. The SCCM is located under the steering column covers (see illustration).

15   To replace the SCCM requires downloading the information from the original SCCM and then reinstalling that information into the replacement SCCM. This requires the use of the proper scanner and diagnostic tools. See your local dealer or qualified independent repair facility for any repair or diagnostics regarding the SCCM operation.

Heated steering wheel module - removal and installation

Caution: *The HSWM (Heated Steering Wheel Module) is mounted behind the driver's knee bolster trim panel. Replacing the HSWM requires programming of the replacement module. This task cannot be carried out without the proper scanning and diagnostic tools. See your local dealer or qualified independent repair facility for proper configuration.*

16   Remove the driver's side knee bolster trim panel (see Chapter 11).

17   Disconnect the electrical connector from the HSWM and remove the fasteners.

18   Installation is the same as removal. Have the HSWM reconfigured by a qualified repair facility.

---

9   Ignition switch and key lock cylinder - replacement

Warning: *The models covered by this manual are equipped with a Supplemental Restraint System (SRS), more commonly known as airbags. Always disconnect the negative battery cable(s) and wait two minutes before working in the vicinity of any airbag system component to avoid the possibility of accidental deployment of the airbag, which could cause personal injury (see Section 26).*

Ignition switch

1   Disconnect the cable(s) from the negative battery terminal(s) (see Chapter 5, Section 3).

2   Remove the steering column covers and the tilt-column handle, if equipped (see Chapter 11).

3   Unplug the ignition switch electrical connector.

4   Press in on the two tabs securing the ignition switch to the steering column.

5   Pull the switch away from the column to remove it.

6   The remainder of the installation is the reverse of removal. Check for proper operation of the ignition switch in the lock, start and accessory positions.

Lock cylinder

7   Disconnect the cable(s) from the negative battery terminal(s) (see Chapter 5, Section 3).

8   Remove the steering wheel covers (see Chapter 11).

9   Disconnect the electrical connector to the PATS (Passive Anti-Theft System) transceiver. Use a small pick (such as a dental pick tool) and raise the two locking tabs far enough to allow the transceiver to be disconnected (see illustration). Then remove the PATS transceiver.

10   Turn the key to the ACC position.

9.9 Removing the PATS transceiver

9.11 To remove the ignition lock cylinder, place the key in the "ACC" position, push in on the release tab with a small Allen wrench or a small punch and pull the cylinder straight out

10.2 Carefully pry the headlight switch assembly from the instrument panel

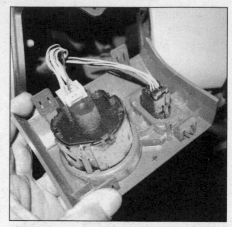

10.3 Turn the switch over and disconnect the electrical leads

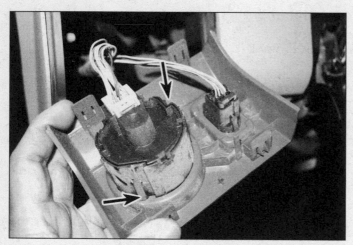

10.4 Press in on these tabs

10.8 Press in on these tabs to remove the selector switch

11   Insert a 1/8-inch punch or Allen wrench into the access hole. Depress the punch while pulling out on the lock cylinder to remove it from the steering column housing (see illustration).

12   To install the lock cylinder, make sure the switch is in the ACC position and align the locking tab while pushing the cylinder into place.

13   Rotate the key back to the Off position. This will allow the retaining pin to extend itself back into the locating hole in the steering column housing.

14   Turn the lock to ensure that operation is correct in all positions.

15   The remainder of installation is the reverse of removal.

---

10   Instrument panel switches - replacement

Warning: *The models covered by this manual*

are equipped with a Supplemental Restraint System (SRS), more commonly known as air-bags. Always disconnect the negative battery cable(s) and wait two minutes before working in the vicinity of any airbag system component to avoid the possibility of accidental deployment of the airbag, which could cause personal injury (see Section 26).

Headlight/fog light switch and panel dimmer switch

1   Disconnect the cable(s) from the negative battery terminal(s) (see Chapter 5, Section 3).

2   Carefully pry the headlight switch bezel plate from the instrument panel (see illustration).

3   Disconnect the electrical connector (see illustration) and remove the switch.

4   Press in on the tabs to remove the switch from the bezel trim plate (see illustration).

Note: *Dash dimmer switch is removed in the same manner.*

5   Installation is the reverse of the removal procedure.

4WD selector switch (MSS - Mode Selector Switch)

6   Disconnect the cable(s) from the negative battery terminal(s) (see Chapter 5).

7   Remove the dash center trim panel (see Chapter 11).

8   Press in on the three locking tabs and push the MSS out of the trim panel (see illustration).

9   Installation is the reverse of the removal procedure.

Rear defogger switch

10   The rear defogger switch is part of the HVAC control module (see Chapter 3) for removal.

Auxiliary power outlet

11   Disconnect the cable(s) from the negative battery terminal(s) (see Chapter 5).

10.13 Disconnect the electrical connector from the backside of the instrument cluster trim panel

11.6 Screw locations

12   Remove the instrument cluster trim panel (see Chapter 11).
13   Disconnect the power outlet electrical connector (see illustration).
14   Remove the mounting screws and lift the assembly from the instrument panel.
15   Installation is the reverse of the removal procedure.

11   Instrument cluster - removal and installation

Warning: *The models covered by this manual are equipped with a Supplemental Restraint System (SRS), more commonly known as air-bags. Always disconnect the negative battery cable(s) and wait two minutes before working in the vicinity of any airbag system component to avoid the possibility of accidental deployment of the airbag, which could cause personal injury (see Section 26).*

1   Disconnect the cable from the negative battery terminal (see Chapter 5).
Warning: *If you are replacing the instrument cluster, the original information stored in the old instrument cluster must be uploaded to a scanner to be reinstalled in the replacement cluster. See the appropriate repair facility that is qualified to make the repair before proceeding.*
2   Remove the upper instrument panel center trim (see Chapter 11).
3   Lower the tilt wheel to the lowest position and pull out the lower steering column trim panel to release the retaining clips.
4   Remove the two push pins that are hiding the gap (on the top edge) between the steering column and dash assembly.
5   Remove the instrument cluster trim panel.
6   Remove the two screws securing the instrument cluster to the dash assembly (see illustration).
7   Pull the instrument cluster out and unplug

the electrical connectors from the backside (see illustration), then remove the cluster from the instrument panel.
8   Installation is the reverse of removal.

12   Wiper motor - check and replacement

Wiper motor circuit check

Note: *Refer to the wiring diagrams for wire colors in the following checks. When checking for voltage, probe a grounded 12-volt test light to each terminal at a connector until it lights; this verifies voltage (power) at the terminal. If the following checks fail to locate the problem, have the system diagnosed by a dealer service department or other properly equipped repair facility.*
Note: *There are some general voltage and ground tests that can be made to check the operation of the wiper system. To test it properly requires the use of a scanner with bi-directional capabilities. If the following outlined generic tests do not lead to a result, take your vehicle to the appropriate repair facility for diagnostics and repairs.*
1   If the wipers work slowly, make sure the battery is in good condition and has a strong charge (see Chapter 5). If the battery is in good condition, remove the wiper motor and operate the wiper arms by hand. Check for binding linkage and pivots. Lubricate or repair the linkage or pivots as necessary. Reinstall the wiper motor. If the wipers still operate slowly, check for loose or corroded connections, especially the ground connection. If all connections look OK, replace the motor.
2   If the wipers fail to operate when activated, check the fuse (see Section 3). If the fuse is OK, connect a jumper wire between the wiper motor's ground terminal and ground, then retest. If the motor works now, repair the ground connection. If the motor still doesn't work, turn the wiper switch to the HI position

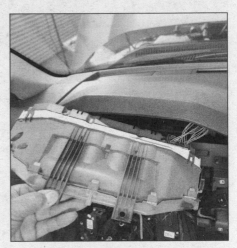

11.7 Disconnect the instrument cluster electrical connectors

and check for voltage at the motor.
3   If there's voltage at the connector, remove the motor and check it off the vehicle with fused jumper wires from the battery. If the motor now works, check for binding linkage (see Step 1). If the motor still doesn't work, replace it. If there's no voltage to the motor, check for voltage at the wiper control relays. If there's voltage at the wiper control relays and no voltage at the wiper motor, have the switch tested. If the switch is OK, the wiper control relay is probably bad. See Section 5 for relay testing.
4   If the interval (delay) function is inoperative, check the continuity of all the wiring between the switch and wiper control module.
5   If the wipers stop at the position they're in when the switch is turned off (fail to park), check for voltage at the park feed wire of the wiper motor connector when the wiper switch is Off but the ignition is On. If no voltage is present, check for an open circuit between the wiper motor and the fuse panel.

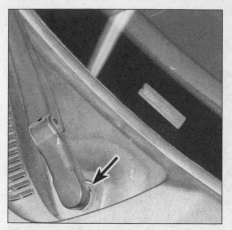

12.7a Wiper arm locking tab location

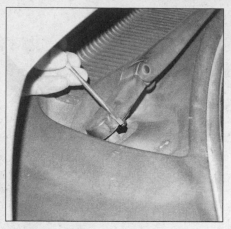

12.7b Release the wiper locking tab using a small screwdriver or pick and . . .

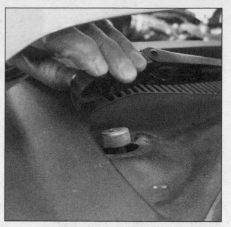

12.7c . . . pull the wiper arm off the wiper motor shaft

12.9 Wiper assembly mounting bolts

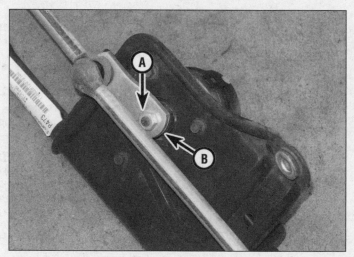

12.10 Crank arm nut and match marks

A   Crank arm nut
B   Match mark the crank arm to the wiper motor shaft and to the bracket before removing the crank arm

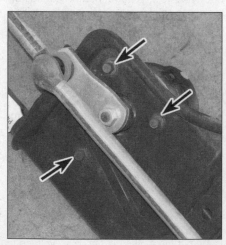

12.11 Remove the fasteners

Wiper motor replacement

6   Disconnect the cable from the negative battery terminal (see Chapter 5).
7   Mark the positions of the wiper arm(s) on the windshield, then remove the wiper arm(s) (see illustrations).
8   Remove the windshield cowl cover (see Chapter 11).
9   Disconnect the wiper motor harness connector and remove the three bolts securing the windshield wiper motor/linkage assembly and remove the assembly (see illustration).
10   Mark the location of the wiper crank arm to the linkage assembly and wiper motor shaft. Remove the wiper motor crank arm nut and separate the crank arm from the motor (see illustration).
11   Remove the fasteenrs securing the motor to the linkage assembly (see illustration).

12   Installation is the reverse of removal. Tighten all the fasteners securely.

13   Radio and speakers - removal and installation

Warning: *The models covered by this manual are equipped with a Supplemental Restraint System (SRS), more commonly known as airbags. Always disconnect the negative battery cable(s) and wait two minutes before working in the vicinity of any airbag system component to avoid the possibility of accidental deployment of the airbag, which could cause personal injury (see Section 26).*

1   Disconnect the cable from the negative battery terminal (see Chapter 5).

13.3 Remove the mounting fasteners

13.4 Disconnect the electrical connectors and antenna

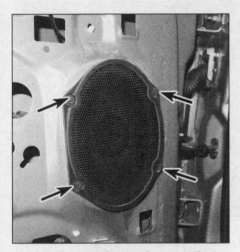

13.7a Remove the speaker mounting fasteners, then . . .

13.7b . . . tilt the speaker out to disconnect the electrical connector and separate the speaker from the door

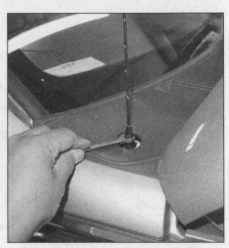

14.2 Remove the antenna from the antenna base

Radio

Warning: *If you are replacing the ACM (Audio Front Control Module) upload the ACM configuration into a scan tool so that it can be reloaded into the replacement unit. This will have to be done by the appropriately equipped repair facility.*

Note: *If equipped with the base AM/FM radio, it does not need to be programmed.*

2   Remove the dash center trim panel (see Chapter 11).

3   Remove the fasteners securing the radio to the dash (see illustration).

4   Pull the unit out far enough to disconnect the electrical connections and antenna lead (see illustration).

5   Installation is the reverse of removal.

Speakers

6   Remove the door trim panel (see Chapter 11).

7   Remove the mounting fasteners (see illustration), withdraw the speaker, unplug the electrical connector (see illustration) and remove the speaker from the vehicle.

8   Installation is the reverse of removal.

---

14   Antenna and cable - removal and installation

Antenna

1   Remove the right side cowling cover (see Chapter 11).

2   Remove the antenna mast (see illustration) and the rubber grommet.

3   Remove the right hand cowl section to gain access to the antenna base mounting screws (see illustration).

4   Lift the antenna and disconnect the antenna cable from the base.

5   Installation is the reverse of removal.

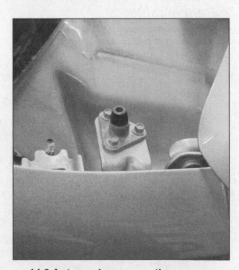

14.3 Antenna base mounting screws

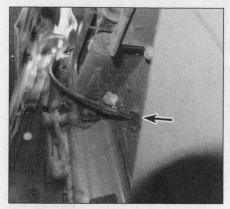

14.9a Location of the antenna cable grommet in the right hand fenderwell

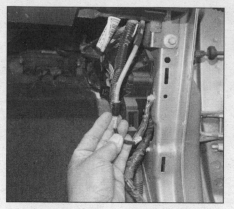

14.9b Disconnect the cable from behind the right kick panel area to remove the cable with the grommet

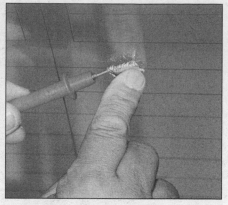

15.4 When measuring the voltage at the rear window defogger grid, wrap a piece of aluminum foil around the positive probe of the voltmeter and press the foil against the wire with your finger

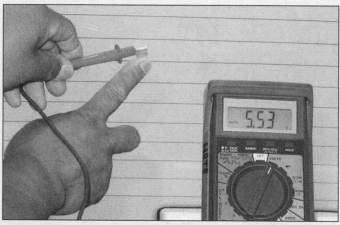

15.5 To determine if a heating element has broken, check the voltage at the center of each element - if the voltage is 5- or 6-volts, the element is unbroken - if the voltage is 10- or 12-volts, the element is broken between the center and the ground side - if there is no voltage, the element is broken between the center and the positive side

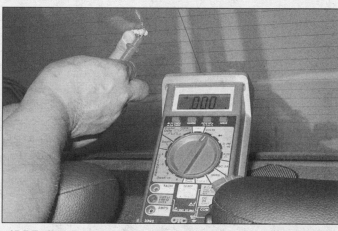

15.7 To find the break, place the voltmeter negative lead against the defogger ground terminal, place the voltmeter positive lead with the foil strip against the heating element at the positive terminal end and slide it toward the negative terminal end - the point at which the voltmeter reading changes abruptly is the point at which the element is broken

Antenna cable

6   Remove the radio (see Section 13) and disconnect the antenna lead from the rear side of the radio.

7   Remove the right side cowling cover (see Chapter 11) and disconnect the antenna cable from the antenna.

8   Lower the glove box completely and remove the right side kick panel to gain access to the antenna cable.

9   Pull the antenna with the firewall grommet intact through the glove box opening and pull the antenna cable from the radio side as well. Remove the cable through the glove box opening (see illustrations).

10   Installation is the reverse of removal.

---

15   Rear window defogger - check and repair

1   The rear window defogger consists of a number of horizontal elements baked onto the glass surface.

2   Small breaks in the element can be repaired without removing the rear window.

Check

3   Turn the ignition switch and defogger system switches to the On position. Using a voltmeter, place the positive probe against the defogger grid positive terminal and the negative probe against the ground terminal. If battery voltage is not indicated, check the fuse, defogger switch and related wiring. If voltage is indicated, but all or part of the defogger doesn't heat, proceed with the following tests.

4   When measuring voltage during the next two tests, wrap a piece of aluminum foil around the tip of the voltmeter positive probe and press the foil against the heating element with your finger (see illustration). Place the negative probe on the defogger grid ground terminal.

5   Check the voltage at the center of each heating element (see illustration). If the voltage is 5- or 6-volts, the element is OK (there is no break). If the voltage is 0-volts, the element is broken between the center of the element and the positive end. If the voltage is 10- to 12-volts the element is broken between the center of the element and ground. Check each heating element.

6   Connect the negative lead to a good body ground. The reading should stay the same. If it doesn't, the ground connection is bad.

7   To find the break, place the voltmeter negative probe against the defogger ground terminal. Place the voltmeter positive probe with the foil strip against the heating element at the positive terminal end and slide it toward the negative terminal end. The point at which the voltmeter deflects from several volts to zero is the point at which the heating element is broken (see illustration).

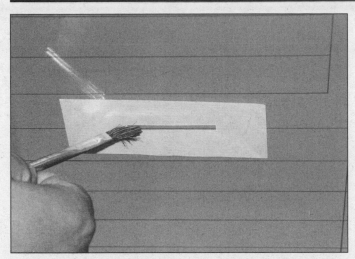

15.13 To use a defogger repair kit, apply masking tape to the inside of the window at the damaged area, then brush on the special conductive coating

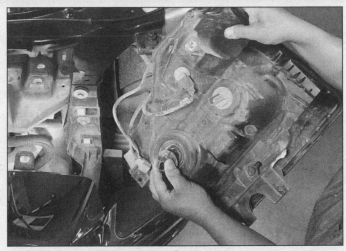

16.2 Disconnect the electrical connector and twist the bulb socket out of the housing

Repair

8   Repair the break in the element using a repair kit specifically recommended for this purpose, available at most auto parts stores. Included in this kit is plastic conductive epoxy.

9   Prior to repairing a break, turn off the system and allow it to cool off for a few minutes.

10   Lightly buff the element area with fine steel wool, then clean it thoroughly with rubbing alcohol.

11   Use masking tape to mask off the area being repaired.

12   Thoroughly mix the epoxy, following the instructions provided with the repair kit.

13   Apply the epoxy material to the slit in the masking tape, overlapping the undamaged area about 3/4-inch on either end (see illustration).

14   Allow the repair to cure for 24 hours before removing the tape and using the system.

17.1 Location of the headlight adjustment screw

16   Headlight bulb - replacement

Warning: *Halogen gas filled bulbs are under pressure and may shatter if the surface is scratched or the bulb is dropped. Wear eye protection and handle the bulbs carefully, grasping only the base whenever possible. Do not touch the surface of the bulb with your fingers because the oil from your skin could cause it to overheat and fail prematurely. If you do touch the bulb surface, clean it with rubbing alcohol.*

1   Remove the headlamp housing (see Section 18).

2   Rotate the headlamp bulb 1/8 of a turn and remove the bulb (see illustration).

Caution: *Always replace the headlamp with the same type of bulb. Some aftermarket*

bulbs that claim to give off a more intense field of light may actually do that. However, the way most of these bulbs accomplish the higher intensity is with a higher current flow. This puts a heavy strain on the factory wiring and connections that were never designed to carry an excessive current as required by some of these aftermarket bulbs. This also includes the HID aftermarket systems. There is a risk changing from the factory systems to any aftermarket lighting system.

3   Installation is the reverse of removal.

17   Headlights - adjustment

Note: *The headlights must be aimed correctly. If adjusted incorrectly they could blind the driver of an oncoming vehicle and cause a serious accident or seriously reduce your ability to see the road. The headlights should be checked for proper aim every 12 months and any time*

a new headlight is installed or front end body work is performed. It should be emphasized that the following procedure is only an interim step which will provide temporary adjustment until the headlights can be adjusted by a properly equipped shop.

1   Identify the exact location of the adjustment screws:

a) *On halogen bulb style headlights, the adjustment screws are located on top of the housing.*

Note: *On 2006 and later models, only the vertical aiming can be adjusted (see illustration).*

b) *On sealed beam style headlights, the inboard (vertical) adjustment screw and the upper (horizontal) adjustment screw are accessible on the exterior of the vehicle.*

2   There are several methods of adjusting the headlights. The simplest method requires masking tape, a blank wall and a level floor.

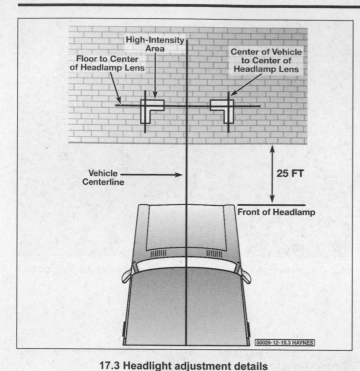

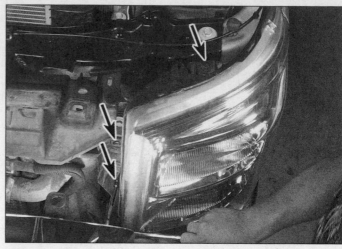

18.2a Two bolts are along the inside edge (upper bolt also shown)

17.3 Headlight adjustment details

3   Position masking tape vertically on the wall in reference to the vehicle centerline and the centerlines of both headlights (see illustration).

4   Position a horizontal tape line in reference to the centerline of all the headlights.

Note: *It may be easier to position the tape on the wall with the vehicle parked only a few inches away.*

5   Adjustment should be made with the vehicle parked 25 feet from the wall, sitting level, the gas tank half-full and no heavy load in the vehicle.

6   Starting with the low beam adjustment, position the high intensity zone so it is two inches below the horizontal line and two inches to the side of the headlight vertical line, away from oncoming traffic. Adjustment is made by turning the top (sealed beam) or inner (aerodynamic type) adjusting screw clockwise to raise the beam and counterclockwise to lower the beam. The adjusting screw on the side should be used in the same manner to move the beam left or right.

7   With the high beams on, the high intensity zone should be vertically centered with the exact center just below the horizontal line.

Note: *It may not be possible to position the headlight aim exactly for both high and low beams. If a compromise must be made, keep in mind that the low beams are the most used and have the greatest effect on safety.*

18 Headlight housing - removal and installation

1   Remove the front grille (see Chapter 11).

2   Remove the four bolts securing the headlight assembly to the core support (see illustrations).

3   Pull the housing from the core support and disconnect the electrical connector (see illustration).

4   Installation is the reverse of removal.

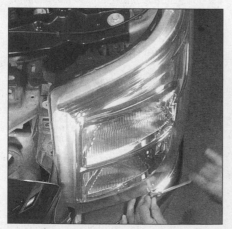

18.2b Remove the lower bolt

18.2c Remove the upper bolt

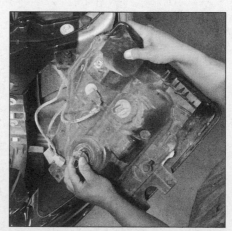

18.3 Disconnect the electrical connections and remove the housing

19   Horn - replacement

1   Remove the front grille (see Chapter 11).
2   Disconnect the electrical connector(s) from the horn(s) (see illustration).
3   Remove the horn mounting bolt(s) and separate the horn(s) from the body.
4   Installation is the reverse of removal.

20   Bulb - replacement

Bulb removal
Fog light

1   Lift and support the front end of the vehicle (if applicable).
2   Locate the fog light bulb socket and disconnect the electrical connector.

19.2 Location of the horns

Bulb removal

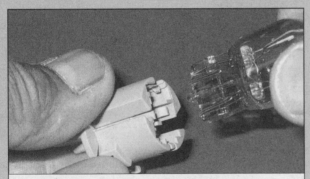

To remove many modern exterior bulbs from their holders, simply pull them out

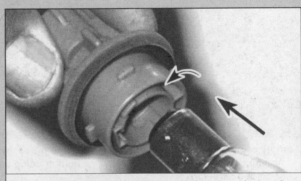

On bulbs with a cylindrical base ("bayonet" bulbs), the socket is spring-loaded; a pair of small posts on the side of the base hold the bulb in place against spring pressure. To remove this type of bulb, push it into the holder, rotate it 1/4-turn counterclockwise, then pull it out

If a bayonet bulb has dual filaments, the posts are staggered, so the bulb can only be installed one way

To remove most overhead interior light bulbs, simply unclip them

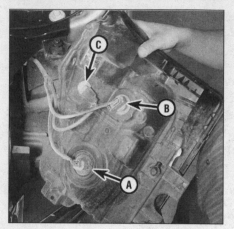

20.6 Bulb locations
A   *Headlamp bulb*
B   *Turn signal/park light bulb*
C   *Park light/marker light bulb*

20.10 Remove the rear taillight housing mounting screws

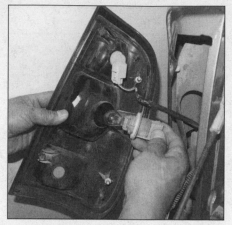

20.11 Remove the taillight housing and replace the defective bulbs

20.14 Carefully pry the license plate light housing out using a flat-bladed screwdriver

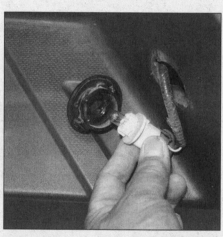

20.15 Remove the bulb holder from the light housing by twisting it 1/4-turn

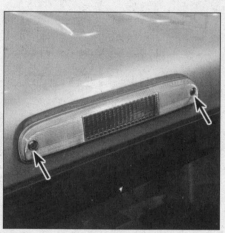

20.17 Location of the mounting screws on the high-mounted brake light assembly

3   Twist the bulb socket 1/8 of a turn counterclockwise to remove.
4   Installation is the reverse of removal.

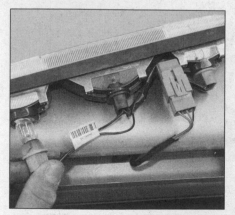

20.18 Remove the defective bulb from the high-mounted brake light assembly

Front turn signal and side marker lights

5   Remove the headlight assembly (see Section 18).
6   Twist the bulb socket 1/8 of a turn counterclockwise to remove (see illustration).
7   The defective bulb can then be removed from the socket and replaced.
8   Installation is the reverse of removal.

Rear turn signal, brake, tail and back-up lights

9   Open the tailgate or liftgate.
10   Remove the retaining screws securing the rear taillight housing, then pull the taillight assembly outward to access the taillight bulbs (see illustration).
11   Twist the bulb socket a quarter turn counterclockwise, then remove the bulb assembly from the housing (see illustration).
12   The defective bulb can then be removed from the socket and replaced.

13   Installation of the taillight housing is the reverse of removal.

License plate light

14   The license plate light bulbs can be accessed from the rear of the bumper (see illustration). Carefully pry the light housing out using a flat-bladed screwdriver, then pull the bulb assembly from the bumper.
15   Twist the bulb holder 1/4-turn to remove it from the housing (see illustration). The defective bulb can then be pulled straight out of the socket and replaced.
16   Installation is the reverse of removal.

High-mounted brake light

17   Remove the lens retaining screws and pull the lamp assembly outward to access the bulbs (see illustration).
18   Twist the bulb socket a quarter turn counterclockwise (see illustration), then remove the bulb assembly from the housing.
19   The defective bulb can then be pulled straight out of the socket and replaced.

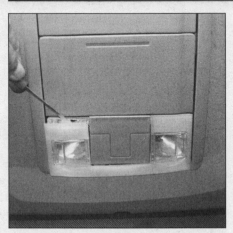

20.20a Carefully pry the lens cover off to access the bulbs

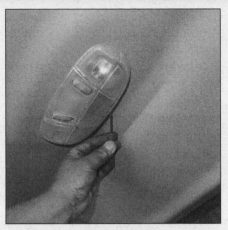

20.20b Overhead entry light/dome light lens can be removed with a small flat screwdriver

20.20c Remove the bulb by pulling the old bulb straight out of the housing

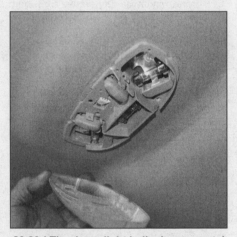

20.20d The dome light bulbs have a metal cap on each end. Remove the bulb and replace with the same type

20.22 Press the clip inside the mirror housing forward as you hold onto the lens

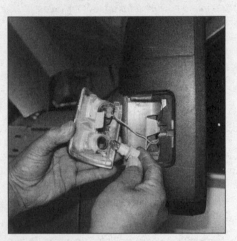

20.25 Once the socket is removed, pull the bulb out of the socket

Dome and cargo lights

20   Using a small screwdriver, remove the lens and replace the bulb (see illustrations).

Outside mirror bulb replacement

21   Rotate mirror glass all the way inward.
22   Using a small flat bladed screwdriver, reach into the outside edge between the mirror and the mirror housing and release the plastic clip by pushing the clip forward (see illustration).
23   Grasp the lens cover as you apply pressure to the clip so that the lens doesn't fall.
24   Rotate the lens outward to free the two front tabs.
25   Replace the bulbs by pulling straight out away from the housing (see illustration).
26   Installation is the reverse of removal

Instrument cluster illumination

27   To gain access to the instrument cluster illumination lights, the instrument cluster will have to be removed (see Section 11). The bulbs are fixed into position inside the instrument cluster. Take the instrument cluster to the appropriate repair facility that is equipped to repair the cluster.

---

21   Mirrors - description

Outside rear view mirrors

Note: *These models are equipped with a Body Control Module (BCM), (which doubles as the main interior fuse box). Several systems are linked to a centralized control module that allows simple and accurate troubleshooting, but only with a professional-grade scan tool. The Body Control Module governs the door locks, the power windows, the ignition lock and security system, the interior lights, the Daytime Running Lights system, the horn, the windshield wipers, the heating/air condi-tioning system and the power mirrors. In the event of malfunction with this system, have the vehicle diagnosed by a dealership service department or other qualified automotive repair facility.*

1   The electric rear view mirrors use two motors to move the glass; one for up and down adjustments and one for left-right adjustments. Some vehicles are equipped with heated (defog), power folding, power telescopic, and memory mirrors as well. These mirrors are an integral part of the power seat memory unit and the navigation systems as well as incorporated with the Body Control Module (BCM). If there is a problem with these systems it is advised to seek out a qualified independent repair facility or your local dealer.
2   The control switch has a selector portion which sends voltage to the left or right side mirror. With the ignition in the ACC position and the engine OFF, roll down the windows

and operate the mirror control switch through all functions (left-right and up-down) for both the left and right side mirrors.

3   Listen carefully for the sound of the electric motors running in the mirrors.

4   If the motors can be heard but the mirror glass doesn't move, there's probably a problem with the drive mechanism inside the mirror. Power mirrors have no user-serviceable parts inside - a defective mirror must be replaced as a unit.

5   If the mirrors don't operate and no sound comes from the mirrors, check for a blown fuse.

6   If the fuses are OK, remove the mirror control switch. Have the switch continuity checked by a dealer service department or other qualified shop.

Note: *If the mirrors are continuously moved 10 times in a row the system will shut down the power to the mirrors for 3 1/2 minutes allowing the motor to cool off. In extreme cases it can keep the power off for up to 30 minutes.*

7   Check the ground connections.

8   If the mirror still doesn't work, remove the mirror and check the wires at the mirror for voltage.

9   If there's not voltage in each switch position, check the circuit between the mirror and control switch for opens and shorts.

10   If the mirror is inoperative, try holding the switch in one of the directions and swing the door open and closed. If there is a break in the door jamb you may occasionally make contact long enough to avoid removing the mirror from the door for further testing. This will also give you some clue as to where the break is rather than testing things in one given position.

Note: *If there's voltage, remove the mirror and test it off the vehicle with jumper wires. Before testing, check the wiring diagram for the correct leads that have to be used for each position. Applying voltage to the wrong leads can damage the mirror drive motors. Replace the mirror if it fails this test.*

Interior mirror

11   The inside mirror removal can be found in Chapter 11.

12   The vehicle may be equipped with one or more of the following interior mirror features.

a) *Basic interior mirror with manual flip lever for dimming.*

b) *Automatic mirror dimming to reduce night time glare.*

c) *Auto dimming with video display.*

d) *Compass*

Note: *The video display is turned on by the navigation system anytime the vehicle is put into reverse. In vehicles with the touch screen, the image is not in the mirror but on the touch screen video display.*

Note: *The interior mirror with the built-in compass has an internal compass module as part of the mirror assembly. It cannot be removed or serviced. For calibration procedures see your owner's manual for details.*

22 Cruise control system - general information

1   The Powertrain Control Module (PCM) controls the cruise control system electronically via the electronic throttle control system. If you have problems with the cruise control system, check for the presence of trouble codes stored in the PCM (see Chapter 6A, Section 2). If that doesn't turn up any problems, have it checked by a dealer service department or other qualified repair shop.

23 Power window system - description and check

Note: *These models are equipped with a Body Control Module (BCM) incorporated with the window system. Several systems are linked to a centralized control module that allows simple and accurate troubleshooting, but only with a professional-grade scan tool. The Body Control Module governs the door locks, the power windows, the ignition lock and security system, the interior lights, the Daytime Running Lights system, the horn, the windshield wipers, the heating/air conditioning system and the power mirrors. In the event of malfunction with this system, have the vehicle diagnosed by a dealership service department or other qualified automotive repair facility.*

1   The power window system operates electric motors, mounted in the doors, which lower and raise the windows. The system consists of the control switches, the motors, regulators, glass mechanisms, the Body Control Module (BCM) and associated wiring.

2   The power windows can be lowered and raised from the master control switch by the driver or by remote switches located at the individual windows. Each window has a separate motor that is reversible. The position of the control switch determines the polarity and therefore the direction of operation.

3   The window motor circuits are protected by a fuse. Each motor is also equipped with an internal circuit breaker; this prevents one stuck window from disabling the whole system.

4   The power window system will only operate when the ignition switch has activated the Retained Accessory Power (RAP) relay. In addition, many models have a window lockout switch at the master control switch which, when activated, disables the switches at the rear windows and, sometimes, the switch at the passenger's window also. Always check these items before troubleshooting a window problem related to any of the other windows besides the driver's window. Window lock-out does not affect the driver's window.

5   These procedures are general in nature, so if you can't find the problem using them, take the vehicle to a dealer service department or to an independent repair facility that specializes in electrical repairs.

6   If the power windows won't operate, always check the fuse and circuit breaker first.

7   If only the rear windows are inoperative, or if the windows only operate from the master control switch, check the rear window lockout switch for continuity in the unlocked position. Replace it if it doesn't have continuity.

8   Check the wiring between the switches and fuse panel for continuity. Repair the wiring, if necessary.

9   If only one window is inoperative from the master control switch, try the other control switch at each individual window.

Note: *This doesn't apply to the driver's door window.*

10   If the same window works from one switch, but not the other, check the switch and/or wiring for continuity.

11   If the switch tests OK, check for a short or open in the circuit between the affected switch and the window motor.

12   If one window is inoperative from both switches, remove the switch panel from the affected door. Check for voltage at the motor (refer to Chapter 11 for door panel removal) while the switch is operated.

Note: *There are voltage and ground signals present at the switch connector. However, on 2010 and later models, there are also two data lines to the BCM that are not true positive/ negative signals. Be sure you are testing the correct leads. Do not apply voltage or ground to any leads you are not sure of. Seek professional help in diagnosing any problems with the window circuits if you are unsure.*

13   If voltage is reaching the motor, disconnect the glass from the regulator. Move the window up and down by hand while checking for binding and damage. Also check for binding and damage to the regulator. If the regulator is not damaged and the window moves up and down smoothly, replace the motor. If there's binding or damage, lubricate, repair or replace parts, as necessary.

14   If voltage isn't reaching the motor, check the wiring in the circuit for continuity between the switches and the BCM, and between the BCM (if applicable) and the motors. You'll need to consult the wiring diagram. If the circuit is equipped with a relay, check that the relay is grounded properly and receiving voltage.

15   Test the windows to confirm proper repairs.

Note: *To verify if the motor is getting the needed voltage or ground, a simple but effective test is to sit in the car with the ignition on, open a door and look at the dome light. Then operate the switch to the faulty window. If you see the dome light dimming slightly this is a good indication that the motor is getting power and is probably a stuck motor or faulty wiring. Do not hold the switch on for very long when a motor is stuck or wiring is in question or more damage may occur. Finally, in some cases, a good rap on the door panel in the general area of the window motor - while the key is on and the window switch is depressed in the direc-*

tion the window needs to move - will free up a stuck motor temporarily. You can damage the door panel or more internal components if you hit it too hard or are too aggressive.

Window express down programming

Note: *Any time the battery or window motor are disconnected, you will need to perform this procedure to reestablish the auto feature.*

16 Lower the window to its lowest position, holding the switch in the Down position for an additional five seconds.

17 Raise the window to its highest position, holding the window switch in the Up position for an additional five seconds.

18 Verify proper operation.

24 Power door lock and keyless entry system - description and check

Note: *These models are equipped with a Body Control Module (BCM). Several systems are linked to a centralized control module that allows simple and accurate troubleshooting, but only with a professional-grade scan tool. The Body Control Module governs the door locks, the power windows, the ignition lock and security system, the interior lights, the Daytime Running Lights system, the horn, the windshield wipers, the heating/air conditioning system and the power mirrors. In the event of malfunction with this system, have the vehicle diagnosed by a dealership service department or other qualified automotive repair facility.*

1 The power door lock system operates the door lock actuators mounted in each door. The system consists of the switches, actuators, lock and unlock relays, Body Control Module (BCM) and associated wiring. Diagnosis can usually be limited to simple checks of the wiring connections and actuators for minor faults that can be easily repaired.

2 Power door lock systems are operated by bi-directional solenoids located in the doors. The lock actuators are mounted as part of the door latch. Remove the door latch for access to the door lock actuator. The lock switches have two operating positions: Lock and Unlock. These switches send a signal to the BCM, which in turn sends a signal to the door lock relays, the relays then send the needed voltage to each of the door lock solenoids.

3 If you are unable to locate the trouble using the following general steps, consult your dealer service department or qualified independent repair shop.

4 Always check the circuit protection first. Some vehicles use a combination of circuit breakers and fuses.

5 Check for voltage at the switches. If no voltage is present, check the fuse first. If the fuse is good then check the wiring between the fuse panel and the switches for an open lead.

6 If voltage is present, test the switch for continuity. Replace it if there's not continuity in both switch positions. There should be a voltage input and, when switch is depressed, voltage should be going out on the appropriate lead. Follow the wiring diagram for the actual wire and position on the switch. To remove the switch, use a flat-bladed trim tool to pry out the door/window switch assembly.

7 If the switch has continuity, check the wiring between the switch, BCM, door lock relay and the door lock solenoid.

8 If all but one of the lock solenoids operates, remove the trim panel from the affected door and check for voltage at the solenoid while the lock switch is operated. One of the wires should have positive voltage in the Lock position; the other lead should have positive voltage in the Unlock position.

9 If the inoperative solenoid is receiving positive voltage on one lead and negative on the other, the solenoid is most likely defective. Check the connections for good contact; if the connection is good, replace the solenoid.

10 If the inoperative solenoid isn't receiving voltage or ground, check for an open or short in the wire between the lock solenoid and the relay. A good method of non-destructive testing is to squeeze the rubber corrugated tubing and search with your fingers for an individual wire. Follow the wire as far as possible and feel for any breaks in the leads.

Note: *It's not uncommon for wires to break in the portion of the harness between the body and door (opening and closing the door fatigues and eventually breaks the wires).*

11 On the models covered by this manual, power door lock system communication goes through the Body Control Module. If the above tests do not pinpoint a problem, take the vehicle to a dealer or qualified shop with the proper scan tool to retrieve trouble codes from the BCM. Replacing of some components may result in programming issues. To avoid replacing good components always test thoroughly before any parts are deemed faulty.

Keyless entry system

12 The keyless entry system consists of a remote control transmitter that sends a coded infrared signal to a receiver, which then operates the door lock system.

13 Replace the battery when the transmitter doesn't operate the locks at a distance of 10 feet. Normal range should be about 65 feet.

14 For more information on key fob battery replacement, programming and additional keys see Section 6.

25 Daytime Running Lights (DRL) - general information

1 The Daytime Running Lights (DRL) system illuminates the headlights whenever the engine is running. The only exception is with the engine running and the parking brake engaged. Once the parking brake is released, the lights will remain on as long as the ignition switch is on, even if the parking brake is later applied.

2 The DRL system supplies reduced power to the headlights so they won't be too bright for daytime use, while prolonging headlight life.

26 Airbag system - general information

General information

1 All models are equipped with a Supplemental Restraint System (SRS), more commonly known as airbags. This system is designed to protect the driver, and the front seat passenger, from serious injury in the event of a head-on or frontal collision. It uses an electronic crash sensor (ESC) (also known as the Restraints Control Module [RCM] on later models) mounted on the center tunnel behind the instrument panel. The airbag assemblies are mounted on the steering wheel and the right side of the passenger's side dash. Beginning with the 2002 model year, seat belt pre-tensioners became available. These are pyrotechnic devices controlled by the Restraints Control Module (RCM) that reduce the slack in the seat belts during an impact of sufficient force to trigger the airbags.

Airbag module

Driver's side

2 The airbag inflator module contains a housing incorporating the cushion (airbag) and inflator unit, mounted in the center of the steering wheel. The inflator assembly is mounted on the back of the housing over a hole through which gas is expelled, inflating the bag almost instantaneously when an electrical signal is sent from the system. A "clockspring" on the steering column under the steering wheel carries this signal to the module.

3 This clockspring assembly can transmit an electrical signal regardless of steering wheel position. The igniter in the airbag converts the electrical signal to heat and ignites the powder, which inflates the bag.

Passenger's side

4 The airbag is mounted above the glove compartment and designated by the letters SRS (Supplemental Restraint System). It consists of an inflator containing an igniter, a reaction housing/airbag assembly and a trim cover.

5 The airbag is considerably larger than the steering wheel-mounted unit and is supported by the steel reaction housing. The trim cover is textured and painted to match the

instrument panel and has a molded seam which splits when the bag inflates (see illustration).

Electronic Crash Sensor (ECS)/ Restraints Control Module (RCM) diagnostic unit

6   This unit supplies the current to the airbag system (and seat belt pre-tensioners, on models so equipped) in the event of the collision, even if battery power is cut off. It checks this system every time the vehicle is started, causing the "SRS" light to go on and then off, if the system is operating properly. If there is a fault in the system, the light will go on and stay on, flash, or the dash will make a beeping sound. If this happens, the vehicle should be taken to your dealer immediately for service.

Servicing components near the SRS system

7   There are times when you need to remove the steering wheel, the instrument cluster, the radio, the heater/air conditioning control assembly or other components that are near airbag components. At these times you'll be working around components and wire harnesses for the SRS system. Do not use electrical test equipment on airbag system wires; it could cause the airbag(s) to deploy. ALWAYS DISABLE THE SRS SYSTEM BEFORE WORKING NEAR THE SRS SYSTEM COMPONENTS OR RELATED WIRING.

Disarming the system and other precautions

Warning: *Failure to follow these precautions could result in accidental deployment of the airbag and personal injury.*

8   Whenever working in the vicinity of the steering wheel, instrument panel or any of the other SRS system components, the system must be disarmed. To disarm the system:

a) *Point the wheels straight ahead and turn the key to the Lock position.*
b) *Disconnect the cable from the negative battery terminal(s). On dual battery systems, be sure to disconnect both negative battery terminals. Refer to Chapter 5 for the disconnecting procedure.*
c) *Wait at least two minutes for the back-up power supply to be depleted.*

Disabling

9   Turn off all accessories on the vehicle and disconnect any memory-saver devices. Turn off the ignition switch.
10   At the interior fuse/relay box under the right side of the instrument panel, remove the fuse marked RCM. On later models the fuse is in the junction box behind the right kick-panel.
11   Turn the ignition switch to On. Observe the airbag warning light on the instrument panel for 30 seconds. If you have removed the correct fuse, the warning light should be

26.5 Passenger air bag location

on steady (no blinking).
12   Turn the ignition switch off.
13   Disconnect the negative battery cable and wait 30 minutes before doing any work around SRS components.

Enabling

14   Turn the ignition switch from Off to On.
15   Reinstall the RCM fuse.
16   Connect the battery cable.
Warning: *When the battery cable is connected, no one should be in front of any airbag module, in case of accidental deployment.*
17   To test if the SRS system is working properly, turn the ignition switch from On to Off for ten seconds, then back to On. The warning light should illuminate for six seconds and go out.
18   The warning light indicates a problem with the SRS system if the light blinks, stays illuminated continuously, or fails to illuminate in one minute. If one of these conditions exists, have your system checked with a scan tool at a dealership.
19   Whenever handling an airbag module, always keep the airbag opening (the trim side) pointed away from your body. Never place the airbag module on a bench or other surface with the airbag opening facing the surface. Always place the airbag module in a safe location with the airbag opening facing up.
20   Never measure the resistance of any SRS component or use any electrical test equipment on any of the wiring or components. An ohmmeter has a built-in battery supply that could accidentally deploy the airbag.
21   Never use electrical welding equipment on a vehicle equipped with an airbag without first disconnecting the airbag electrical connectors, located under the steering column near the combination switch connector (driver's airbag) (see Chapter 10) and behind the glovebox (passenger's airbag). On models with seat belt pre-tensioners, the pre-tensioner electrical connectors are located behind the B-pillar trim panels and/or under the seats, depending on the model.
22   Never dispose of a live airbag module or seat belt pre-tensioner. Return it to a dealer service department or other qualified repair shop for safe deployment and disposal.

26.25 Passenger airbag module fasteners

Airbag module removal and installation

Driver's side airbag module and clockspring

23   Refer to Chapter 10, *Steering wheel - removal and installation*, for the driver's side airbag module and clockspring removal and installation procedures.

Passenger's side airbag module

24   Disarm the airbag system as described previously in this Section.
25   Lower the glovebox door fully to gain access to the airbag by pressing the glovebox door tabs and pushing down, then unplug the electrical connector and remove the airbag module mounting nuts and bolts (see illustration). Be sure to heed the precautions outlined previously in this Section.
26   Carefully slide a flat-bladed screwdriver under the right bottom edge of the airbag panel and lift up. Separate the panel from the clips.
27   Installation is the reverse of the removal procedure. Tighten the airbag module mounting screws securely.

27   Wiring diagrams - general information

1   Since it isn't possible to include all wiring diagrams for every year covered by this manual, the following diagrams are those that are typical and most commonly needed.
2   Prior to troubleshooting any circuits, check the fuse and circuit breakers (if equipped) to make sure they're in good condition. Make sure the battery is properly charged and check the cable connections (see Chapter 1).
3   When checking a circuit, make sure that all connectors are clean, with no broken or loose terminals. When unplugging a connector, do not pull on the wires. Pull only on the connector housings themselves.

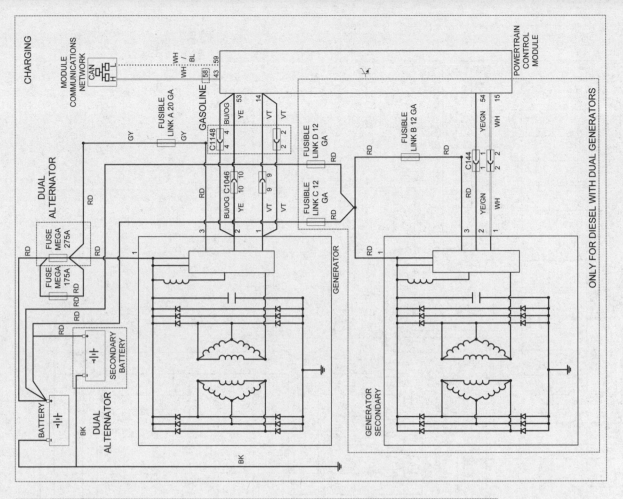

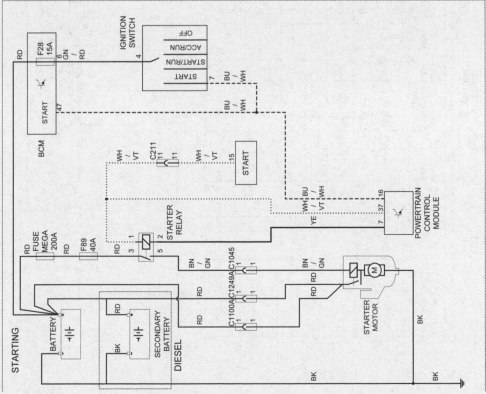

Starting and charging systems

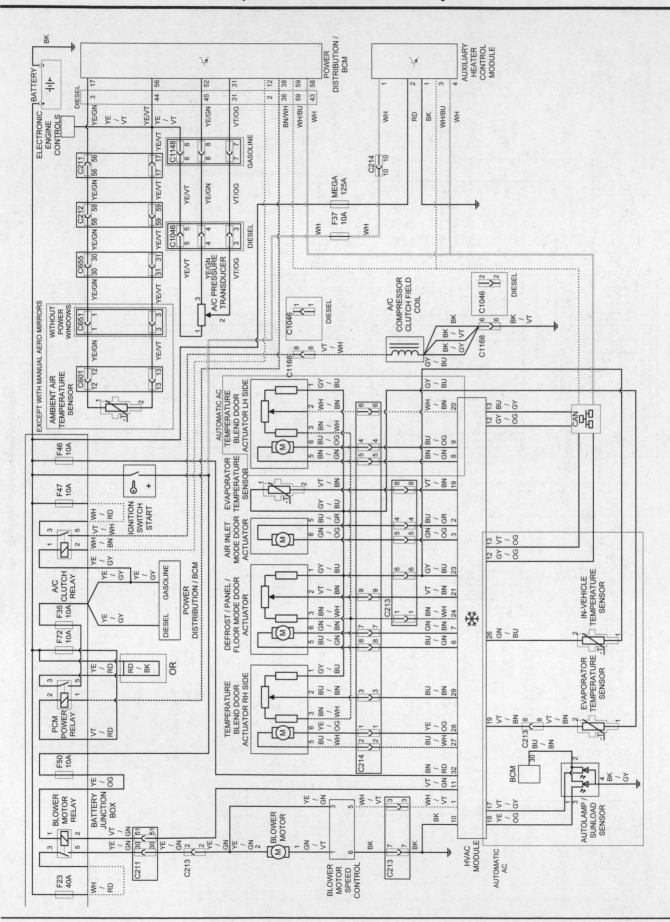

Heating and air conditioning systems

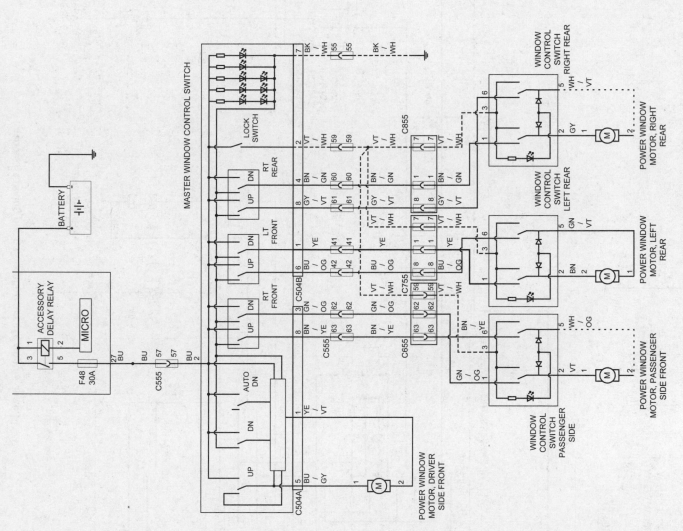

Power window system (Crew Cab models)

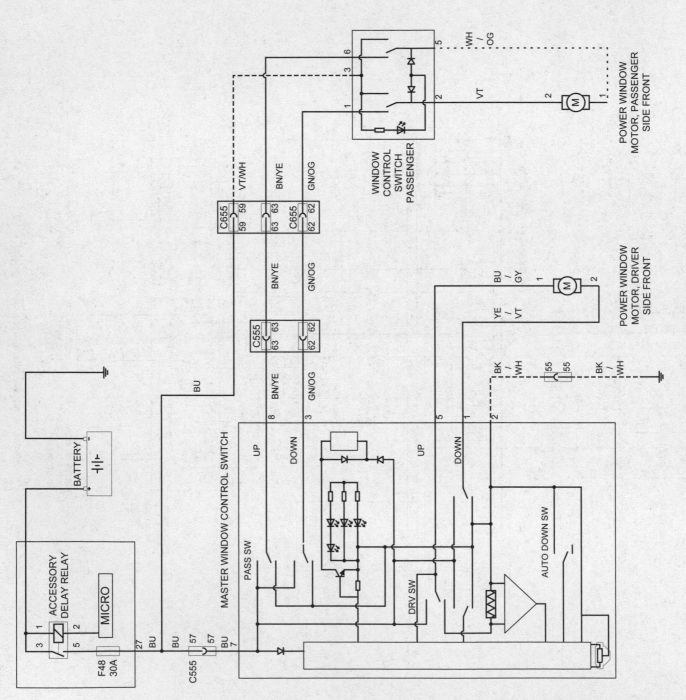

Power window system (regular cab and Supercab models)

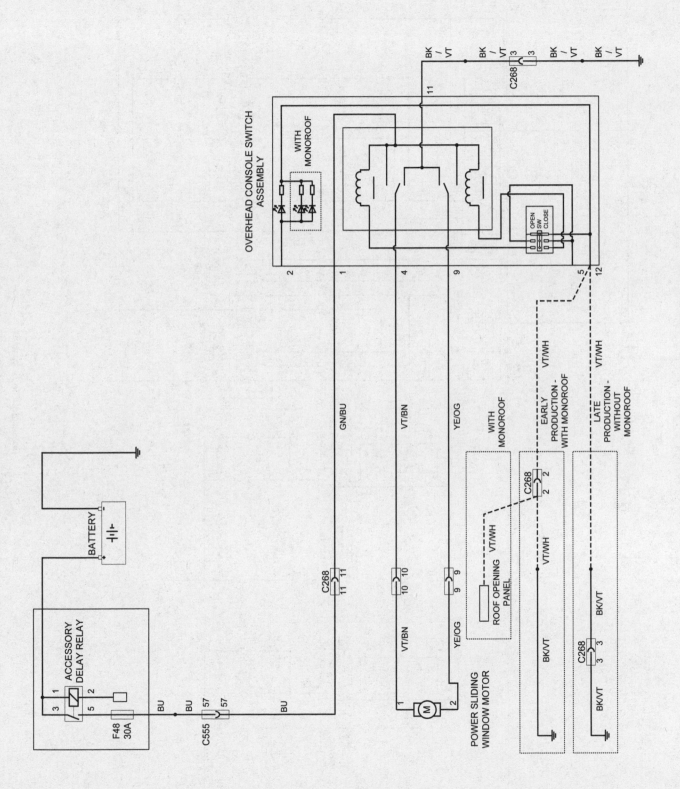
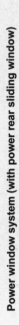

Power window system (with power rear sliding window)

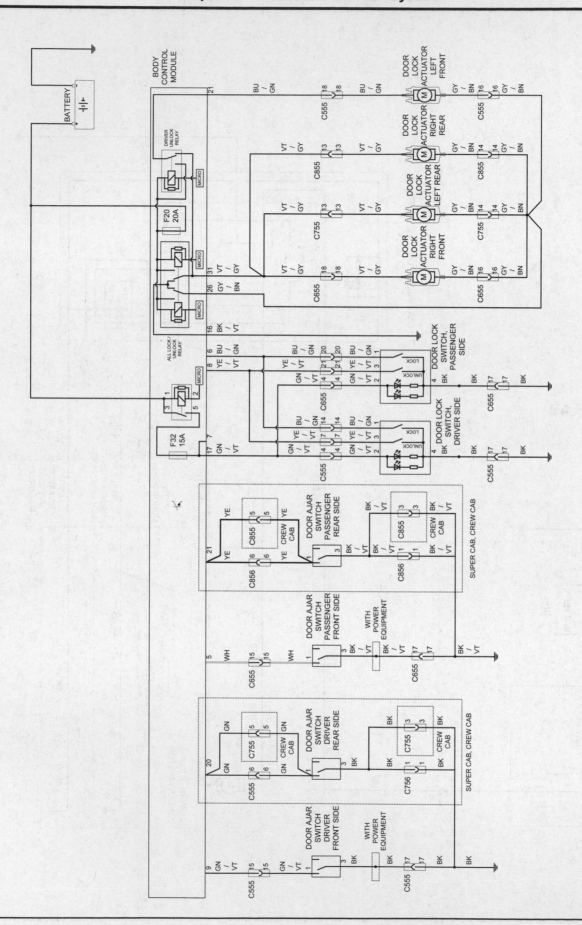

Power door lock system

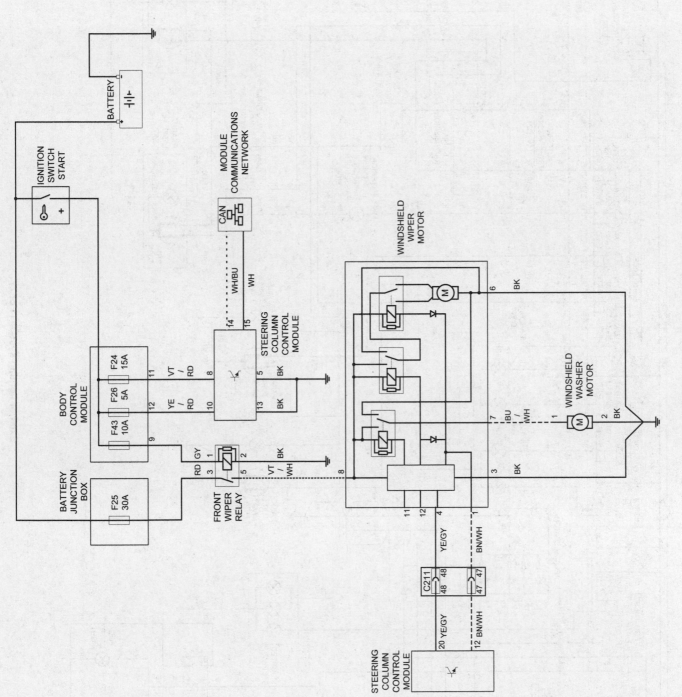

Windshield wiper and washer system

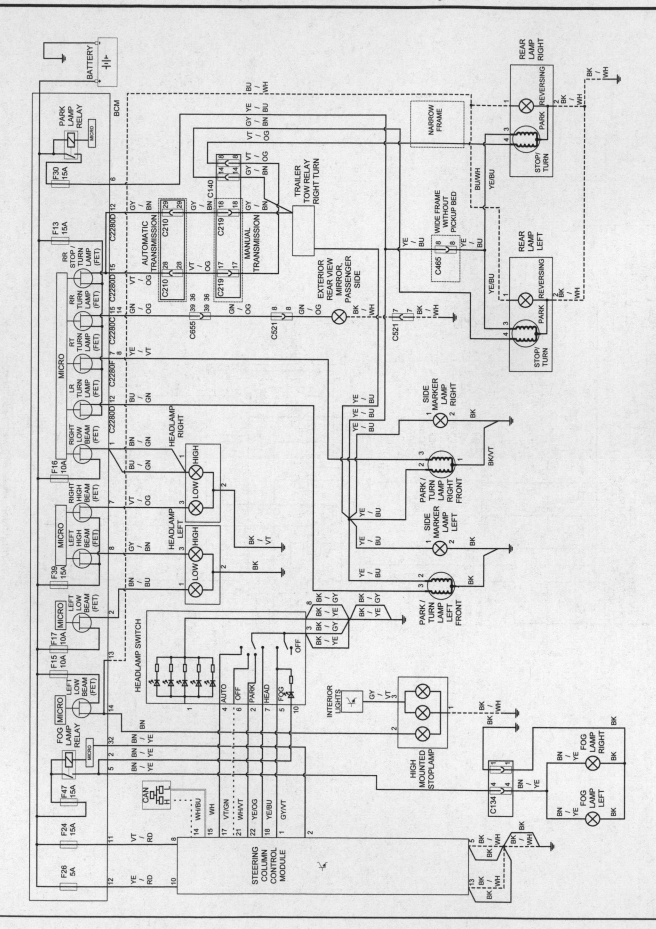

Exterior lighting system (1 of 2)

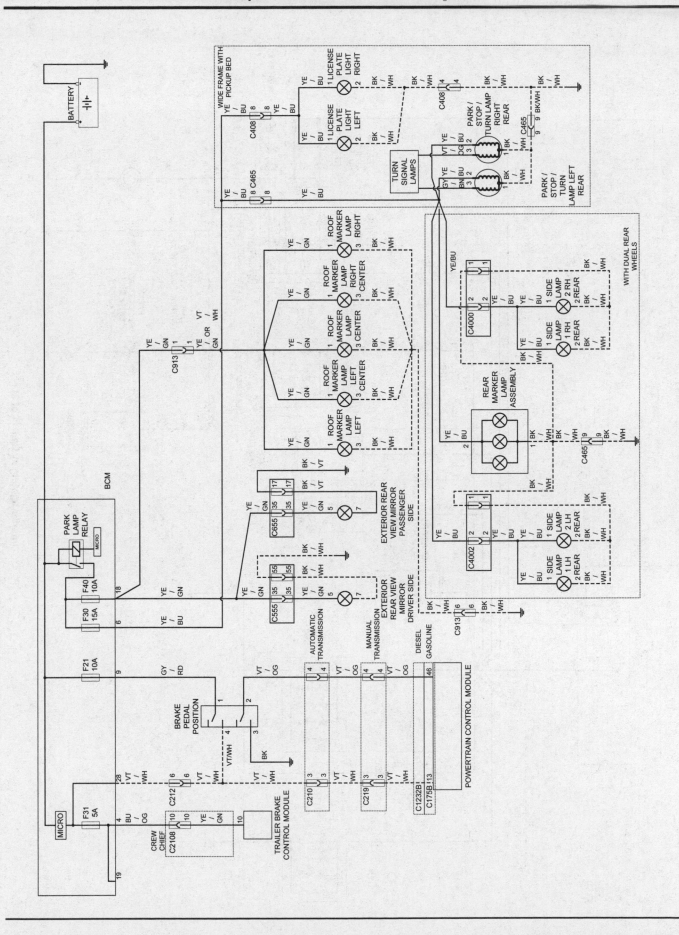

Exterior lighting system (2 of 2)

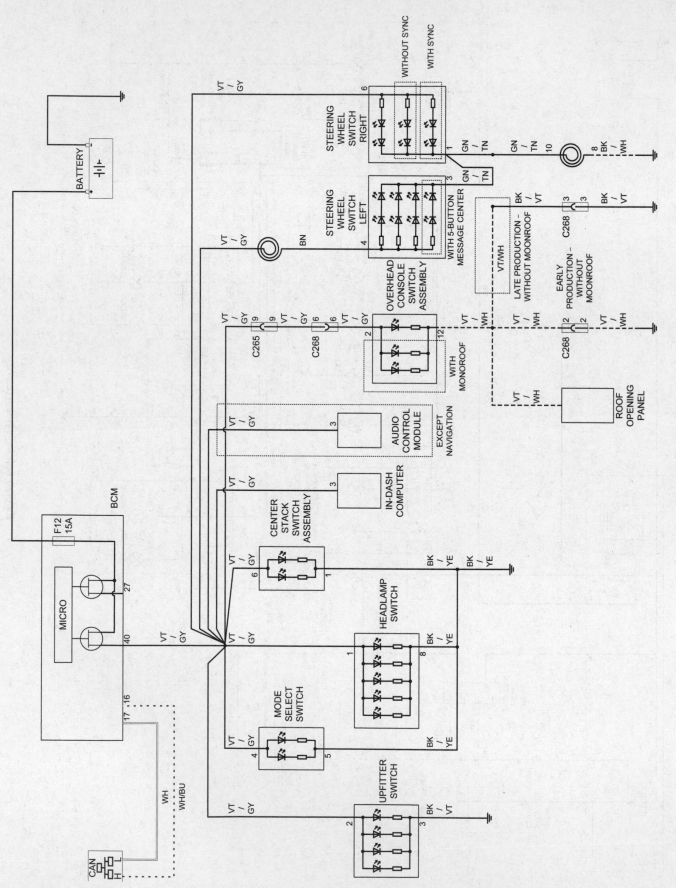

Interior lighting system (1 of 2)

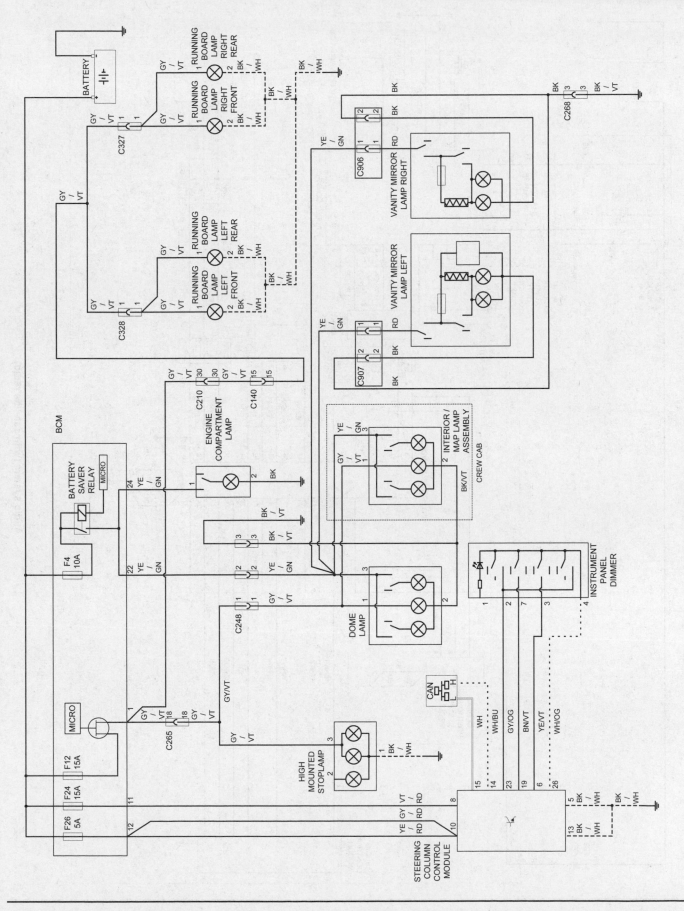

Interior lighting system (2 of 2)

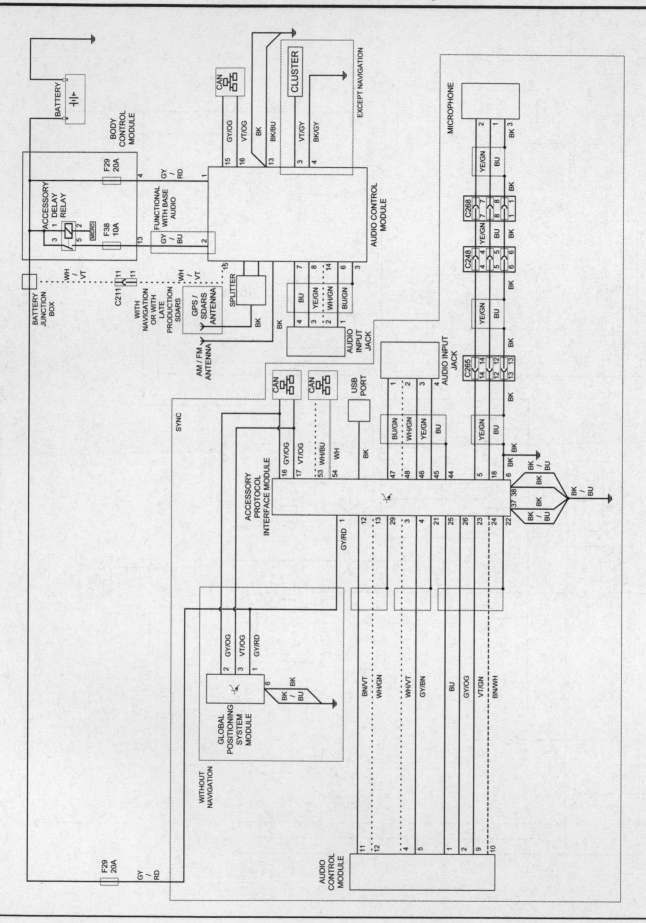

Audio system (with satellite radio)

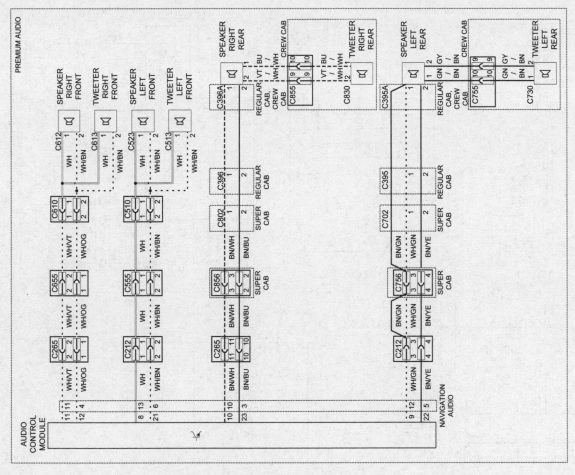

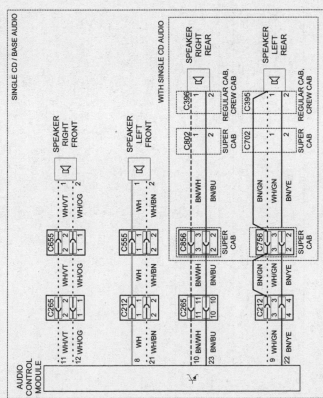

Audio system (base and premium, without navigation)

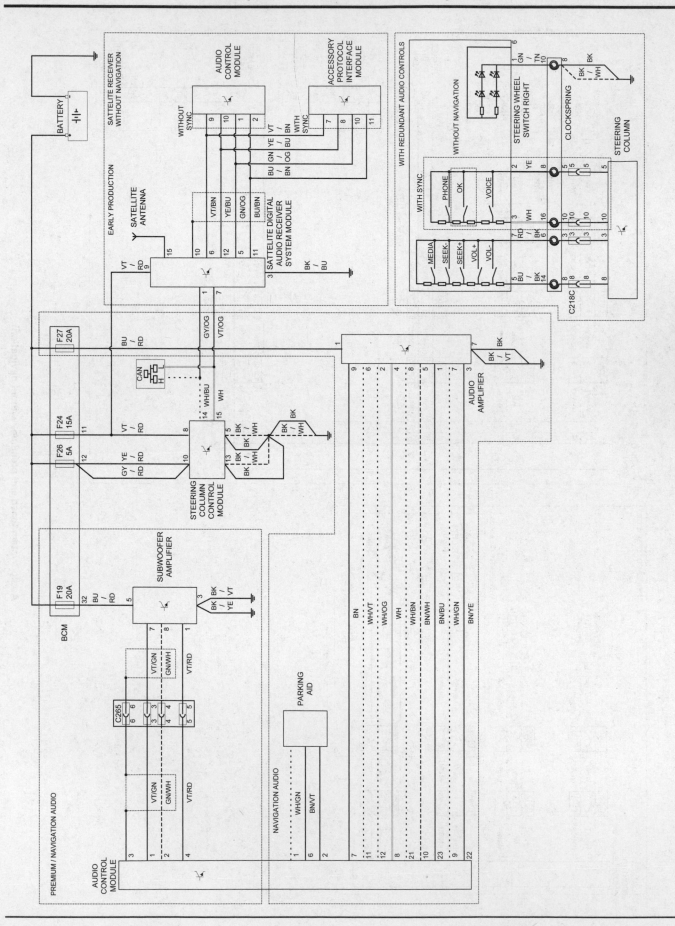

Audio system (premium, with navigation)

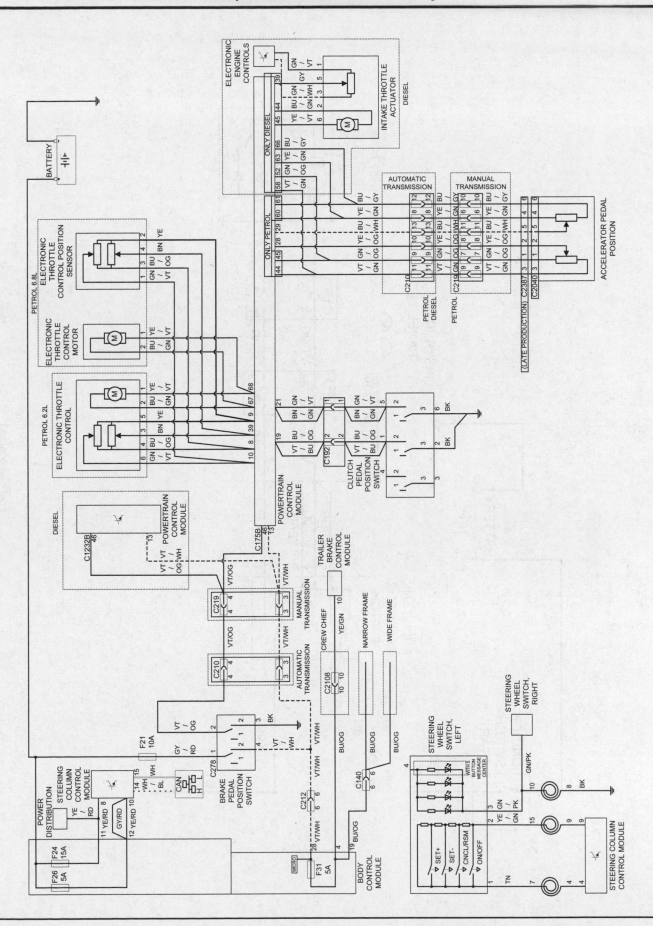

Cruise control circuit

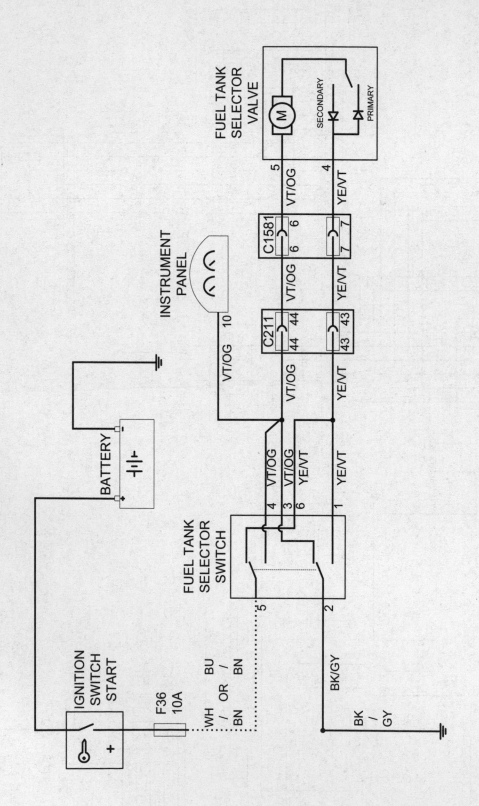
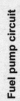

Fuel pump circuit

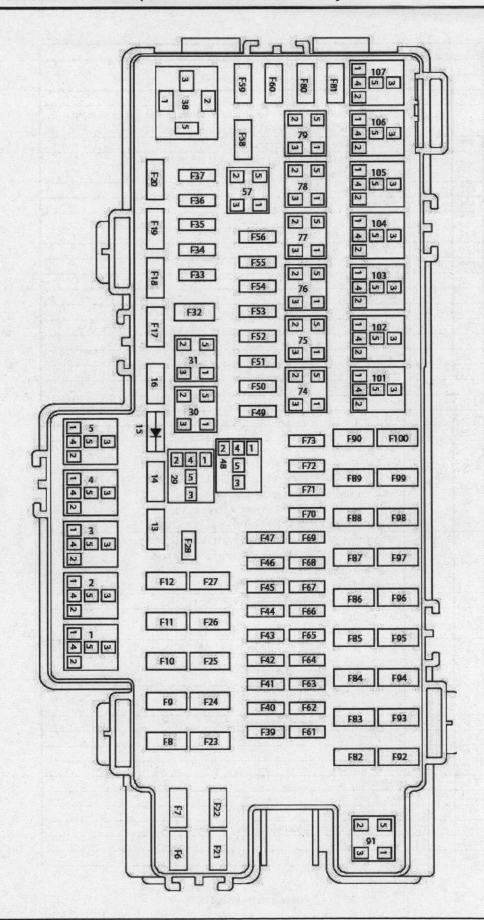

Fuses and relays (1 of 5)

BATTERY JUNCTION BOX (BJB)		
Fuse	Current (A)	Circuits protected
F6	-	Not used
F7	40	Rear window defrost relay - without rear heated power sliding window. Rear window defrost relay - with rear heated power sliding window
F8	50	Passenger side front seat control switch
F9	30	Driver side front seat control switch - without memory, Driver Seat Module (DSM) - with memory
F10	-	Not used
F11	-	Not used
F12	-	Not used
F17	15	Exterior rear view mirrors - heating element
F18	-	Not used
F19	-	Not used
F20	-	Not used
F21	-	Not used
F22	30	Trailer Brake Control (TBC) module, Telematics module
F23	40	Telematics module
F24	-	Not used
F25	30	Front wiper relay
F26	30	Trailer tow relay, parking lamp
F27	25	Reductant heater relay - Diesel
F28	-	Not used
F32	-	Not used
F33	15	Powertrain Control Module (PCM)
F34	20	Heated oxygen sensors, Mass Air Flow / Intake Air Temperature (MAF/IAT) sensor, Variable Camshaft Timing (VCT) solenoid 1 - 6.2L, Variable Camshaft Timing (VCT) solenoid 2 - 6.2L, Evaporative emission (EVAP) canister purge valve, Intake Manifold Tuning Valve (IMTV) - 6.8L (GASOLINE)
F34	15	Turbocharger actuator, EGR cooler bypass valve solenoid, Turbocharger (TC) wastegate regulating valve solenoid, Reductant pump assembly, Reductant Purge Valve (RDPGV) (DIESEL)
F35	10	A/C clutch relay, Cooling fan - 6.2L, 6.7L Diesel
F36	20	Coil on Plugs, Ignition transformer capacitors (GASOLINE)
F36	15	Mass Air Flow / Intake Air Temperature (MAF/IAT) sensor, Glow Plug Control Module (GPCM), NOx sensor module (DIESEL)
F37	10	Fuel pressure control valve - Diesel, Fuel volume control valve - Diesel
F39	10	Constant Vacuum Hublock (CVH) solenoid
F40	15	Transfer Case Control Module (TCCM)
F41	-	Not used
F42	-	Not used
F43	-	Not used
F44	-	Not used
F45	10	Run / Start relay
F46	10	Transmission Control Module (TCM) - Diesel
F47	10	A/C clutch relay
F49	10	Video camera

F50	10	Blower motor relay
F51	-	Not used
F52	10	Powertrain Control Module (PCM), Transmission Control Module (TCM) - Diesel
F53	10	Transfer Case Control Module (TCCM)
F54	10	Anti-lock Brake System (ABS) module
F55	10	Rear window defrost relay, Trailer tow relay, battery charge
F56	20	Body Control Module (BCM) - F33, F34, F35, F36, F37, F41, F42
F58	-	Not used

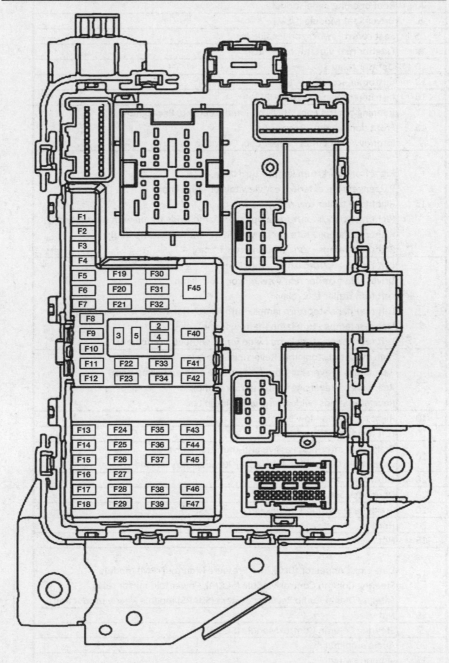

BODY CONTROL MODULE (BCM)		
Fuse	Current (A)	Circuits protected
F1	30	Not used
F2	15	Upfitter relay 4
F3	30	Not used
F4	10	Telescoping exterior rear view mirror switch, Tool link reader Engine compartment lamp, Front dome lamp - Regular Cab, Interior map / lamps - Super Cab, Crew Cab, Vanity mirror lamps - Super Cab, Crew Cab
F5	20	Roof opening panel module
F6	5	Driver Seat Module (DSM)
F7	7.5	Seat control switch, driver side front
F8	10	Exterior rear view mirror switch
F9	10	Upfitter relay 3
F10	10	Customer access
F11	10	Instrument Panel Cluster (IPC)
F12	15	Running board lamps, High mounted stoplamp, Backlighting, Front dome lamp - Regular Cab, Interior map / lamps - Super Cab, Crew Cab
F13	15	Right front park/turn lamps - turn lamp, Passenger side exterior rear view mirror - turn lamp, Right turn trailer tow relay, Right rear park/stop/turn lamps - turn lamp (wide frame with pickup bed), Right rear lamps - turn lamp (wide frame without pickup bed), Right rear lamps - turn lamp (narrow frame)
F14	15	Left front park/turn lamps - turn lamp, Driver side exterior rear view mirror - turn lamp, Left turn trailer tow relay, Left rear park/stop/turn lamps - turn lamp (wide frame with pickup bed), Left rear lamps - turn lamp (narrow frame), Left rear lamps - turn lamp (wide frame without pickup bed)
F15	15	High mounted stoplamp, Reversing lamp relay, Rear lamps - reversing lamp (wide frame without pickup bed), Rear lamps - reversing lamp (narrow frame), Reversing lamps (wide frame with pickup bed
F16	10	Right headlamp - low beam
F17	10	Left headlamp - low beam
F18	10	Brake shift interlock, Passive anti-theft transceiver, Powertrain Control Module (PCM), Keyless entry keypad
F19	20	Subwoofer amplifier
F20	20	Door latch actuators
F21	10	Brake Pedal Position (BPP) switch
F22	20	Horn
F23	15	Not used
F24	15	Data Link Connector (DLC), Tire Pressure Monitor (TPM) module, Steering Column Control Module (SCCM), Power fold mirror relay, Satellite Digital Audio Receiver System (SDARS) module - early production
F25	15	Not used
F26	5	Steering Column Control Module (SCCM)
F27	20	Audio amplifier
F28	15	Ignition switch

Fuses and relays (4 of 5)

F29	20	Audio Control Module (ACM), In-dash computer, Accessory Protocol Interface Module (APIM) - with SYNC, Global Position System Module (GPSM) - SYNC without navigation
F30	15	Parking lamp trailer tow relay, Side marker lamps, Front park/turn lamps - parking lamp, License plate lamps (wide frame with pickup bed), Rear park/stop/turn lamps - parking lamp (wide frame with pickup bed), Rear lamps - parking lamp (narrow frame), Rear lamps - parking lamp (wide frame without pickup bed), Rear marker lamp assembly - parking lamp (dual rear wheels), Side lamps (dual rear wheels)
F31	5	Trailer Brake Control (TBC) module
F32	15	Door lock switches, Roof opening panel module, Auto-dimming interior mirror, DC/AC inverter module
F33	10	Restraints Control Module (RCM)
F34	10	Not used
F35	5	Parking Aid Module (PAM), Trailer Brake Control (TBC) module, Tow haul switch, Telematics module
F36	10	Fuel tank selector switch
F37	10	Auxiliary Heater Control Module (AHCM)
F38	10	Audio Control Module (ACM), In-dash computer
F39	15	Headlamps - high beam
F40	10	Exterior rear view mirrors - park lamps, Roof marker lamps
F41	7.5	Center stack switch assembly
F42	5	Not used
F43	10	Front wiper relay
F44	10	Upfitter switch
F45	5	Not used
F46	10	HVAC module
F47	15	Fog lamps
F48	30	Master window control switch, Passenger window control switch, Overhead console switch assembly - with power sliding rear window

WIRE COLOR CODE INDEX

BK – BLACK	———
BN – BROWN	———
BU – BLUE	———
GN – GREEN	———
GY – GREY	———
LB – LIGHT BLUE	———
LG – LIGHT GREEN	———
RD – RED	———
OG – ORANGE	———
PK – PINK	———
YE – YELLOW	———
WH – WHITE	———

Notes

Index

G

H

Notes

Haynes Automotive Manuals

ACURA
- **12020** Integra '86 thru '89 & **Legend** '86 thru '90
- **12021** Integra '90 thru '93 & **Legend** '91 thru '95
- Integra '94 thru '00 - see *HONDA Civic (42025)*
- MDX '01 thru '07 - see *HONDA Pilot (42037)*
- **12050** Acura TL all models '99 thru '08

AMC
- Jeep CJ - see *JEEP (50020)*
- **14020** Mid-size models '70 thru '83
- **14025** (Renault) Alliance & Encore '83 thru '87

AUDI
- **15020** 4000 all models '80 thru '87
- **15025** 5000 all models '77 thru '83
- **15026** 5000 all models '84 thru '88
- Audi A4 '96 thru '01 - see *VW Passat (96023)*
- **15030** Audi A4 '02 thru '08

AUSTIN-HEALEY
- Sprite - see *MG Midget (66015)*

BMW
- **18020** 3/5 Series '82 thru '92
- **18021** 3-Series incl. Z3 models '92 thru '98
- **18022** 3-Series incl. Z4 models '99 thru '05
- **18023** 3-Series '06 thru '10
- **18025** 320i all 4 cyl models '75 thru '83
- **18050** 1500 thru 2002 except Turbo '59 thru '77

BUICK
- **19010** Buick Century '97 thru '05
- Century (front-wheel drive) - see *GM (38005)*
- **19020** Buick, Oldsmobile & Pontiac Full-size (Front-wheel drive) '85 thru '05
  Buick Electra, LeSabre and Park Avenue; Oldsmobile Delta 88 Royale, Ninety Eight and Regency; Pontiac Bonneville
- **19025** Buick, Oldsmobile & Pontiac Full-size (Rear wheel drive) '70 thru '90
  Buick Estate, Electra, LeSabre, Limited, Oldsmobile Custom Cruiser, Delta 88, Ninety-eight, Pontiac Bonneville, Catalina, Grandville, Parisienne
- **19030** Mid-size Regal & Century all rear-drive models with V6, V8 and Turbo '74 thru '87
- Regal - see *GENERAL MOTORS (38010)*
- Riviera - see *GENERAL MOTORS (38030)*
- Roadmaster - see *CHEVROLET (24046)*
- Skyhawk - see *GENERAL MOTORS (38015)*
- Skylark - see *GM (38020, 38025)*
- Somerset - see *GENERAL MOTORS (38025)*

CADILLAC
- **21015** CTS & CTS-V '03 thru '12
- **21030** Cadillac Rear Wheel Drive '70 thru '93
- Cimarron - see *GENERAL MOTORS (38015)*
- DeVille - see *GM (38031 & 38032)*
- Eldorado - see *GM (38030 & 38031)*
- Fleetwood - see *GM (38031)*
- Seville - see *GM (38030, 38031 & 38032)*

CHEVROLET
- **10305** Chevrolet Engine Overhaul Manual
- **24010** Astro & GMC Safari Mini-vans '85 thru '05
- **24015** Camaro V8 all models '70 thru '81
- **24016** Camaro all models '82 thru '92
- **24017** Camaro & Firebird '93 thru '02
- Cavalier - see *GENERAL MOTORS (38016)*
- Celebrity - see *GENERAL MOTORS (38005)*
- **24020** Chevelle, Malibu & El Camino '69 thru '87
- **24024** Chevette & Pontiac T1000 '76 thru '87
- Citation - see *GENERAL MOTORS (38020)*
- **24027** Colorado & GMC Canyon '04 thru '10
- **24032** Corsica/Beretta all models '87 thru '96
- **24040** Corvette all V8 models '68 thru '82
- **24041** Corvette all models '84 thru '96
- **24045** Full-size Sedans Caprice, Impala, Biscayne, Bel Air & Wagons '69 thru '90
- **24046** Impala SS & Caprice and Buick Roadmaster '91 thru '96
- Impala '00 thru '05 - see *LUMINA (24048)*
- **24047** Impala & Monte Carlo all models '06 thru '11
- Lumina '90 thru '94 - see *GM (38010)*
- **24048** Lumina & Monte Carlo '95 thru '05
- Lumina APV - see *GM (38035)*
- **24050** Luv Pick-up all 2WD & 4WD '72 thru '82
- Malibu '97 thru '00 - see *GM (38026)*
- **24055** Monte Carlo all models '70 thru '88
- Monte Carlo '95 thru '01 - see *LUMINA (24048)*
- **24059** Nova all V8 models '69 thru '79
- **24060** Nova and Geo Prizm '85 thru '92
- **24064** Pick-ups '67 thru '87 - Chevrolet & GMC
- **24065** Pick-ups '88 thru '98 - Chevrolet & GMC

- **24066** Pick-ups '99 thru '06 - Chevrolet & GMC
- **24067** Chevrolet Silverado & GMC Sierra '07 thru '12
- **24070** S-10 & S-15 Pick-ups '82 thru '93, Blazer & Jimmy '83 thru '94,
- **24071** S-10 & Sonoma Pick-ups '94 thru '04, including Blazer, Jimmy & Hombre
- **24072** Chevrolet TrailBlazer, GMC Envoy & Oldsmobile Bravada '02 thru '09
- **24075** Sprint '85 thru '88 & Geo Metro '89 thru '01
- **24080** Vans - Chevrolet & GMC '68 thru '96
- **24081** Chevrolet Express & GMC Savana Full-size Vans '96 thru '10

CHRYSLER
- **10310** Chrysler Engine Overhaul Manual
- **25015** Chrysler Cirrus, Dodge Stratus, Plymouth Breeze '95 thru '00
- **25020** Full-size Front-Wheel Drive '88 thru '93
- K-Cars - see *DODGE Aries (30008)*
- Laser - see *DODGE Daytona (30030)*
- **25025** Chrysler LHS, Concorde, New Yorker, Dodge Intrepid, Eagle Vision, '93 thru '97
- **25026** Chrysler LHS, Concorde, 300M, Dodge Intrepid, '98 thru '04
- **25027** Chrysler 300, Dodge Charger & Magnum '05 thru '09
- **25030** Chrysler & Plymouth Mid-size front wheel drive '82 thru '95
- Rear-wheel Drive - see *Dodge (30050)*
- **25035** PT Cruiser all models '01 thru '10
- **25040** Chrysler Sebring '95 thru '06, Dodge Stratus '01 thru '06, Dodge Avenger '95 thru '00

DATSUN
- **28005** 200SX all models '80 thru '83
- **28007** B-210 all models '73 thru '78
- **28009** 210 all models '79 thru '82
- **28012** 240Z, 260Z & 280Z Coupe '70 thru '78
- **28014** 280ZX Coupe & 2+2 '79 thru '83
- 300ZX - see *NISSAN (72010)*
- **28018** 510 & PL521 Pick-up '68 thru '73
- **28020** 510 all models '78 thru '81
- **28022** 620 Series Pick-up all models '73 thru '79
- 720 Series Pick-up - see *NISSAN (72030)*
- **28025** 810/Maxima all gasoline models '77 thru '84

DODGE
- 400 & 600 - see *CHRYSLER (25030)*
- **30008** Aries & Plymouth Reliant '81 thru '89
- **30010** Caravan & Plymouth Voyager '84 thru '95
- **30011** Caravan & Plymouth Voyager '96 thru '02
- **30012** Challenger/Plymouth Saporro '78 thru '83
- **30013** Caravan, Chrysler Voyager, Town & Country '03 thru '07
- **30016** Colt & Plymouth Champ '78 thru '87
- **30020** Dakota Pick-ups all models '87 thru '96
- **30021** Durango '98 & '99, Dakota '97 thru '99
- **30022** Durango '00 thru '03 Dakota '00 thru '04
- **30023** Durango '04 thru '09, Dakota '05 thru '11
- **30025** Dart, Demon, Plymouth Barracuda, Duster & Valiant 6 cyl models '67 thru '76
- **30030** Daytona & Chrysler Laser '84 thru '89
- Intrepid - see *CHRYSLER (25025, 25026)*
- **30034** Neon all models '95 thru '99
- **30035** Omni & Plymouth Horizon '78 thru '90
- **30036** Dodge and Plymouth Neon '00 thru '05
- **30040** Pick-ups all full-size models '74 thru '93
- **30041** Pick-ups all full-size models '94 thru '01
- **30042** Pick-ups full-size models '02 thru '08
- **30045** Ram 50/D50 Pick-ups & Raider and Plymouth Arrow Pick-ups '79 thru '93
- **30050** Dodge/Plymouth/Chrysler RWD '71 thru '89
- **30055** Shadow & Plymouth Sundance '87 thru '94
- **30060** Spirit & Plymouth Acclaim '89 thru '95
- **30065** Vans - Dodge & Plymouth '71 thru '03

EAGLE
- Talon - see *MITSUBISHI (68030, 68031)*
- Vision - see *CHRYSLER (25025)*

FIAT
- **34010** 124 Sport Coupe & Spider '68 thru '78
- **34025** X1/9 all models '74 thru '80

FORD
- **10320** Ford Engine Overhaul Manual
- **10355** Ford Automatic Transmission Overhaul
- **11500** Mustang '64-1/2 thru '70 Restoration Guide
- **36004** Aerostar Mini-vans '86 thru '97
- **36006** Contour & Mercury Mystique '95 thru '00
- **36008** Courier Pick-up '72 thru '82
- **36012** Crown Victoria & Mercury Grand Marquis '88 thru '10
- **36016** Escort/Mercury Lynx all models '81 thru '90
- **36020** Escort/Mercury Tracer '91 thru '02

- **36022** Escape & Mazda Tribute '01 thru '11
- **36024** Explorer & Mazda Navajo '91 thru '01
- **36025** Explorer/Mercury Mountaineer '02 thru '10
- **36028** Fairmont & Mercury Zephyr '78 thru '83
- **36030** Festiva & Aspire '88 thru '97
- **36032** Fiesta all models '77 thru '80
- **36034** Focus all models '00 thru '11
- **36036** Ford & Mercury Full-size '75 thru '87
- **36044** Ford & Mercury Mid-size '75 thru '86
- **36045** Fusion & Mercury Milan '06 thru '10
- **36048** Mustang V8 all models '64-1/2 thru '73
- **36049** Mustang II 4 cyl, V6 & V8 models '74 thru '78
- **36050** Mustang & Mercury Capri '79 thru '93
- **36051** Mustang all models '94 thru '04
- **36052** Mustang '05 thru '10
- **36054** Pick-ups & Bronco '73 thru '79
- **36058** Pick-ups & Bronco '80 thru '96
- **36059** F-150 & Expedition '97 thru '09, F-250 '97 thru '99 & Lincoln Navigator '98 thru '09
- **36060** Super Duty Pick-ups, Excursion '99 thru '10
- **36061** F-150 full-size '04 thru '10
- **36062** Pinto & Mercury Bobcat '75 thru '80
- **36066** Probe all models '89 thru '92
- Probe '93 thru '97 - see *MAZDA 626 (61042)*
- **36070** Ranger/Bronco II gasoline models '83 thru '92
- **36071** Ranger '93 thru '10 & Mazda Pick-ups '94 thru '09
- **36074** Taurus & Mercury Sable '86 thru '95
- **36075** Taurus & Mercury Sable '96 thru '05
- **36078** Tempo & Mercury Topaz '84 thru '94
- **36082** Thunderbird/Mercury Cougar '83 thru '88
- **36086** Thunderbird/Mercury Cougar '89 thru '97
- **36090** Vans all V8 Econoline models '69 thru '91
- **36094** Vans full size '92 thru '10
- **36097** Windstar Mini-van '95 thru '07

GENERAL MOTORS
- **10360** GM Automatic Transmission Overhaul
- **38005** Buick Century, Chevrolet Celebrity, Oldsmobile Cutlass Ciera & Pontiac 6000 all models '82 thru '96
- **38010** Buick Regal, Chevrolet Lumina, Oldsmobile Cutlass Supreme & Pontiac Grand Prix (FWD) '88 thru '07
- **38015** Buick Skyhawk, Cadillac Cimarron, Chevrolet Cavalier, Oldsmobile Firenza & Pontiac J-2000 & Sunbird '82 thru '94
- **38016** Chevrolet Cavalier & Pontiac Sunfire '95 thru '05
- **38017** Chevrolet Cobalt & Pontiac G5 '05 thru '11
- **38020** Chevrolet Citation, Olds Omega, Pontiac Phoenix '80 thru '85
- **38025** Buick Skylark & Somerset, Oldsmobile Achieva & Calais and Pontiac Grand Am all models '85 thru '98
- **38026** Chevrolet Malibu, Olds Alero & Cutlass, Pontiac Grand Am '97 thru '03
- **38027** Chevrolet Malibu '04 thru '10
- **38030** Cadillac Eldorado, Seville, Oldsmobile Toronado, Buick Riviera '71 thru '85
- **38031** Cadillac Eldorado & Seville, DeVille, Fleetwood & Olds Toronado, Buick Riviera '86 thru '93
- **38032** Cadillac DeVille '94 thru '05 & Seville '92 thru '04 Cadillac DTS '06 thru '10
- **38035** Chevrolet Lumina APV, Olds Silhouette & Pontiac Trans Sport all models '90 thru '96
- **38036** Chevrolet Venture, Olds Silhouette, Pontiac Trans Sport & Montana '97 thru '05
- General Motors Full-size Rear-wheel Drive - see *BUICK (19025)*
- **38040** Chevrolet Equinox '05 thru '09 Pontiac Torrent '06 thru '09
- **38070** Chevrolet HHR '06 thru '11

GEO
- Metro - see *CHEVROLET Sprint (24075)*
- Prizm - '85 thru '92 see *CHEVY (24060)*, '93 thru '02 see *TOYOTA Corolla (92036)*
- **40030** Storm all models '90 thru '93
- Tracker - see *SUZUKI Samurai (90010)*

GMC
- Vans & Pick-ups - see *CHEVROLET*

HONDA
- **42010** Accord CVCC all models '76 thru '83
- **42011** Accord all models '84 thru '89
- **42012** Accord all models '90 thru '93
- **42013** Accord all models '94 thru '97
- **42014** Accord all models '98 thru '02
- **42015** Accord '03 thru '07
- **42020** Civic 1200 all models '73 thru '79
- **42021** Civic 1300 & 1500 CVCC '80 thru '83
- **42022** Civic 1500 CVCC all models '75 thru '79

(Continued on other side)

Haynes North America, Inc., 859 Lawrence Drive, Newbury Park, CA 91320-1514 • (805) 498-6703 • http://www.haynes.com

Haynes Automotive Manuals (continued)

NOTE: If you do not see a listing for your vehicle, consult your local Haynes dealer for the latest product information.

42023 **Civic** all models '84 thru '91
42024 **Civic & del Sol** '92 thru '95
42025 **Civic** '96 thru '00, **CR-V** '97 thru '01,
  Acura Integra '94 thru '00
42026 **Civic** '01 thru '10, **CR-V** '02 thru '09
42035 **Odyssey** all models '99 thru '10
  Passport - see ISUZU Rodeo (47017)
42037 **Honda Pilot** '03 thru '07, **Acura MDX** '01 thru '07
42040 **Prelude CVCC** all models '79 thru '89

HYUNDAI
43010 **Elantra** all models '96 thru '10
43015 **Excel & Accent** all models '86 thru '09
43050 **Santa Fe** all models '01 thru '06
43055 **Sonata** all models '99 thru '08

INFINITI
  G35 '03 thru '08 - see NISSAN 350Z (72011)

ISUZU
  Hombre - see CHEVROLET S-10 (24071)
47017 **Rodeo, Amigo & Honda Passport** '89 thru '02
47020 **Trooper & Pick-up** '81 thru '93

JAGUAR
49010 **XJ6** all 6 cyl models '68 thru '86
49011 **XJ6** all models '88 thru '94
49015 **XJ12 & XJS** all 12 cyl models '72 thru '85

JEEP
50010 **Cherokee, Comanche & Wagoneer Limited**
  all models '84 thru '01
50020 **CJ** all models '49 thru '86
50025 **Grand Cherokee** all models '93 thru '04
50026 **Grand Cherokee** '05 thru '09
50029 **Grand Wagoneer & Pick-up** '72 thru '91
  Grand Wagoneer '84 thru '91, Cherokee &
  Wagoneer '72 thru '83, Pick-up '72 thru '88
50030 **Wrangler** all models '87 thru '11
50035 **Liberty** '02 thru '07

KIA
54050 **Optima** '01 thru '10
54070 **Sephia** '94 thru '01, **Spectra** '00 thru '09,
  Sportage '05 thru '10

LEXUS
  ES 300/330 - see TOYOTA Camry (92007) (92008)
  RX 330 - see TOYOTA Highlander (92095)

LINCOLN
  Navigator - see FORD Pick-up (36059)
59010 **Rear-Wheel Drive** all models '70 thru '10

MAZDA
61010 **GLC Hatchback** (rear-wheel drive) '77 thru '83
61011 **GLC** (front-wheel drive) '81 thru '85
61012 **Mazda3** '04 thru '11
61015 **323 & Protogé** '90 thru '03
61016 **MX-5 Miata** '90 thru '09
61020 **MPV** all models '89 thru '98
  Navajo - see Ford Explorer (36024)
61030 **Pick-ups** '72 thru '93
  Pick-ups '94 thru '00 - see Ford Ranger (36071)
61035 **RX-7** all models '79 thru '85
61036 **RX-7** all models '86 thru '91
61040 **626** (rear-wheel drive) all models '79 thru '82
61041 **626/MX-6** (front-wheel drive) '83 thru '92
61042 **626, MX-6/Ford Probe** '93 thru '02
61043 **Mazda6** '03 thru '11

MERCEDES-BENZ
63012 **123 Series Diesel** '76 thru '85
63015 **190 Series** four-cyl gas models, '84 thru '88
63020 **230/250/280** 6 cyl sohc models '68 thru '72
63025 **280** 123 Series gasoline models '77 thru '81
63030 **350 & 450** all models '71 thru '80
63040 **C-Class:** C230/C240/C280/C320/C350 '01 thru '07

MERCURY
64200 **Villager & Nissan Quest** '93 thru '01
  All other titles, see FORD Listing.

MG
66010 **MGB** Roadster & GT Coupe '62 thru '80
66015 **MG Midget, Austin Healey Sprite** '58 thru '80

MINI
67020 **Mini** '02 thru '11

MITSUBISHI
68020 **Cordia, Tredia, Galant, Precis &
  Mirage** '83 thru '93
68030 **Eclipse, Eagle Talon & Ply. Laser** '90 thru '94
68031 **Eclipse** '95 thru '05, **Eagle Talon** '95 thru '98
68035 **Galant** '94 thru '10
68040 **Pick-up** '83 thru '96 & **Montero** '83 thru '93

NISSAN
72010 **300ZX** all models including Turbo '84 thru '89
72011 **350Z & Infiniti G35** all models '03 thru '08
72015 **Altima** all models '93 thru '06
72016 **Altima** '07 thru '10
72020 **Maxima** all models '85 thru '92
72021 **Maxima** all models '93 thru '04
72025 **Murano** '03 thru '10
72030 **Pick-ups** '80 thru '97 **Pathfinder** '87 thru '95
72031 **Frontier Pick-up, Xterra, Pathfinder** '96 thru '04
72032 **Frontier & Xterra** '05 thru '11
72040 **Pulsar** all models '83 thru '86
  Quest - see MERCURY Villager (64200)
72050 **Sentra** all models '82 thru '94
72051 **Sentra & 200SX** all models '95 thru '06
72060 **Stanza** all models '82 thru '90
72070 **Titan pick-ups** '04 thru '10 **Armada** '05 thru '10

OLDSMOBILE
73015 **Cutlass** V6 & V8 gas models '74 thru '88
  *For other OLDSMOBILE titles, see BUICK,
  CHEVROLET or GENERAL MOTORS listing.*

PLYMOUTH
  For PLYMOUTH titles, see DODGE listing.

PONTIAC
79008 **Fiero** all models '84 thru '88
79018 **Firebird** V8 models except Turbo '70 thru '81
79019 **Firebird** all models '82 thru '92
79025 **G6** all models '05 thru '09
79040 **Mid-size Rear-wheel Drive** '70 thru '87
  Vibe '03 thru '11 - see TOYOTA Matrix (92060)
  *For other PONTIAC titles, see BUICK,
  CHEVROLET or GENERAL MOTORS listing.*

PORSCHE
80020 **911** except Turbo & Carrera 4 '65 thru '89
80025 **914** all 4 cyl models '69 thru '76
80030 **924** all models including Turbo '76 thru '82
80035 **944** all models including Turbo '83 thru '89

RENAULT
  Alliance & Encore - see AMC (14020)

SAAB
84010 **900** all models including Turbo '79 thru '88

SATURN
87010 **Saturn** all S-series models '91 thru '02
87011 **Saturn Ion** '03 thru '07
87020 **Saturn** all L-series models '00 thru '04
87040 **Saturn VUE** '02 thru '07

SUBARU
89002 **1100, 1300, 1400 & 1600** '71 thru '79
89003 **1600 & 1800** 2WD & 4WD '80 thru '94
89100 **Legacy** all models '90 thru '99
89101 **Legacy & Forester** '00 thru '06

SUZUKI
90010 **Samurai/Sidekick & Geo Tracker** '86 thru '01

TOYOTA
92005 **Camry** all models '83 thru '91
92006 **Camry** all models '92 thru '96
92007 **Camry, Avalon, Solara, Lexus ES 300** '97 thru '01
92008 **Toyota Camry, Avalon and Solara and
  Lexus ES 300/330** all models '02 thru '06
92009 **Camry** '07 thru '11
92015 **Celica Rear Wheel Drive** '71 thru '85
92020 **Celica Front Wheel Drive** '86 thru '99
92025 **Celica Supra** all models '79 thru '92
92030 **Corolla** all models '75 thru '79
92032 **Corolla** all rear wheel drive models '80 thru '87
92035 **Corolla** all front wheel drive models '84 thru '92
92036 **Corolla & Geo Prizm** '93 thru '02
92037 **Corolla** models '03 thru '11
92040 **Corolla Tercel** all models '80 thru '82
92045 **Corona** all models '74 thru '82
92050 **Cressida** all models '78 thru '82
92055 **Land Cruiser FJ**40, 43, 45, 55 '68 thru '82
92056 **Land Cruiser** FJ60, 62, 80, FZJ80 '80 thru '96
92060 **Matrix & Pontiac Vibe** '03 thru '11
92065 **MR2** all models '85 thru '87
92070 **Pick-up** all models '69 thru '78
92075 **Pick-up** all models '79 thru '95
92076 **Tacoma, 4Runner, & T100** '93 thru '04
92077 **Tacoma** all models '05 thru '09
92078 **Tundra** '00 thru '06 & **Sequoia** '01 thru '07
92079 **4Runner** all models '03 thru '09
92080 **Previa** all models '91 thru '95
92081 **Prius** all models '01 thru '08
92082 **RAV4** all models '96 thru '10
92085 **Tercel** all models '87 thru '94
92090 **Sienna** all models '98 thru '09
92095 **Highlander & Lexus RX-330** '99 thru '07

TRIUMPH
94007 **Spitfire** all models '62 thru '81
94010 **TR7** all models '75 thru '81

VW
96008 **Beetle & Karmann Ghia** '54 thru '79
96009 **New Beetle** '98 thru '11
96016 **Rabbit, Jetta, Scirocco & Pick-up** gas
  models '75 thru '92 & Convertible '80 thru '92
96017 **Golf, GTI & Jetta** '93 thru '98, **Cabrio** '95 thru '02
96018 **Golf, GTI, Jetta** '99 thru '05
96019 **Jetta, Rabbit, GTI & Golf** '05 thru '11
96020 **Rabbit, Jetta & Pick-up** diesel '77 thru '84
96023 **Passat** '98 thru '05, **Audi A4** '96 thru '01
96030 **Transporter 1600** all models '68 thru '79
96035 **Transporter 1700, 1800 & 2000** '72 thru '79
96040 **Type 3 1500 & 1600** all models '63 thru '73
96045 **Vanagon** all air-cooled models '80 thru '83

VOLVO
97010 **120, 130 Series & 1800 Sports** '61 thru '73
97015 **140 Series** all models '66 thru '74
97020 **240 Series** all models '76 thru '93
97040 **740 & 760 Series** all models '82 thru '88
97050 **850 Series** all models '93 thru '97

TECHBOOK MANUALS
10205 **Automotive Computer Codes**
10206 **OBD-II & Electronic Engine Management**
10210 **Automotive Emissions Control Manual**
10215 **Fuel Injection Manual** '78 thru '85
10220 **Fuel Injection Manual** '86 thru '99
10225 **Holley Carburetor Manual**
10230 **Rochester Carburetor Manual**
10240 **Weber/Zenith/Stromberg/SU Carburetors**
10305 **Chevrolet Engine Overhaul Manual**
10310 **Chrysler Engine Overhaul Manual**
10320 **Ford Engine Overhaul Manual**
10330 **GM and Ford Diesel Engine Repair Manual**
10333 **Engine Performance Manual**
10340 **Small Engine Repair Manual**, 5 HP & Less
10341 **Small Engine Repair Manual**, 5.5 - 20 HP
10345 **Suspension, Steering & Driveline Manual**
10355 **Ford Automatic Transmission Overhaul**
10360 **GM Automatic Transmission Overhaul**
10405 **Automotive Body Repair & Painting**
10410 **Automotive Brake Manual**
10411 **Automotive Anti-lock Brake (ABS) Systems**
10415 **Automotive Detaiing Manual**
10420 **Automotive Electrical Manual**
10425 **Automotive Heating & Air Conditioning**
10430 **Automotive Reference Manual & Dictionary**
10435 **Automotive Tools Manual**
10440 **Used Car Buying Guide**
10445 **Welding Manual**
10450 **ATV Basics**
10452 **Scooters 50cc to 250cc**

SPANISH MANUALS
98903 **Reparación de Carrocería & Pintura**
98904 **Manual de Carburador Modelos
  Holley & Rochester**
98905 **Códigos Automotrices de la Computadora**
98906 **OBD-II & Sistemas de Control Electrónico
  del Motor**
98910 **Frenos Automotriz**
98913 **Electricidad Automotriz**
98915 **Inyección de Combustible** '86 al '99
99040 **Chevrolet & GMC Camionetas** '67 al '87
99041 **Chevrolet & GMC Camionetas** '88 al '98
99042 **Chevrolet & GMC Camionetas
  Cerradas** '68 al '95
99043 **Chevrolet/GMC Camionetas** '94 al '04
99048 **Chevrolet/GMC Camionetas** '99 al '06
99055 **Dodge Caravan & Plymouth Voyager** '84 al '95
99075 **Ford Camionetas y Bronco** '80 al '94
99076 **Ford F-150** '97 al '09
99077 **Ford Camionetas Cerradas** '69 al '91
99088 **Ford Modelos de Tamaño Mediano** '75 al '86
99089 **Ford Camionetas Ranger** '93 al '10
99091 **Ford Taurus & Mercury Sable** '86 al '95
99095 **GM Modelos de Tamaño Grande** '70 al '90
99100 **GM Modelos de Tamaño Mediano** '70 al '88
99106 **Jeep Cherokee, Wagoneer & Comanche**
  '84 al '00
99110 **Nissan Camioneta** '80 al '96, **Pathfinder** '87 al '95
99118 **Nissan Sentra** '82 al '94
99125 **Toyota Camionetas y 4Runner** '79 al '95

Over 100 Haynes
motorcycle manuals
also available

7-12

Haynes North America, Inc., 859 Lawrence Drive, Newbury Park, CA 91320-1514 • (805) 498-6703 • http://www.haynes.com